Unquenchable ~

Unquenchable ~

AMERICA'S WATER CRISIS AND WHAT TO DO ABOUT IT

Robert Glennon

June 14, 2012

To Susan –
Thanks for hosting me!
Let's confront the crisis
and avert a catastrophe!
Robert Glennon

Island Press / Shearwater Books
Washington • Covelo • London

Published by Island Press

Library of Congress Cataloging-in-Publication data.

Glennon, Robert Jerome, 1944-
Unquenchable : America's water crisis and what to do about it / by Robert Glennon.

Includes bibliographical references and index.
ISBN-13: 978-1-59726-436-5 (cloth : alk. paper)
ISBN-10: 1-59726-436-9 (cloth : alk. paper)
1. Water-supply—United States. 2. Droughts—United States.
3. Water consumption—United States--Forecasting. I. Title.
TD223.G578 2009
363.6'10973—dc22

The paperback edition carries the 13-digit ISBN: 978-1-59726-816-5 and the 10-digit ISBN: 978-1-59726-816-X

Printed on recycled, acid-free paper

Design by Joyce C. Weston

Manufactured in the United States of America

10 9 8 7 6 5 4 3

To Karen

Spirit of the water
Give us all the courage and the grace
To make genius of this tragedy unfolding
The genius to save this place.

—Joni Mitchell

Contents

Introduction

"There is no lack of water in the Mojave Desert unless you try to establish a city where no city should be."

—Edward Abbey

"WHAT HAPPENS IN Vegas stays in Vegas," teases the city's risqué advertising slogan as it invites visitors to lose their inhibitions, violate their moral principles, forget about their spouses, and ignore their credit card balances. A metropolitan area of 1.8 million people in the Mojave Desert, Sin City encourages irresponsible behavior about everything from sex to water.

During the 1980s and 1990s, garishness characterized development in Las Vegas. Extravagant homeowners built Bavarian châteaus with Italian marble and gilded fixtures. Excess was first a goal, then a standard. Ornate and lavish fountains, rococo entrances, manicured emerald green lawns, lush landscapes, swimming pools, and spas dominated many subdivisions. Lake Las Vegas, a subdivision seventeen miles east of the Strip, boasts a 320-acre privately owned lake, three golf courses, and homes built for only the most upscale tastes. Celine Dion maintains a home at Lake Las Vegas and for five years traveled via helicopter to and from her nightly show at Caesars Palace. The fastest-growing city in the United States, Las Vegas welcomed growth with open arms. But, in 2001, the city of illusion and fantasy stumbled upon a stark reality. It had run out of water.

DURING THE 1920S, no one worried about Nevada needing water. With a population of roughly 5,000, Las Vegas was merely a whistle-stop on the Union Pacific Railroad. Its prospects for growth seemed minuscule until the construction of Hoover Dam provided a water supply—Lake Mead—and hydroelectric power to run air conditioners. In the 1930s, Benjamin Siegelbaum, a.k.a. Bugsy Siegel, part of the East Coast mob's bootlegging and gambling operations, arrived in Vegas. As a hit man, Bugsy had been sent to Los Angeles, where he killed a police informant and escaped conviction. But rather than head back East, Bugsy then turned his attention to Las Vegas, which he envisioned as a gambling mecca run by the mob. Gambling had been legal in Nevada since 1931, when construction workers poured into town to build Hoover Dam. With funding from his New York bosses, Bugsy oversaw construction of the Flamingo casino, which opened in 1946. When the Flamingo lost money, the mob suspected Bugsy of skimming off the top and had him killed in Beverly Hills in 1947. But the seed had been planted. Meyer Lansky and other figures in organized crime soon recognized Las Vegas' gambling potential and an opportunity to launder money. The mob went on to open several casinos on the Strip in the 1950s, including the Sahara, the Sands, and the Tropicana. In addition to gambling, these casinos also featured entertainment by celebrities, including Elvis Presley, Frank Sinatra, Dean Martin, and Bing Crosby.

Today, as if in homage to the sacred gift from Lake Mead, water features at casinos on the Strip create the illusion that the city of Las Vegas has an abundance of water. In the late 1990s, when Steve Wynn developed the Bellagio hotel and casino, an architect and an artist created a $40 million water feature with an eight-acre pond that holds 27 million gallons of water. With a computer-choreographed musical score and light show, it sends water as high as 244 feet through more than 1,200 individually controlled water jets.

Since 1989, the Mirage Hotel and Casino has featured a volcano set on a three-acre pool; the volcano erupts hourly, sending smoke

Figure I.1. Bellagio fountain.
Photograph courtesy of Myles Davidson.

and fire spray 100 feet into the night sky. At the entrance, terraced waterfalls and a rain forest greet tourists as though they've arrived on an island in the South Pacific rather than at a casino in the Mojave Desert. Every evening, the Treasure Island Hotel and Casino uses a lagoon to stage a naval battle between pirates and the British Royal Navy, accompanied by fireworks and scantily clad women—the latter a departure from historical fidelity. Down the Strip, at the Mandalay Bay Resort and Casino, a 1.6-million-gallon "wave pool" provides six-foot-high waves for surfers. But my favorite, for kitsch, is the Venetian Resort Hotel Casino, where, fittingly, tourists ride gondolas poled by gondoliers in striped shirts. Unlike the canals in the real Venice, these are indoors, on the second floor of the casino, beneath a roof painted to resemble someone's sense of a Venetian sky.

An engineering marvel, Las Vegas has repeatedly demonstrated an uncanny and unmatched ability to reinvent itself. In the 1950s, the mob transformed a sleepy cow town into a shady operation, financed with Teamsters Union pension funds, that laundered money and promoted prostitution. Casinos offered all-you-can-eat buffets, cheap booze, free lounge acts, and low-cost, dingy rooms to draw hordes of down-and-out transients hoping for a little change in their luck. The cleanup process began with the arrival of Howard Hughes in 1966, when he bought the Desert Inn casino and then the Sands. He paid $14.6 million for the Sands and 183 acres of prime real estate that eventually became the Hughes Center. At the time, this acquisition confirmed most people's suspicions that the reclusive billionaire was certifiably insane. When he told the managers of the casinos that he expected the restaurants to make money and not merely serve as loss leaders to attract gamblers, that confirmed his reputation.

In the 1980s, Steve Wynn ushered in the next phase in Las Vegas by building flamboyant, glitzy hotels on the sites of dull old casinos. At an unheard-of construction cost of $630 million, the 3,000-room Mirage opened in 1989, with French restaurants and museum-quality artwork on display. Elegance had arrived in Las Vegas. Imitators quickly enlarged Caesars Palace and built the 4,000-room Excalibur and 5,000-room MGM Grand. All four projects baffled naysayers when they quickly turned hefty profits.

In the 1990s, as competition from Atlantic City and Indian reservations began to eat into profits, Las Vegas embarked on another expansion. Hoping to attract families to the Strip, the city created Disneyland-like attractions with amusement-park rides, video arcades, dragons, and water-park features. This effort fizzled; by 2007, only 10 percent of visitors arrived with children under the age of twenty-one. Las Vegas stopped trying to be something it wasn't and returned to its naughty roots.

In 1998, Steve Wynn set a new standard for opulence when he opened the $1.6 billion Bellagio, which stressed elegance, sophistication, and even high culture. The Venetian followed suit in 1999 with more than 4,000 luxury suites (even the regular rooms have a three-room bathroom area with thirteen light fixtures); seventeen restaurants featuring famous chefs such as Thomas Keller, Emeril Lagasse, and Wolfgang Puck; an opulent spa; high-end designer shops; 1.5 million square feet of meeting space; a 1,720-seat theater where Blue Man Group performs; and the Guggenheim Hermitage Museum, with original works by Picasso, Rubens, and Van Gogh. The era of the full-service, five-diamond, internationally recognized destination resort had descended on Las Vegas. Yet, in 2005, Wynn again raised the stakes when he opened Wynn Las Vegas, a 2,800-room casino-hotel that cost $2.8 billion—a whopping $1 million per room. Others have already called the bet and raised him. Current expansion plans shock even Wynn, who describes them as "the most outrageous, over-the-top expansion ever."

By 2005, Las Vegas had fifteen of the twenty largest hotels in the *world*, yet several are expanding, including the Venetian, which in 2008 added a 3,200-suite tower of hotel rooms and condominiums, making it the world's largest hotel, with 7,200 rooms. The granddaddy of them all, an MGM Mirage project called CityCenter, will cost $9.3 billion. The costliest privately financed development in American history, CityCenter will cover seventy-six acres on the Strip and include seven towers as high as sixty-one stories. Most developments have a construction trailer for supervisors; CityCenter has a $40 million two-building complex. The concrete required to build CityCenter could make a four-foot-wide sidewalk from Las Vegas to New York City and back again. At completion, CityCenter's population may reach 70,000.

Surprisingly, of the seven towers, only the ARIA Resort and Casino will have gaming. The others will feature condominium-hotels or simply residences. To many people around the United States, it will come as a shock to discover that some people actually want to spend serious money to buy a condo on the Las Vegas Strip. Lots of people. And for lots of money. The Mandarin Oriental, one of the seven towers within CityCenter, offers 227 condominiums on the upper floors, oversized hotel rooms and suites on the lower floors, a private owner's lobby, and a sky lounge at the very top. It began selling condos in January 2007. With units starting at $1.5

Figure I.2. Artist's rendering of CityCenter.
Sketch courtesy of Gordon Absher, MGM Mirage.

million and reaching an eye-popping $10 million, 90 percent of the condos sold in the first fourteen days. While the company refuses to release a profile of its buyers, more than a third come from outside the United States. What's driving them, along with our fellow Americans, to plunk down so much money for a piece of the Las Vegas Strip? How can Las Vegas possibly need more hotel rooms?

Las Vegas has become the premier convention destination in the United States. The city already has 151,000 hotel rooms and more than 9 million square feet of meeting space. And developers aren't finished: 11,000 hotel rooms are under construction, with another 35,000 on the drawing board. In 2006, an astonishing 39 million tourists visited Las Vegas. That's more than the combined population of New York, Pennsylvania, and New Jersey. (Disney World, the most popular entertainment destination on the earth, attracts 25 million visitors a year.) Las Vegas hosts conventions, fairs, expositions, rodeos, conferences, and trade shows. In 2006, it hosted 23,825 events, an average of 460 per week. Perhaps the best known is the annual Consumer Electronics Show, which draws roughly 150,000 people to see the latest electronic gizmos and hear Bill Gates spar with Google's founders. It seems as though every group in the country meets in Las Vegas, including funeral directors, celebrity impersonators, and pizzeria operators, who learn about the latest crusts, ingredients, and toppings. The World of Concrete Exposition attracted 85,000 people, but Magic International topped that with 120,000.

By no means are all gatherings large groups. The Alpaca Owners and Breeders Association had 700 delegates and the National Network of Embroidery Professionals 2,000. It's hard to tell whether the size of the gathering is an accurate gauge of public interest in the subject, but the Evangelical Lutheran Church in America managed to draw only 100 people, while the Adult Entertainment Expo attracted 37,000. Thanks to conventions such as these, Las Vegas' hotel room occupancy rate exceeded 93 percent in 2007, compared with a nationwide average of 63 percent.

But it's not just the numbers of people who descend upon Vegas, it's what they do while they're there: consume. Las Vegas visitors still gamble, but an increasing number are searching for other leisure-time activities, especially fine dining, top-drawer entertainment, indulgent spas, haute couture boutiques, and impeccable accommodations with every conceivable amenity. And let's not ignore the club scene, the latest example of Las Vegas reinventing itself. Groups of friends, whether sorority sisters from USC or lawyers from San Francisco, travel to Las Vegas for the weekend to go clubbing at night and recuperate at fashionable pool parties during the day. From all over the country, people descend on Las Vegas for bachelor parties, bridal showers, and divorce parties. One currently hot venue, known as Rehab, rents private cabanas on Sunday afternoons for $2,000 to $5,000. On a recent Sunday, forty people were on the waiting list. To compete, casinos have begun to offer what they call "European sunbathing." The Mirage's top-optional pool club is called Bare.

The epitome of trendiness is Tao Las Vegas. In 2006, its first full year of operation, the restaurant did a startling $55 million in business, making it the highest-grossing restaurant in the United States.

Photographs of fashionable celebrities visiting Tao Las Vegas regularly adorn the pages of *People* magazine. Its nightclub attracts those who need only a single name, such as Bono and Madonna, as well as Paris Hilton, Chelsea Clinton, and Tom Brady. For the opportunity to hobnob with such personalities, the club charges $300 to $5,000 for bottle service. When asked what you get for $5,000, co-owner Noah Tepperberg replied, "A table on the dance floor on a busy holiday weekend."

LAS VEGAS HAS GROWN AS though water weren't a problem, as if the city weren't in a desert. Meanwhile, it has exhausted its rights to Colorado River water from Lake Mead. Even though Sin City's tourists don't dwell on the city's dwindling water supply, Patricia Mulroy thinks about it every day. She's the general manager of the Southern Nevada Water Authority. An anomaly in a male-dominated, good-old-boy water world, Pat Mulroy more than holds her own. She is a physically fit, energetic, well-coiffed woman in her fifties who wears tasteful jewelry, silk blouses, and finely tailored jackets. Not exactly a pushover, Mulroy is known for her fiery disposition, her take-no-prisoners attitude, and her propensity to dismiss those who disagree with her as "brain-dead." Oh, yes, and she has unapologetically defended water use in Las Vegas, particularly on the Strip.

Mulroy has been scrambling to find additional water since she became general manager in 1989. She began by having the Water Authority file claims to groundwater located in aquifers scattered around the state, but these claims lay dormant until 2003. Then she began to look for water outside Nevada because she had run out of in-state options. Unlike its neighbors, California and Arizona, Nevada has few farmers who could fallow their land during drought emergencies.

So Mulroy has her hands full. Las Vegas is expected to add another 1.2 million people, equivalent to a city the size of Dallas, by 2020. "These people are going to come," she notes matter-of-factly.

"It is our job as the authority to make sure that there's an adequate water supply for them. And that is a delicate jigsaw puzzle."

Mulroy is desperate to find new sources of water. In 2005, she approached the cities of San Diego and Tijuana with an offer to build them treatment plants to desalinate Pacific Ocean water in exchange for a portion of their Colorado River allocation. Mulroy notes that if Las Vegas had its own coastal property, "we would be building our own desalter. But we don't." Instead, she sees the Colorado River as Las Vegas' best bet.

In 2007, when the Colorado River Basin states (California, Nevada, Arizona, New Mexico, Colorado, Utah, and Wyoming) finally accepted sweeping revisions to the allocation of Colorado River water that gave greater flexibility to Nevada, Mulroy exclaimed, "This is it!" One change allows Nevada to fund water conservation projects in other states and use the conserved water. Mulroy has moved aggressively to take advantage of these opportunities. For example, during the closing hours of the 109th Congress, she partnered with Nevada senator Harry Reid to persuade Congress to construct a new reservoir in Southern California. The water collected in this reservoir would come from managing the Colorado River system more efficiently.

It would work this way. If a farmer in Southern California's Imperial Valley orders water from the U.S. Bureau of Reclamation, that water is released from Hoover Dam. It takes three days for the water to reach the intake point for the canal that will carry it to the Imperial Valley. If it rains in the meantime, the farmer no longer needs to irrigate his fields, and the water is not diverted from the river to the canal. When that happens, the water flows down the Colorado River, unused by U.S. or Mexican farmers, into the Sea of Cortés, where it nourishes the Colorado River Delta estuary. But from the perspective of municipal water providers, that water is wasted. So Mulroy and Reid devised something called the Drop 2 Storage Reservoir, a structure designed to capture the water the farmer no longer needs and keep it from flowing into the Sea of Cortés. The new reservoir will be located in southeastern California, about twenty-five miles west of the Colorado River, alongside the All-American Canal. Mulroy agreed to fund it, to the tune of $172 million, in exchange for rights to 40,000 acre-feet of water per year for seven years. (An acre-foot is roughly 325,000 gallons.)

When I suggested to Mulroy that she seemed willing to sign a blank check in order to get water, she did not disagree. "Money goes a long way toward solving Colorado River water problems," she said. Rita Maguire, the former head of the Arizona Department of Water Resources, agrees. "They have more money than water . . . and they're using that money to bring more water."

As a seven-year drought in the Colorado River Basin continued into 2006, Mulroy turned her attention to the dormant applications for groundwater in central and eastern Nevada. The Southern Nevada Water Authority spent tens of millions of dollars purchasing ranches and water rights in Spring and Snake valleys, which extend into Utah. The Water Authority eventually hopes to transport 91,000 acre-feet of water to Las Vegas through a pipeline estimated to cost $2 billion.

The pipeline project involves a clash of lifestyles. Spring and Snake valleys, dotted with cattle ranches and alfalfa farms, cover a region bigger than Connecticut but have a population of only 10,000. And while some ranchers were ready to cash out and profit from Las Vegas' offers, others were not, including Dean Baker and his sons, who have been ranching in Snake Valley for fifty years.

Baker rejected Las Vegas' $20 million offer for his family's 12,000-acre cattle operation. "If it was just about the money, we'd sell to the Southern Nevada Water Authority." "But," he continues, "my family, particularly the boys, love farming. They enjoy the livestock, they like raising their family here and the children, so it's the values of here and not the money." Baker fears that Las Vegas will simply pump the water anyway, from neighboring ranches and other locations within the watershed, lowering the water table and ultimately destroying his ranch, which depends on a shallow water table for irrigation.

More than 830 parties filed protests to the Water Authority's applications for groundwater in the valleys, including the Church of Jesus Christ of Latter-day Saints, which owns a 4,000-acre cattle ranch in Spring Valley. The collision of interests pits Sin City's showgirls and slot machines against the Mormon Church, whose members are forbidden to smoke cigarettes, drink alcohol, or gamble. "Gluttony, glitter, girls, and gambling are what Las Vegas is all about," grumbles Cecil Garland, an eighty-one-year-old rancher in Callao, Utah. "What it's all about here is children, cattle, country, and church."

But Pat Mulroy counters Garland's concerns: "Ninety percent of Nevada's water goes to agriculture and generates a total of 6,000 jobs, which is less than the Mirage Hotel generates," Mulroy argues. "The West was settled by the federal government as an agrarian economy, but it isn't that anymore. The West is becoming an urban area."

Mulroy's plans to acquire additional water are interesting, but maybe, as critics suggest, Las Vegas should get its own house in order before plundering other areas. In fact, Pat Mulroy has moved aggressively to do exactly that. With one caveat: limiting growth is not an option.

In 2001, Mulroy had an epiphany. The Colorado River Basin states had, ironically, spent the 1990s fighting over how to divide up *surplus* water. Nevada entered into a ten-year agreement with Arizona to "bank" 1.2 million acre-feet of Colorado River water in Arizona's underground storage aquifers at a cost to Nevada of $330 million. But the situation deteriorated as the Colorado River drought worsened, and Arizona told Nevada that it no longer had excess water for Las Vegas. "Overnight, the world changed," said Mulroy. "We were out of water."

Mulroy responded with an about-face that resembled a religious conversion. The unrepentant apologist for growth remade herself as an apostle of water conservation. The combative advocate morphed into a kinder, gentler conciliator. She saw that the daunting task she faces will not be solved by legal, technical, or engineering solutions alone. The solutions are political and cultural. After twenty years at the helm of the Water Authority, this "European city girl," as she describes herself, has learned that "nothing is more emotional and nothing is more political than water in the West."

When Mulroy arrived in Las Vegas in the late 1980s, she found a city whose residents were, as she puts it, "living under the illusion that they were in a subtropical climate." The city's per capita water consumption of 350 gallons per day was double that of New York City, which gets ten times the rainfall of Las Vegas. Changing cultural assumptions about water use has proved a daunting task. In 2002, the Water Authority, faced with a worsening drought, implemented an ambitious conservation program targeting outdoor water use, which accounts for 70 percent of Las Vegas' water usage. The authority banned water-thirsty grass in front yards, limited the number of hours and days people could water, imposed water budgets on golf courses, and instituted fines for people and businesses that didn't comply. Quite innovatively, it encouraged homeowners to rip out lawns and install water-smart landscaping by offering them $2 for each square foot of grass that they remove. The program has been spectacularly successful. Mulroy proudly notes, "We've removed 80 million square feet of turf."

The Water Authority has also sponsored a publicity campaign to persuade citizens to conserve. My favorite, an ad that's making the rounds on the Internet, involves a public service announcement run on Las Vegas television stations. In it, a little old lady hobbles down a sidewalk, aided by a cane, past a lawn with sprinklers watering the sidewalk as well as the lawn. She walks to the front door and rings the bell. A young man answers, looks puzzled, because the woman is a total stranger, and asks if he can help her. In response, she gives him a swift, hard kick in the groin. As he doubles over in pain, a voice-over threatens, "We warned you not to waste water."

Mulroy faces challenges to conservation nonetheless, especially from the overuse of water to wash cars and to fill fountains. An exasperated Mulroy cautions, "Don't ever get between senior citizens and their car washing schedule." As a compromise, seniors can still wash their cars, but they must have a shutoff nozzle on the hose. When she proposed limits on fountains, to Mulroy's complete amazement, "people went nuts." She mockingly asks, "Are you going to stop banking with Wells Fargo because they shut their fountain off in front of the bank building?"

Las Vegas has also imposed tiered water rates to encourage conservation, but the rate increases were quite modest. The average Las Vegas household uses 17,000 gallons in a typical summer month but pays only $37, or about two cents per 10 gallons. Las Vegas residents use more water per person than do residents of other western cities, such as Tucson and Albuquerque, that charge sharply higher rates. Mulroy is not convinced that higher rates would stimulate greater conservation. "Look at what's happening with gasoline: people are not using less gas as a result of price hikes." In contrast to Mulroy's claim, sales of fuel-efficient cars, such as the Toyota Prius, are way up. Nevertheless, she deserves great credit for the progress the city has made. Between 2002 and 2006, Las Vegas slashed its water demand by more than 18 billion gallons a year, even though its population grew by 330,000.

Okay, so she's taken on the seniors and the lawn lovers, but what about the Strip? How can Pat Mulroy justify the casinos' fountains, ponds, volcanoes, rain forests, and canals? Quite easily, actually. The biggest misconception about Las Vegas' water use is that the hotels waste water, says Mulroy. "The entire Las Vegas Strip uses three percent of our water. And they are the economic driver in Nevada, bar none. It's just that their water use is visible and looks wasteful."

Mulroy pitched water conservation to casino owners and developers as a business issue. If they didn't conserve, the city would run out of water. No water, no tourists. The transformation began with the Mirage and Treasure Island casinos in the late 1980s. She remembers Steve Wynn saying, "Pat, I just have to have a water feature. I just have to. Don't tell me I can't do it. Just tell me how I can do it." So she challenged him: "Are you willing to tap the contaminated groundwater supply and double plumb the hotels?" He was. Wynn spent the extra money required to build a state-of-the-art reverse osmosis wastewater treatment system in the parking structure beneath Treasure Island. The dual plumbing system captures water from the casinos' sinks and showers. After treatment, it's used for the pirate lagoon and volcano.

Wynn set the standard for innovative water use on the Strip. Since then, the other casinos have followed suit by recycling their water, using low-flow fixtures, and installing drip irrigation. In the city of fantasy, shower aerators create the illusion of abundant water pressure while reducing water use. Each aerator saves 6,000 gallons per year. Some water-saving practices probably escape the attention of most tourists. On-demand water heaters magically provide hot water the moment hotel guests turn on a faucet, rather than taking upward of a minute as gallons of water go down the drain. With reused water and recirculating pumps, the Strip's water features cater to tourists' demand for a surreal world. The tourists don't seem to mind. "Everything that hits the sewer system is recycled," Mulroy boasts, "unlike the situation in San Diego, which just dumps it into the ocean. We recycle 100 percent of our water." The fountains at Bellagio use water from wells beneath the Bellagio footprint that previously irrigated a golf course. Quite remarkably, the 3,933-room Bellagio uses less water than the golf course did. The largest consumptive use of water is to run the resorts' immense air-conditioning systems.

Innovative water use planned for CityCenter includes water features that will be sealed to reduce evaporation. Ten percent of CityCenter's electricity will come from its own on-site heat and power cogeneration plant, which will capture the heat by-product of producing energy and reuse it to heat water, thus reducing the amount of water needed to produce electricity.

Despite these grand conservation successes, Pat Mulroy is not yet finished searching for other ways to "augment" her supply, a euphemism for water importation. Las Vegas footed the bill for the consulting firms CH2M HILL and Black & Veatch to explore all viable options for securing more water. Their report, released in 2008, explored some "crazy" ideas, says Kay Brothers, Mulroy's chief deputy, such as shipping water from Alaska and building a pipeline to get water from the Columbia River, on the border of Washington and Oregon.

But the one that most fascinates Mulroy involves importing Mississippi River water. When I spoke with her in her nicely appointed corner office at the Southern Nevada Water Authority, she became very animated about the future. Asking me to look forward twenty years and to think outside the box, she focused on the junction of the Mississippi and Ohio rivers. Each year, some 436,000 million acre-feet of water flow by. Just imagine, she told me, a pipeline that diverted 6 million acre-feet per year and transported it west through Missouri, Kansas, and Colorado, then south across New Mexico and onto the Navajo Nation, then west across Arizona and into Nevada, where it would augment Las Vegas' supply. This proposal took my breath away.

In years past, such grandiose schemes have occasionally surfaced, usually dreamt up by wild-eyed government engineers, idly doodling designs while buried in obscure Washington, DC, offices. No one ever took them seriously. But Pat Mulroy is a major player, perhaps the most important figure in western water and an oft-rumored gubernatorial candidate. She gets things done. Mulroy argues that such a project could replenish the Ogallala Aquifer, help the Navajo Nation, reduce the flood threat to New Orleans, and, not coincidentally, provide water to Las Vegas. The day after our meeting, David Donnelly, an engineer for one of Las Vegas' consulting firms, floated this idea at a conference about the Colorado River. He conceded that "the institutional issues are significant," engineer-speak for a nightmare of environmental permitting, a political bloodbath with the other states, sustained protests from environmentalists, daunting engineering challenges, and an initial price tag of $11 billion. Among other obstacles is a physical one: the Rocky Mountains. The pipeline would need to move the water more than 6,000 feet in elevation—more than a mile straight up in the air—to get it over the Rockies.

As Las Vegas remakes itself again, it remains to be seen whether CityCenter paves the way for a new urban West that will grow up rather than out. Pat Mulroy sure hopes so. "I love high-rises," she says. To Mulroy, their small footprint and minimal outdoor landscaping "are much more sustainable for southern Nevada." Las Vegas is both attractive and repellent at the same time, even to the same people. Its rapacious, self-conscious ideology of growth, its love of all things glittery and banal, its boundless 24/7 energy, and its seductive allure ensure it will always have its boosters and detractors. For those of us watching from afar, it's a sober lesson of a city that has run out of water yet charges ahead full of dreams for the future that would make Bugsy proud.

"WHEN THE WELL'S DRY, we know the worth of water," observed Benjamin Franklin in 1774. But he was wrong. In the United States, we utterly fail to appreciate the value of water, even as we are running out. We Americans are spoiled. When we turn on the tap, out comes a limitless quantity of high-quality water for less money than we pay for our cell phone service or cable television. But as we'll see, what is happening in Vegas is not staying in Vegas. It's becoming a national epidemic.

Ignorance is bliss when it comes to water. In almost every state in the country, a landowner can drill a domestic well anywhere, anytime—no questions asked. Many states don't even require permits for commercial wells unless the pumping will exceed 100,000 gallons a day (that's 36 million gallons annually). For each well. We know so little about this pumping that the federal government cannot even estimate the total number of these wells across the country.

In many agricultural regions where the government does know the number of wells, such as California's Central Valley, it is still clueless as to how much water farmers pump out of those wells, because they're unmetered.

Water is a valuable, exhaustible resource, but as Las Vegas did until just a few years ago, we treat it as valueless and inexhaustible. Just as the energy crisis brought to the nation's consciousness an acute awareness of energy consumption, global warming, and carbon footprints, so too the impending national water crisis will inspire us to rethink how and why we use water.

My aim in this book is to explore the crisis and to stimulate that rethinking. Part of the problem is that water shortages in many parts of the country, lacking the exhibitionist tendencies of Las Vegas, are often hidden. This book will illustrate the true dimensions of the crisis and offer solutions to it. Alas, the dimensions are immense.

Water lubricates the American economy just as oil does. It is intimately linked to energy because it takes water to make energy, and it takes energy to divert, pump, move, and cleanse water. Water plays a critical role in virtually every segment of the economy, from heavy industry to food production, from making semiconductors to providing Internet service. A prosperous future depends on a secure and reliable water supply. And we don't have it. To be sure, water still flows from taps, but we're draining our reserves like gamblers at the craps table.

We tend to look at Las Vegas and think it's a unique case, perhaps a cautionary tale but barely relevant to where the rest of us live. But the truth is, when it comes to water, Vegas offers us a glimpse of our own future. The evidence is everywhere—though if it is noticed, it is forgotten with the next drenching rain. Consider the following events that have occurred since 2007:

- Colorado farmers watched their crops wither because of a lack of irrigation water.
- Atlanta, Georgia, came within three months of running out, so

it banned watering lawns, washing cars, and filling swimming pools.
- Orme, Tennessee, did run out and was forced to truck water in from Alabama.
- Scientists at the Scripps Institution of Oceanography predicted that Lake Mead, which supplies water to Los Angeles and Phoenix, could dry up by 2021.
- Hundreds of workers lost their jobs at Bowater, a South Carolina paper company, because low river flows prevented the plant from discharging its wastewater.
- Lack of adequate water prompted the Nuclear Regulatory Commission to rebuff Southern Nuclear Operating Company's request to build two new reactors in Georgia.
- Water shortages caused California farmers to cut the tops off hundreds of healthy, mature avocado trees in a desperate attempt to keep them alive.
- Lake Superior, the earth's largest freshwater body, was too shallow to float fully loaded cargo ships.
- Decimated salmon runs prompted cancellation of the commercial fishing season off the coasts of California and Oregon.
- A lack of adequate water led regulators in Idaho, Arizona, and Montana to deny permits for new coal-fired power plants.
- In Riverside County, California, water shortages forced a water district to put on hold seven proposed commercial and residential developments.

To understand the depth of the water crisis, consider that more than thirty-five of the lower forty-eight states are fighting with their neighbors over water.

Our existing supplies are stretched to the limit, yet demographers expect the U.S. population to grow by 120 million by midcentury. Before the crisis becomes a catastrophe, we must embark in a fundamentally new direction. Business as usual just won't cut it. We have traditionally engineered our way out of water shortages by building dams, diverting rivers, and drilling wells. But proposals for new dams engender immense political and environmental opposition, diversions have already dried up many rivers and reduced the flow in others to a trickle, and groundwater tables are plummeting around the United States. Meanwhile, the environment suffers as excessive water use causes springs, creeks, rivers, and wetlands to go dry, salt water to contaminate potable supplies, the ground to collapse, and sinkholes to appear. Even lakes are not immune. Dozens in Florida have already gone dry.

Are there alternatives to business as usual? Some dreamers offer grandiose plans that include seeding clouds and towing icebergs from Alaska, but these are not viable options. We can expand the supply by reusing municipal effluent and by desalinating ocean water, but neither of these choices is a panacea. On the demand side, we can encourage water conservation. In some water-wasteful regions, conservation has great potential; however, many water-stressed communities have already implemented ambitious conservation programs but need to reduce demand even more. The reality is that reusing, desalinating, and conserving water may help to alleviate our crisis but will not solve it. We must find other ways to free up water. Las Vegas has pioneered very expensive solutions, but they can succeed only by taking water from other places. Is this sustainable?

In his 2005 book *Collapse*, Pulitzer Prize–winning author Jared Diamond describes how flourishing societies have precipitously collapsed. Examining spatially and temporally diverse cultures, such as those of Easter Island in the South Pacific, Norse settlements in Scandinavia, and the Anasazi in North America, Diamond finds a disturbing pattern, one that resembles contemporary conditions in the United States. As these societies grew and flourished, they mismanaged natural resources, eventually stretching the resources' carrying capacity to the breaking point. Still, the societies continued on in their customary practices, assuming that what they were familiar with was the norm. Then something happened—environmental damage, climate change, hostile neighbors, loss of trading partners, or the culture's own response to its environmental problems—to change the familiar, but it was too late for the society to correct course and avert a catastrophe. With the Anasazi, a growing population depended on ever-increasing use of water and firewood. When a sustained drought hit in the twelfth century and lasted more than fifty years, the society collapsed.

We, however, still control our destiny. The United States is entering an era of water reallocation, when water for new uses will come from existing users who have incentives to use less. Sounds good, but how will this happen? One possible approach is for the government to target wasteful practices by simply prohibiting current water users from using so much. However, heavy-handed government mandates would generate bitter political controversy and endless litigation. What we can do, yet haven't done, in the United States is encourage water conservation by using price signals and market forces. Pricing water appropriately would stimulate all users to reexamine their uses and decide for themselves, on the basis of their own pocketbooks, which uses to curtail and which to continue. The government should encourage a voluntary reallocation of water between current and new users. The alternative is to fight over the water. Which do we prefer?

Water nourishes our bodies and our souls. Our lives are impoverished without the sight, sound, smell, and touch of bubbling brooks, cascading waterfalls, and quiet ponds. The terrifying future depicted in science fiction doomsday novels conspicuously features barren landscapes. Our future needn't be so bleak. Our water crisis should occasion grave concern but not panic. We have solutions available; now we need a national commitment to pursue them.

Part One

The Crisis ~

CHAPTER 1

Atlanta's Prayer for Water

"Water sustains all."
—Thales of Miletus, 600 BC

IN OCTOBER 2007, Atlanta's watershed commissioner, Rob Hunter, issued a dire warning: if it didn't rain, Atlanta would run out of water in four months. His was the optimistic estimate. Other federal and state officials predicted that Lake Lanier, the principal water supply for almost 5 million people in Metro Atlanta, could go dry in three months. A sustained two-year drought had dropped the lake's level by fifteen feet, leaving docks and boathouses high and dry, exposing tree stumps not seen since the lake was first filled, fifty years earlier, and creating red mudflats below what used to be swimming beaches. And Lake Lanier is not some dinky puddle: its surface covers 38,000 acres, almost twice the size of Manhattan. Lanier is one of America's favorite lakes—more than 7.5 million people a year enjoy boating, fishing, waterskiing, and jet skiing as they patronize its marinas, water parks, and resorts. Vacation homes crowd the 692 miles of shoreline.

As Lanier's waters shrank, Georgia announced a Level 4 drought emergency and banned all outdoor watering except for agricultural and "essential" business uses. Governor Sonny Perdue ordered

Figure 1.1. Lake Lanier, fall 2007.
Photograph courtesy of Robert Elzey.

North Georgia businesses and utilities to cut water use by 10 percent. Atlanta's mayor, Shirley Franklin, begged of her constituents, "This is a not a test. Please, please, please do not use water unnecessarily." Concerned about water supplies, the Paulding County Board of Commissioners, in Metro Atlanta, imposed an indefinite ban on new rezoning requests. The development community went bonkers. Michael Paris, head of a pro-growth organization, argued that telling developers to stop building would be like telling Metro Atlanta families to stop having children because there's no more water. "That's how silly [halting development] is," said Paris.

The restrictions caught the business community by surprise. Sam Williams, president of the Metro Atlanta Chamber of Commerce, called the drought the top threat to Atlanta's economy and warned it could be a "dress rehearsal" for what the future holds. The Coca-Cola Company has a major presence in Atlanta, as does PepsiCo, with a Gatorade plant that is the largest water user in Atlanta. Bruce A. Karas, Coca-Cola's vice president for sustainability, sounded an alarm. "We're very concerned. Water is our main ingredient. As a company, we look at areas where we expect water abundance and water scarcity, and we know water is scarce in the Southwest. It's very surprising to us that the Southeast is in a water shortage."

In Georgia, as in the rest of the United States, water is essential not only to human life. It is vital to the entire economy, not just for companies such as Coca-Cola and Kellogg but also for less obvious industrial operations, including automobile manufacturers; steel plants; copper, gold, and coal mines; defense industries; and semiconductor manufacturers, as well as for the energy sector, which includes oil and gas companies, refineries, power plants, and hydroelectric generating facilities. We may worry loudly about the price of oil, but water is the real lubricant of the American economy.

This is most evident in high-growth states such as Georgia. The largest state east of the Mississippi River, Georgia is blessed with extraordinary water resources: 70,000 miles of streams, 400,000 acres of lakes, 4.5 million acres of wetlands, an additional 384,000 acres of tidal wetlands, 854 square miles of estuaries, and 100 miles of coastline, all nourished by an average annual rainfall of forty-nine inches. Georgia's problem comes not from a lack of water but from uneven distribution of the water. Most of it is in the southern part of the state, whereas Atlanta (and most of the state's population) is in the northern part.

As one of the fastest-growing states in the country, with a population approaching 9 million, Georgia expects to be home to another 2 million people by 2015. Population in Metro Atlanta is expected to increase by 50 percent by 2030. To grasp these statistics, consider Atlanta's traffic congestion woes. Metro Atlanta has sprawled in all directions, gobbling up fifty-five acres a day, with the city itself in the middle, constricting the movement of commuters from one side to another. Commuting time has grown faster than in any other city. Traveling ten miles can take forty-five minutes, and some commutes involve fifty miles each way. A portion of Interstate 75 that skirts Atlanta is currently fifteen lanes wide, putting it at the top of the Federal Highway Administration's list of America's biggest highways, yet expansion plans will widen that section. Once completed, it will have twenty-three lanes and be 388 feet wide, more than the length of a football field.

As Georgia's population climbed from 3.4 million in 1950 to 8.2 million in 2000, the state's use of water rose from 150 million to 1.3 billion gallons per day. And in southern Georgia, an additional and unexpected change in water use has come from the rise of irrigated agriculture. Irrigation, once confined to the arid West, where artificial watering supplements natural rainfall, has spread as farmers in the Midwest, East, and Southeast have discovered that applying more water to their fields than Mother Nature provides can boost crop production. Center-pivot systems, those big circles that airplane passengers can see from 35,000 feet, have proliferated in central and southern Georgia. As a result, agricultural water use, mostly from groundwater, grew twelvefold between 1950 and 1980 and has doubled since then, to 1.1 billion gallons per day. A spike in acreage planted with corn, driven by the ethanol boom, is further increasing agricultural water use. Georgia's first ethanol refinery, operated by First United Ethanol, will require roughly 500 million gallons of water per year.

Georgia has also become a second-home haven for retirees from the North and for others who are increasingly choosing Georgia over Florida. Seventy-five miles east of Atlanta is Lake Oconee, an enormous reservoir created in 1979 by Georgia Power to serve a hydroelectric plant. The area is booming, with more than 100 subdivisions and developments featuring gated communities with lushly watered golf courses. Homes, from one-bedroom condominiums to immense estates, line the 374 miles of shoreline. The major developers have plans for thousands of new homes.

The turf industry satisfies Georgia homeowners' demand for impeccably manicured lawns. Indeed, much of the increase in agricultural irrigation comes from more than two dozen turf farms that grow grass for Atlanta's suburbs. With typical American impatience, new homeowners don't want to plant grass seed when they can have mature lawns rolled out in sheets before they move in. As the drought in North Georgia continues and as watering bans take their toll on existing lawns, one perverse result will be the use of more water to grow turf. As Georgia's state geologist, Jim Kennedy, explained, sod production will increase "because when your lawn dies, you will need to replace it." Georgia developers also find imaginative ways to use water, including building artificial lakes in subdivisions to induce waterskiing enthusiasts to buy homes. "Atlanta's unrestrained growth and cavalier attitude to water use has got to be on the table," complained Sally Bethea, executive director of Upper Chattahoochee Riverkeeper, an organization dedicated to protecting the river.

But rather than confront the problem at its core, as the drought worsened, Governor Sonny Perdue played the blame game, claiming that Atlanta's water woes were due to the U.S. Army Corps of Engineers releasing too much water from Lake Lanier to protect three endangered species—two types of mussel and the Gulf sturgeon—that live downstream in the Apalachicola River in Florida. On top of the natural drought, Perdue asserted, "we are mired in a man-made disaster of federal bureaucracy." The governor asked a federal court to issue an injunction to halt the releases from Lanier. He also appealed to President George W. Bush to grant emergency drought relief and to use his powers to exempt Georgia from the Endangered Species Act. Meanwhile, Georgia's United States senators, Johnny Isakson and Saxby Chambliss, as well as members of its congressional delegation, introduced legislation to create a temporary exemption from the act. "Blaming the endangered fish and mussels for our water woes," a University of Georgia ecology professor responded, "is as silly and misdirected as blaming the sick canary for shutting down the mine."

Things had become so dire by November 2007 that Governor Perdue tried another tack. Outside the Georgia Capitol, he led several hundred ministers, legislators, landscapers, and office workers in prayers for rain. Holding Bibles and crucifixes, the group linked arms and sang "What a Mighty God We Serve" and "Amazing Grace." The governor repented, "Oh, Father, we acknowledge our wastefulness," and promised that Georgians would do better in conserving water. The governor, not one to take chances, timed the prayer service to coincide with weather forecasters' predictions of the first rain-bearing front in months.

Georgia's downstream neighbors are angry about its water use. Florida's governor, Charlie Crist, and Alabama's governor, Bob Riley, asked President Bush to reject Governor Perdue's request for emergency relief. Riley called any reduction in releases from Lake Lanier "a radical step that would ignore the vital downstream interests of Alabama."

The three states have been locked in a bitter fight over water since 1990, when Florida and Alabama filed federal lawsuits to stop Metro Atlanta from taking more water from Lake Lanier. In 1997, the three states reached an agreement, known as the Apalachicola–Chattahoochee–Flint River Basin Interstate Compact (ACF), which was nothing more than a mandate for them to develop a formula for allocating the waters of the three rivers by 1998. That never happened. The negotiations collapsed because Georgia had no incentive to curb its water use. As the upstream state, it has physical control over the water. If Georgia holds back water for its own use, there is nothing—short of war or litigation—that Florida and Alabama can do about it. Thus, in 2004, amid much finger-pointing, Florida and Alabama revived their 1990 lawsuit and the three states went back to court. The three governors seemed to reach a truce in November 2007, when they agreed to settle their differences by February 15. This agreement prompted one cynic to jest, "Of what year?" Indeed, a week after the governors' meeting, Florida backed away from the truce in the tristate water war.

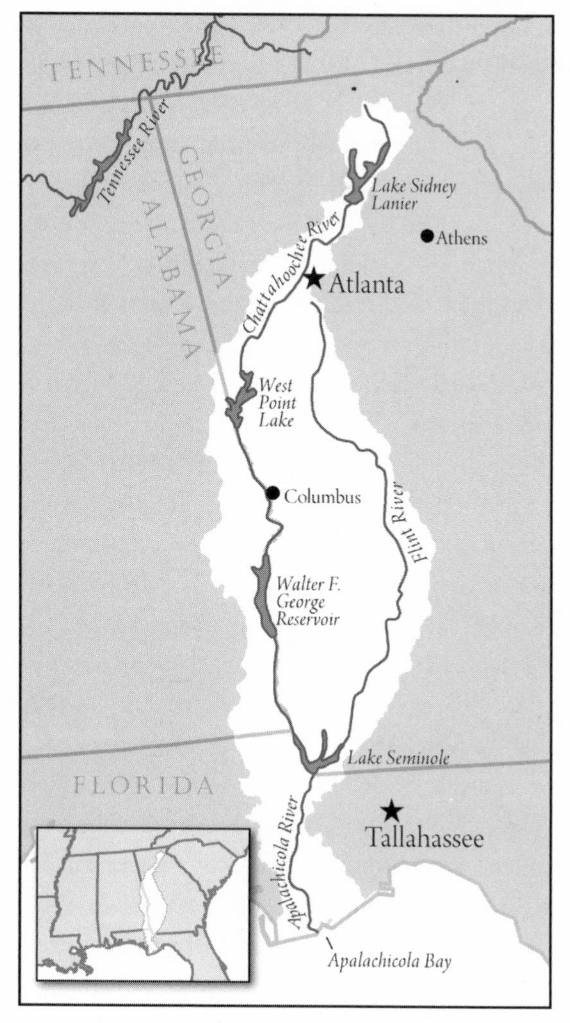

Figure 1.2. Apalachicola–Chattahoochee–Flint River Basin.

When the U.S. Army Corps of Engineers decided, in November 2007, that it could temporarily reduce releases from Lake Lanier without jeopardizing the endangered species, Florida commenced new legal action. In February 2008, the U.S. Court of Appeals sided with Florida and Alabama in ruling that Georgia could not withdraw as much water as it wanted from Lake Lanier.

From Florida's vantage point, greater diversions from Lake

Lanier mean lower flows downstream in the Apalachicola River. The fresh water plays a critical role in sustaining Florida's $134 million commercial oyster industry. Michael Sole, secretary of the Florida Department of Environmental Protection, complained that cutting river flows would cause a "catastrophic collapse of the oyster industry in Apalachicola Bay" and "displace the entire economy of the Bay region." Florida's concerns did not make much of an impression on Jackie Joseph of the Lake Lanier Association, who shrugged and said, "We have a $5.5 billion economy around this lake."

Such attitudes have alienated people outside Metro Atlanta and even other Georgians. The *Valdosta Daily Times* editorialized: "Atlanta is a greedy, poorly designed behemoth of a city incapable of hearing the word 'no' and dealing with it. Atlanta's politicians can't bring themselves to tell their greedy constituents complaining about the low flows in their toilets this week that perhaps if they didn't have six bathrooms, it might ease the situation a bit."

Georgia has also squared off against Alabama over the Alabama-Coosa-Tallapoosa River Basin (ACT). Metro Atlanta gets a substantial amount of its water from Allatoona Lake, which impounds the waters of the Etowah River. The Etowah River eventually becomes part of the Coosa River before it flows into Alabama. From Alabama's perspective, water flows in the Coosa River are critical to its downstream businesses, including the Joseph M. Farley Nuclear Plant, which supplies 20 percent of Alabama Power's electricity. Georgia-Pacific Corporation paper mills and Gulf Power hydroelectric plants would also suffer from lower flows. As with the ACF negotiations, ACT negotiations haven't produced a solution.

Eventually, the United States Supreme Court may step in. It has sole jurisdiction over disputes between states, and, in a set of rulings that mostly involved western rivers, it has divided up interstate rivers. Most ominously from Georgia's perspective, the Supreme Court has ruled that it is "essentially irrelevant" whether the headwaters of a river begin in one state or another. It has not permitted upstream states to hoard water that their downstream neighbors need. A state's record of water conservation and efficiency may affect the Court's allocation. On this front, Georgia is vulnerable.

In a separate battle raging on the other side of the state, Savannah, Georgia, has incurred the wrath of Hilton Head, South Carolina. David Baize, assistant chief of South Carolina's Bureau of Water, notes that "the whole coastal region, including Savannah, is just the definition of sprawl." Driving this growth is the Port of Savannah, the fourth-busiest and the fastest-growing container terminal in the United States. Target, IKEA, and Heineken recently opened distribution centers to take advantage of the port. Baize's agency is charged with issuing permits for groundwater wells in South Carolina, including for the Hilton Head area, a popular tourist destination. Before Savannah's increase in development, groundwater flowed through the Upper Floridan Aquifer and discharged into the ocean at Port Royal Sound in South Carolina. The Savannah area uses six times the groundwater that Hilton Head does, and its collective pumping has reversed the direction of the flow of groundwater. The gradient created by the pumping has caused salt water to migrate laterally and contaminate freshwater wells.

Richard Cyr, general manager of the Hilton Head Public Service District, notes that the utility had to shut down five of its twelve wells because chloride levels, as a result of saltwater migration, exceeded the maximum federal contaminant level. Because Savannah's wells are in a deeper part of the aquifer, which the salt water has yet to reach, Savannah has thus far avoided Hilton Head's fate. "We're the damaged party in this," says Baize. "It is our wells that are being lost and it is our aquifer that's being contaminated." South Carolina considered litigation but decided that the money would be better spent seeking alternative water sources. But this water won't come cheaply. According to Cyr, the two utilities that serve Hilton Head have spent $90 million constructing a reverse osmosis saltwater desalination plant to offset their loss of water from the Floridan Aquifer.

In 2008, reduced flow in the Savannah River led the Nuclear Regulatory Commission's Atomic Safety and Licensing Board to withhold approval of Southern Nuclear Operating Company's request to build two new reactors that would withdraw as much as 83 million gallons per day.

Even Tennessee is unhappy with Georgia's plans for finding more water. After Atlanta's mayor, Shirley Franklin, suggested piping in water from the Tennessee River, Tennessee's governor, Phil Bredesen, responded: "I would have a real problem with a wholesale transfer of water out of the Tennessee watershed." If Georgia lawmakers have their way, that will soon happen. In 2008, the Georgia legislature passed a resolution claiming that an "erroneous" survey in 1818 mistakenly located the border between Georgia and Tennessee approximately one mile south of where it should be. If the border is moved northward, a bend in the Tennessee River would become part of Georgia, giving Georgia unfettered access to the river. On hearing of Georgia's desire to alter the boundary nearly two centuries after the survey, Governor Bredesen responded, "This is a joke, right?" Not in Georgia.

Even before the drought commenced, Georgia needed more water. Plans were already under way to divert an additional 126 billion gallons from the Chattahoochee and Etowah rivers and to build four more reservoirs and expand others, which collectively would hold an additional 33 billion gallons. Glenn Richardson, speaker of the Georgia House of Representatives, pledged his full support. "Frankly, we should have been doing this before now," he said. But given that each and every one of these gallons would travel downstream if it weren't captured by Georgia, it is unlikely that Florida and Alabama will agree with the speaker. These increased withdrawals will severely compromise commercial and sport fisheries, recreation, wildlife habitat, and water quality. Water pollution, a serious problem in the Chattahoochee River near Atlanta, will worsen as increased withdrawals reduce the river's dilution capacity.

As Georgia confronts this severe drought, what is most striking is what the state is not doing: it is not restricting new uses. Georgia continues to approve water permits on a first-come, first-served basis. And that's only when permits are actually required. By state law, there is no need for a permit to divert water from a river or to pump water from a well unless the use exceeds 100,000 gallons a day. The consequence is a booming well-drilling business during one of the Southeast's worst droughts on record. "We could run seven days a week if we wanted to," observes Wes Watson, owner of a small company that drills private and commercial wells. In Watkinsville, Dan Elder, another well driller, used to get three to five calls a week from prospective clients. Now he's getting that many calls a day, and he has a six-week backlog.

Even as conservation restrictions have kicked in, some Georgians want to enjoy the same lush landscapes and green lawns they did before the drought. As a result, the RainHarvest Company, a suburban Atlanta business that installs systems to capture rain, has seen its business quadruple during the drought. One of the company's founders, Paul Morgan, philosophically observes that people "don't want to change their lifestyle." Other Georgians have found a way to maintain green lawns even without Morgan's help. In Atlanta's northeastern suburbs, U-Spray, a do-it-yourself garden center, has suffered declines in the sales of its outdoor products, with one exception: lawn paint. Sales have doubled for paint to make brown lawns green.

This reminds me of a *New Yorker* cartoon that shows a handsome young man gazing at his reflection in a pond as his beautiful young female companion plaintively inquires, "Narcissus, is there someone else?" In fact, Georgians are no more self-absorbed than the rest of us. Well, maybe some of them are. Consider Chris G. Carlos, a member of a prominent Atlanta family who used 440,000 gallons of water in September 2007 at his 14,000-square-foot home on four acres in the Atlanta Country Club. That's enough water to fill the average backyard pool fifty-eight times. After the *Atlanta Journal-Constitution* broke the story, Carlos repented: "I honestly didn't recognize the extent of my water use and regret I didn't act sooner." Yet November 2007 water meter readings on six consecutive days by the

Cobb County Water System found that Carlos was still using roughly 2,000 gallons per day. That's almost ten times what the average Cobb County household uses. But Carlos broke no law, thanks to an exemption for watering by licensed professional landscapers. Nor was using so much water that expensive. His water bills averaged $1,200 a month that year, hardly a big deal for a wealthy investor.

Governor Perdue and most other Georgians seem to believe that once the heavens open, whether from divine intervention or on their own, Georgia's water woes will be over. To be sure, the 2007 drought has had very serious consequences, but Todd Rasmussen, professor of hydrology and water resources at the University of Georgia, points out that "this drought is not particularly different from previous ones." In eight years since 1950, flow levels in northern Georgia rivers have been as low or lower than the 2007 flows. If that's so, why is 2007 so different? It is different because the demand for water has risen, observes Rasmussen, from "burgeoning communities, increased center-pivot irrigation by agriculture, and substantial growth in thermo-electric and nuclear power production." In past droughts, there was enough water to satisfy human demands—but not this time. Nature is not to blame; we are. Unless Georgia alters its water use, Rasmussen predicts, Metro Atlanta will again face a similar predicament. The question is not whether a crisis will occur but when.

The story of Georgia's water plight illustrates both how much we take water for granted and how critical water is to our economic well-being. But if the past is any guide to the future, once rainfall returns to Georgia and slackens the drought, Metro Atlanta will forget all about it . . . until next time.

By May 2008, winter and spring storms had added two feet to Lake Lanier's level, but the lake remained thirteen feet below normal and only a few inches from the lowest level ever recorded for that time of year. The area remained in "extreme" drought, which, according to forecasters, only monumental storms would end. Yet

Governor Perdue relaxed the restrictions on watering plants and filling swimming pools imposed by his own Environmental Protection Division. A three-month, 15 percent drop in water usage persuaded the governor that Georgians had embraced a "culture of conservation," so he could tell Georgia's children, "Swim, kids, swim." The hydro-illogical cycle captures this human capacity to ignore reality.

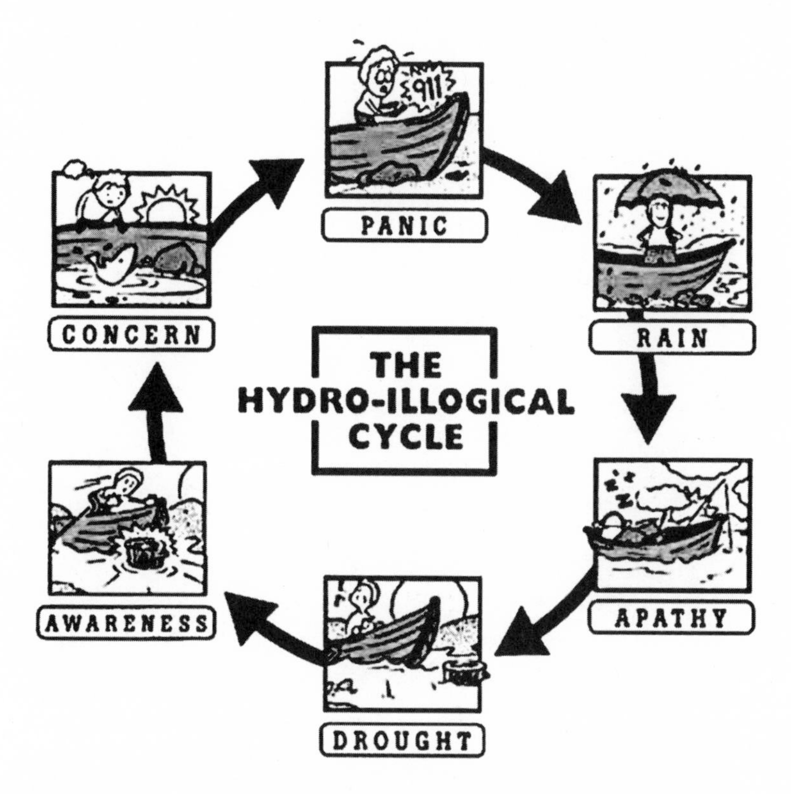

Figure 1.3. Cartoon of the hydro-illogical cycle. Courtesy of Martha Witaker.

CHAPTER 2

Wealth and the Culture of Water Consumption

"If the human body is 60 percent water, why am I only two percent interested?"

—Stephen Colbert

GEORGIA HAS ITS SHARE OF outsized water demands, but the hands-down winner of "the most outrageous use of water in Georgia award" is Stone Mountain Park, a theme park in an Atlanta suburb that attracts 4 million visitors a year. Stone Mountain, a granite dome, became the South's answer to Mount Rushmore, with three figures of Confederate heroes of the Civil War—Stonewall Jackson, Robert E. Lee, and Jefferson Davis—carved into the stone. The second Ku Klux Klan began atop Stone Mountain in 1915. In January 2007, long after the drought began, the park announced the opening of a major new attraction: Coca-Cola® Snow Mountain, featuring deep snow covering a 400-foot tubing hill and a 13,000-square-foot play area. Stone Mountain Park's vice president and general manager, Gerald Rakestraw, explained the marketing angle behind this new attraction. "Research indicated that nearly 70 percent of Atlanta area kids had never experienced real snow, and only 40 percent of Atlanta families have experienced snow as a family unit. Snow Mountain seemed like a great opportunity to address that gap and give Atlanta families a first-of-its-kind snow attraction close to home." The park planned to make more than 200 tons of fresh snow daily. Anthony Esparza, vice president of guest experience, design, and development for Herschend Family Entertainment, which operates Stone Mountain Park, elaborated: "This is the first time in the U.S. an outdoor snow park of this scope has been created in this type of climate."

The park, which invested almost $5 million in planning the attraction and purchasing snowmaking equipment, scheduled the grand opening for the fall of 2007. When Governor Sonny Perdue banned all nonessential outdoor water use on September 28, Stone Mountain management was undeterred. It began making snow on October 1, an eighty-one-degree day. It initially defended the snowmaking as an essential part of its business, and therefore exempt from the state ban, but halted the process after the *Atlanta Journal-Constitution* ridiculed the project. This folly consumed 1.2 million gallons of water, or enough for 106,000 showers.

WHEN IT COMES TO WATER, more than with any other commodity, we throw common sense out the window. We consume water as if it had no value, and we consume it in the most ridiculous ways imaginable.

It's impossible to talk about consumption without talking about population, and I'll begin by acknowledging the elephant in the room. Population growth is the driving force behind all demands for new water and every environmental problem in the United States. Between 2000 and 2007, the population of the United States surged from 285 million to 300 million, with the Southwest leading the way. California's Department of Finance expects the state to add 600,000 new residents per year, increasing its population from 38 million to 60 million by midcentury. To put this population growth in perspective, consider that California adds one new resident per minute and that California's growth rate lags behind that of both Nevada and Arizona. For the country as a whole, the U.S. Census

Bureau projects that our population will reach 420 million by 2050. In the next forty-three years, we will add 120 million people, the equivalent of one person every 11.3 seconds. In the time it took you to read this paragraph, three or four more people joined our ranks.

Now let's consider the regional water consequences of these demographic patterns, a concept pithily captured when I tell you, "I moved to the Sun Belt in 1985." So have lots of other people, as well as to Florida, Georgia, and North Carolina. In 2006, North Carolina bumped New Jersey from the list of the ten most populous states, and New York's population declined for the first time since the 1970s.

Michigan, where I lived for eleven years, is suffering, as everyone knows, from the humiliation of Toyota capturing the number one spot and Daimler-Benz giving Chrysler away at rummage sale prices to the sad spectacle, in 2008, of the chief executive officers of General Motors, Ford, and Chrysler begging Congress for a bailout. Behind these multinational economic shifts are stories of countless Michiganders who've lost their jobs and sometimes their homes. Many of them have joined me in Arizona or moved to Nevada, California, Texas, Florida, Georgia, North Carolina, or Colorado—eight of the ten fastest-growing states in the country. Each of these states has profound water problems. People are migrating from where the water is to where it isn't, or where there isn't enough.

These shifting demographics are critical to understanding a surprising fact: total water consumption in the United States actually went down slightly between 1980 and 2000. Part of the reason was the decline of the manufacturing sector of the nation's economy. But the Clean Water Act also played a key role by requiring industrial users to clean up their discharges. Whether the industry was steel or power generation, many companies concluded that the easiest way to reduce discharges was to use less water in their industrial applications.

The general observation that water use declined during these years obscures the localized reality of water shortages. As the joke goes, if Bill Gates walks into a bar, the average patron becomes a millionaire. Huge spikes in water use in response to regional population growth overwhelm slightly lower overall use. Moreover, since 2000, the nation has witnessed continued population growth, unprecedented droughts, and a surge in well drilling. Other demands for water come from cultural changes, environmental concerns, and economic opportunities. We are such a wealthy society that some cultural changes lead to indulgence.

In 2007, Grohe, a top-end European manufacturer of plumbing supplies, came out with a new line of shower gizmos. Its Aquatower 1000® shower system combines a handheld shower head on a bar to adjust the height, steam heads for a sauna effect, and ten body jets. Suggested retail price is $1,500. Not to be outdone, Kohler introduced its WaterHaven® door assembly, which sprays at least seven heads of water, none of which get the female catalog model's hair wet. But my favorites are Kohler's WaterHaven custom shower tower and its BodySpa ten-jet tower, which sell for more than $4,000 and $6,000, respectively, and deliver as much as eighty gallons per minute.

This trend toward lavishly appointed showers and baths, with "power shower and soaking tubs," has caught on in Phoenix. A 2006 *Phoenix Home & Garden* story featured one interior designer who, utterly without shame, admitted to catering to clients who want "car wash" body sprays. But these indulgences pose plumbing and electrical challenges. With a bathtub that holds eighty gallons and a shower with ten heads, each of which delivers 7.5 gallons per minute, a ten-minute shower empties 750 gallons of water down the drain. Rather than installing the customary 40-gallon hot water tank, builders typically install multiple 100-gallon tanks to service the master bath, the kitchen and laundry, and the guest suites. Homeowners are apparently unwilling to delay their gratification while water heats up, so they gladly pay high electricity bills to keep hundreds of gallons of hot water at the ready. This is why we in Tucson look down on people in Phoenix.

But Phoenix has not cornered the market on gluttony. In Montana, developers are carving up old ranches into mini-ranchettes for rich out-of-state tourists who want 6,000-, 8,000-, or even 10,000-square-foot homes, which they may use only a few weeks during the year. Builders have excavated areas in front of hundreds of these mansions for reflecting ponds. The owners drill wells to pump groundwater to fill the ponds, thereby reducing the flow in Montana's world-famous trout rivers. When water evaporates off the surface of the ponds, the owners must pump even more groundwater. Montana Fish, Wildlife, and Parks is helpless to stop this proliferation. Domestic wells are exempt from regulation in Montana and every other state, save New Mexico, on the theory that a domestic well uses so little water that it's not worth regulating. Although this is true for single wells, cumulatively domestic wells consume large quantities of water, and their omission from regulation is a sizable loophole. A dispute in Cheyenne, Wyoming, in 2007 illustrates the point. Dr. Wayne Lewis decided to construct a pond 300 feet by 157 feet on his property. A hydrologist with Wyoming's Ground Water Division calculated that the pond would lose 1 million gallons per year to evaporation, but the agency had no discretion to deny the permit for the well.

In addition to luxurious showers and wasteful ponds, subdivision developers across the nation have found a new lure for vacation home buyers: water parks. In Delaware, the Peninsula, a 1,400-unit development, has a 14,000-square-foot wave lagoon. Other water parks have recently been built in Telluride, Colorado; Kissimmee, Florida; Cortland, New York; the Wisconsin Dells; Pigeon Forge, Tennessee; and Reno, Nevada.

On a grander scale, consider Waveyard in Mesa, Arizona, which broke ground in 2007. This adventure park, with a price tag that could reach $500 million, will feature an 80,000-square-foot indoor water park, fly-fishing ponds, a sand beach, and a scuba diving and snorkeling lagoon—complete with fiber-optic lights for night divers. A separate outdoor wave pool will allow guests to enjoy swimming, surfing, and boogie boarding, while a man-made white-water river will highlight kayaking and rafting. At the outset, the project will require 30 million to 50 million gallons of water just to fill the recreational features. Then it will take between 60 million and 100 million gallons each year to replenish water lost from evaporation and splashing.

Recreation is driving other new demands for water. As a teenager, I learned to water-ski on a lake in New Hampshire. It's great fun. Most waterskiing is done on natural lakes or reservoirs behind dams. But a recent phenomenon, seen now from California to Florida, involves waterskiing on artificial lakes. Sunset Landing near Clermont, Florida, built specifically for people who water-ski, has three man-made lakes.

Not to be outdone, kayakers and white-water rafters have laid claim to water for their sport as well. California's American River, one of the country's most popular white-water rivers, often lacks sufficient water for rafting and kayaking because the Sacramento Municipal Utility District controls releases from Chili Bar Dam. The flow levels depend on the utility's need to generate hydropower. For years, recreationists pushed for a change. They got a window of opportunity when the fifty-year-old federal license to run the dam came up for renewal. In 2007, negotiations led to an agreement that ensures minimum flows for white-water enthusiasts on weekends and most summer days and that mandates increased flows for rainbow trout to accommodate recreational anglers. The cost to the utility comes in the form of reduced hydroelectric revenues and less water for its customers. The utility will need to replace that lost energy and water from someplace else.

In Golden, Colorado, the city spent $342,000 to create Clear Creek Whitewater Park, which features a series of kayaking courses that let kayakers perform multiple tricks in a streambed previously disturbed by mining. In Washington, the cities of Spokane and Yakima have followed suit, hoping to attract some tourist dollars. (Good luck, I'm thinking, as kayakers are notoriously tightfisted.) But if the facility can attract commercial white-water rafting companies, revenues can pop. That's what communities on the Arkansas River in

Colorado have found out. At least a dozen white-water parks are open, and another seven are under construction. The Colorado legislature, mindful of the economic impact of tourism revenue, has approved "recreational in-channel diversion" water rights. Keeping adequate water in rivers for white-water recreation does not involve a consumptive use of water, in contrast to diversions to irrigate fields. But it does mean there is less water available for other uses, precisely because it must be left in the river.

Some white-water parks are off-channel. In Charlotte, North Carolina, the $25 million U.S. National Whitewater Center, designated as an official training site by the United States Olympic Committee, opened a three-channel course with concrete walls in 2006. Giant pumps recirculate the water between collection pools at the top and bottom. This recycling system does not involve consumption of a substantial amount of water, but it does demand electricity to run the pumps. A lot of electricity—between $750,000 and $1 million worth per year.

Finally, snow-challenged ski resorts have expanded their snowmaking operations. It takes 12,000 gallons of water to blanket an acre of slope in an inch of snow. A decent-sized ski run covers thirty acres or more, and it needs at least a foot of snow to be skiable. That single run requires 4 million gallons of water—enough to fill 300 swimming pools. When a large ski area turns to making snow, it consumes 10,000 gallons of water a minute. So much energy is required to pump this water uphill and to blow it across slopes that some ski resorts have built their own power plants.

But the epitome of a luxury item is bottled water. What a cultural phenomenon it's become. Until recently, few people drank bottled water in the United States, except in Italian restaurants or if they lived in Berkeley. Today, it's ubiquitous. In classrooms, students have bottles of water on their desks next to their laptops. At the gym, bottled water is omnipresent. In fact, the latest generation of treadmills and elliptical trainers has built-in water-bottle holders. Then there's the movie theater, where the price of bottled water rivals the outrageous cost of popcorn. Indeed, bottled water now sells for more money than milk, oil, gasoline (even at $4 a gallon), or, perplexingly, things made with water, such as Coca-Cola. Imagine being chief executive officer of the Coca-Cola Company, sitting in Atlanta with the syrup formula, which you guard as though it were a state secret, and someone comes along and says, "I have a better idea. Forget the syrup."

Driven by health concerns about calories and caffeine, by misgivings about the quality of municipal tap water, and by the convenience of single-serving containers, the market for bottled water has exploded. In 2007, Americans consumed 9 billion gallons of bottled water, almost 28 gallons per capita. By 2010, the volume of bottled water sold is expected to exceed that of soda.

Still, bottled water consumption is a drop in the proverbial bucket of our overall water use. So why is it a problem? One reason is the water and energy needed to manufacture and dispose of the bottles. Jon Mooallem, writing in the *New York Times Magazine* in 2007, observed that "Americans will throw out more than two million tons of PET bottles this year." PET, or polyethylene terephthalate, is a petroleum-based product that is difficult to recycle. In 2005, approximately 18 million barrels of crude oil were required to produce the PET bottles we threw away. Eliminating those bottles would be the equivalent of taking 100,000 cars off the road.

It takes a lot of water to produce bottled water. Elizabeth Royte, author of *Bottlemania: How Water Went on Sale and Why We Bought It*, notes that manufacture of the bottles requires twice as much water as the bottles ultimately contain. Moreover, the filtering process uses "between three and nine gallons of water for every filtered gallon that ends up on the shelf."

In an astonishing marketing coup, the bottled water industry persuaded Americans in 2006 to spend $11 billion on a commodity that most Americans can easily, safely, and cheaply get from their kitchen faucets. Snake-oil salesmen would be envious. P. T. Barnum and bottlers of water both know that there's a sucker born every minute. Taste tests find most people equally split between bottled water and tap water; chemical analyses find most tap water to be as safe as, or safer than, the bottled variety.

The best spoof of the bottled water craze is by Penn & Teller in their Emmy-nominated documentary series, *Bulls**t!*, which airs on the premium channel Showtime. The comedy duo specializes in debunking popular fads and misconceptions. In the bottled water episode, they set up hidden cameras in a trendy California restaurant, created a "water list," much like a fancy French wine list, and used a "water steward" rather than a wine steward to introduce patrons to the establishment's bottled water options. The gullible patrons bought the routine hook, line, and sinker. Despite labels on the bottles suggesting that the water came from Alaska or Yosemite or wherever, the camera shows the water steward filling all bottles from a garden hose at the rear of the restaurant. Yet patrons waxed eloquent about the purity, the cleanness, and the arctic flavor of the specialty bottles, favorably contrasting the bottled variety with tap water.

The backlash has begun. Since 2002, citizens' groups around the United States, such as H_2O for ME in Maine, Water for All and the McCloud Grassroots Committee in California, Michigan Citizens for Water Conservation, Save Our Springs in Florida, Corporate Accountability International in Massachusetts, Save Our Groundwater in New Hampshire, Waterkeepers of Wisconsin, Concerned Citizens of Newport (also in Wisconsin), WaterFirst in Vermont, and other groups without formal names in New York and Texas, have engaged in grassroots organizing and sometimes litigation to thwart new or expanded bottled water plants. In almost every state, the target of the groups' opposition is Nestlé Waters North America, a wholly owned subsidiary of the Switzerland-based food giant Nestlé. That Nestlé bottles water may surprise even bottled water devotees, but Nestlé is the largest bottler of water in the United States. The company has gone on a buying spree and acquired fourteen regional bottling operations, including Arrowhead, Calistoga, Ice Mountain, Deer Park, Zephyrhills, Ozarka, and Poland Spring.

Nestlé has chosen a different marketing strategy for its water from that used by PepsiCo and the Coca-Cola Company, which filter municipal water supplies to produce Aquafina and Dasani, respectively. Nestlé believes that Americans will find greater cachet in, and therefore pay a premium for, water labeled "spring" rather than "artesian," "natural," "mineral," or any of the other categories approved by the U.S. Food and Drug Administration. FDA rules reasonably require wells that pump "spring" water to be located next to the spring. But pumping 500 or 600 gallons of water every minute of every hour of every day in the year—hundreds of millions of gallons—in such close proximity has devastating consequences on the spring's water level; on the birds, animals, and riparian habitats that depend on the spring; and on the river downstream from the spring. Nearby residents and towns may find that their wells dry up.

The perversity of Nestlé's marketing strategy is that, if the company moved its wells a couple of miles away from the springs, it would protect the springs but still get water with the same chemical composition. But then it couldn't market the water as "spring" water. The potentially devastating environmental consequences of Nestlé's strategy—as well as the ease with which it could be remedied—have driven citizens across the nation to resist Nestlé's proposed plants.

So has ideology. To Public Citizen and other groups, water is a public resource essential to life. They believe that a private corporation should not be allowed to bottle, export, and profit from water. It's a moral issue to some, including several religious groups. In 2006, the United Church of Canada, with 1 million members, pressed for a boycott of bottled water unless other sources of safe water are unavailable because "bottling and selling water undermines . . . the use of a public good and public responsibility to provide water."

Even gourmet chefs have weighed in against bottled water. Alice Waters—the culinary genius behind Berkeley's Chez Panisse restaurant, the inspiration for nouveau American cuisine, and the pioneer behind the "eat local, eat fresh" concept—banned bottled water from her flagship restaurant in 2007. That same year, San Francisco's mayor, Gavin Newsom, ordered city departments to stop buying bottled water. Employees can now drink from water coolers that dispense filtered tap water. Officials in Minneapolis, Salt Lake City, and New York City quickly followed suit. In 2007, Chicago's mayor, Richard M. Daley, slapped a nickel-a-container tax on bottled water. Also in 2007, Napa, California, rebuffed Crystal Geyser's request to tap into the city's aquifer. Mayor Jill Techel explained, "We shouldn't be mining our groundwater and shipping it out of our jurisdiction." In 2008, Mayor Newsom asked San Francisco restaurants to stop selling bottled water and to serve tap water instead.

I was first to point the finger at Nestlé, in my 2002 book, *Water Follies*, yet I don't object to people drinking bottled water. I sometimes do so myself, though I assiduously try to avoid "spring" water. Every physician will confirm that we don't drink enough water. Goodness knows, water is better for us than soda—liquid candy, as nutritionists call it—or beer, although many men and some women would disagree. Before consumers began drinking bottled water, they often chose another beverage, usually soda, fruit juice, or a sports drink such as Gatorade. It's a tad misleading, then, to compare the cost of bottled water with that of tap water. A better comparison is with the cost of these other processed beverages. Additionally, this comparison sheds light on the water bottle disposal problem, for these alternative beverages also come in containers, usually plastic (sometimes PET) or aluminum. While aluminum is easily recyclable, making it consumes a lot of energy—and water.

The bottled water industry is fighting back. Nestlé has rolled out Eco-Shape, a 12.5-gram plastic bottle that holds a half liter. It's almost 15 percent lighter than the standard bottle and takes less energy to produce. Nestlé's chief executive officer, Kim E. Jeffrey, points out that bottles used for carbonated beverages are twice as heavy, and Gatorade's bottles are almost three times as heavy. Fiji Water Company has promised to buy verified carbon offsets that will exceed its CO_2 emissions from transporting water thousands of miles from the South Pacific. But these steps are no consolation to

Michael J. Brune, executive director of the Rainforest Action Network. "Bottled water," he scolds, "is a business that is fundamentally, inherently and inalterably unconscionable."

The latest bottled water marketing campaigns try to distinguish bottled water from tap water. A flurry of "enhanced water" brands promise more than just hydration. Consuming these beverages, their manufacturers claim, will make you smarter, boost your energy level, and provide essential vitamins. A July 2007 full-page ad in *People* magazine for Kellogg's Special K_2O Protein Water promoted its five grams of protein and fifty calories: "You get your water, you get your protein, you get your sweet, you get a mid-morning smile that could last you 'til lunch." Janelle Belter, a Canadian nutritionist, observes, "You could just as easily eat a handful of almonds." Actress Jennifer Aniston, with her face plastered on magazine pages and billboards, has become the celebrity spokesdrinker for Glacéau's "vapor distilled" Smartwater with "certified organic essence." To nutritionist Belter, most enhanced waters "are nothing more than a flavored water with a really good marketing campaign." Nonetheless, consumers are buying in, and the market for enhanced waters grew by 35 percent in 2006. As for myself, I can't wait to see the Penn & Teller sequel.

The most subtle, but perhaps the most serious, effect of the bottled water craze is that the marketing of bottled water may erode public confidence in and support for our public water supply systems. Turning to bottled water or even filtering tap water, as the city of San Francisco is doing, implies that there is a quality problem with municipal supplies. In the backlash against bottled water, two markets are booming: water filters and reusable containers that offer guilt-ridden consumers an alternative to disposable water bottles. Domestic water-filtering systems, some costing thousands of dollars, are posting record sales. In third-world countries, such as India, plagued with contaminated water supplies, filtering water can save lives or at least prevent serious health problems. But in the United

States we expect our municipal water providers to deliver safe, wholesome water to our household taps. If affluent Americans turn away from this bedrock premise by relying on bottled water and water-filtering systems, it undercuts our democratic commitment to safe water for all.

CHAPTER 3

Our Thirst for Energy

"The frog does not drink up the pond in which he lives."
—American Indian proverb

IN AN OLD *Saturday Night Live* comedy sketch, Phil Hartman promoted a "Jiffy Pop Air Bag" that saved lives because, on impact, it inflated full of popcorn. But in 2008, Americans are far from laughing as they hope that corn, distilled into ethanol, will save us from global warming. A big push in this direction came when George W. Bush, in his 2007 State of the Union Address, called for a mandatory requirement that fuel producers supply 35 billion gallons of renewable fuels by 2017. That's a fivefold increase in ten years. Ethanol producers were delirious with joy, though they tried not to cackle too loudly.

A major beneficiary of the ethanol boom is Archer Daniels Midland, the agribusiness giant. The company is notoriously tight-lipped about almost everything, including how much money it makes from ethanol. For an October 2006 *Fortune* magazine profile of its chief executive officer, Patricia A. Woertz, "ADM declined comment." *Fortune* instead did a back-of-the-envelope calculation and concluded that "ADM earned a minimum of $610 million pretax—and probably more—from ethanol production during the 2006 fiscal year." The future looks bright for ethanol producers, thanks to a fifty-one-cent-per-gallon tax credit for ethanol *and* a fifty-four-cent-per-gallon tariff on sugarcane-based ethanol imported from Brazil, corn-based ethanol's chief renewable fuels competitor. These tax policies call to mind P. J. O'Rourke's sarcastic comment that the farm sector of the American economy is the last vestige of Soviet communism.

Although the term "six-dollar corn" doesn't mean much to most of us, it's music to the ears of corn farmers. The price of corn tripled between 2006 and 2008. Corn sold at $6 per bushel changes farmers' lives. A farmer in Minnesota who made $27,000 raising 1,000 acres of corn in 2004 made a profit of $270,000 in 2007. But to hog and beef producers, who use corn for feed, six-dollar corn makes it difficult to make ends meet. To the Coca-Cola Company and other beverage producers that sweeten their drinks with corn syrup, six-dollar corn spells competitive trouble. And American consumers have experienced higher food prices. As Michael Pollan brilliantly demonstrates in *The Omnivore's Dilemma*, corn is an ingredient in almost everything we eat, including beef, poultry, and anything that comes in a can, a bottle, or a jar. Higher prices for livestock feed and food ingredients squeeze companies as diverse as the Campbell Soup Company, Hormel Foods, the J. M. Smucker Company, and Tyson Foods.

So what *is* ethanol? It's alcohol, but not the stuff you'd drink, unless you're in serious need of rehab. Recently constructed ethanol plants use a dry milling process that begins with grinding the corn kernels into flour, or meal, and then slurrying the meal with water to form a mash. Added enzymes convert the corn's starch into a single sugar, dextrose. A high-temperature cooker controls the mash's bacteria. Once the mash is cooled, yeast is added to begin the fermentation process that converts the sugar into ethanol. A conventional distilling process then concentrates the ethanol to 190 proof. The process is about the same as that for making moonshine, but ethanol distillers add 5 percent gasoline to render it undrinkable and to avoid beverage alcohol taxes.

In 2007, American farmers planted 94 *million* acres of corn (more than the landmass of New York, Pennsylvania, and Ohio combined), a 15 percent increase over 2006 and the most corn planted since World War II. Ethanol production used 3.2 billion bushels of the 13.2 billion bushels of corn grown in 2007, a 52 percent increase over 2006.

Using corn-based ethanol as an energy source is a fool's bargain. What most people don't know about ethanol made from corn is that it may contain less energy than is required to produce it. How, then, is it a form of renewable energy? Economists and physicists measure the amount of energy required to produce energy using a standard called the energy return on investment, or EROI. The U.S. Department of Agriculture, the U.S. Department of Energy, and the Renewable Fuels Association, ethanol's trade group, insist that ethanol's EROI is positive, about 1.6, but that is far less than the EROI of gasoline, which is approximately 5. Critics such as David Pimentel of Cornell University and Tad Patzek of the University of California, Berkeley, maintain that ethanol's EROI is actually negative. "Ethanol is a boondoggle," says Pimentel, "because it takes more energy in the form of fossil fuels to make ethanol than we get out of it." Even assuming that ethanol's EROI is positive, the debate over ethanol has neglected another variable: water.

Ethanol refineries consume enormous quantities of water in making the mash slurry, as well as during the evaporative cooling process. Even though modern ethanol plants have sophisticated water treatment techniques that allow water to be recycled, the Institute for Agriculture and Trade Policy calculates that ethanol plants consume more than four gallons of water for every gallon of ethanol produced. To put this figure into perspective, consider the following story.

In June 2006, Granite Falls Energy held a grand opening celebration for an ethanol refinery in southwestern Minnesota that had come on line the previous November; the refinery was designed to produce 40 million gallons of ethanol per year. The day's festivities included a speech by Minnesota's governor, Tim Pawlenty, who wants Minnesota to lead the nation in the production of ethanol. It's the future, believes Pawlenty, "a transformational technology that is a pure positive. . . . We need to think like Bill Gates, inspire like

Tony Robbins and promote like Oprah Winfrey." Despite the governor's optimism, I suspect that neither he nor the other speakers at the grand opening realized that the plant was already running out of water.

Granite Falls Energy needs almost 400 gallons of water per minute. That's 160 million gallons of water per year for one modest-sized ethanol plant. The company initially drilled a well and began pumping groundwater. But neighbors three miles away, including homeowners Bay and Roger Reinke, soon found that their more modest domestic wells had gone dry and their pump motors had burned out. Angry well owners such as the Reinkes forced Granite Falls Energy to turn next to diverting water from the Minnesota River. The Minnesota Department of Natural Resources approved this diversion for the time being, but if a drought hits—which happens about one year in ten—the state will order Granite Falls Energy to halt the diversion.

Minnesota, the Land of Ten Thousand Lakes, does not have enough water for the new and proposed ethanol plants. In that state, sixteen plants are already in operation, four of these plants are expanding, and eight new facilities are proposed. In 2005, Cargill, a privately held agribusiness company, wanted to build a refinery in Pipestone with a capacity of 100 million gallons per year, but the local water company couldn't supply enough water. Elsewhere in Minnesota, plants are straining water resources in Winthrop, Windom, and Marshall. In Marshall, the water utility drilled a well field twenty miles east of town in order to pipe in water for an Archer Daniels Midland plant. If this can happen in a water-rich state such as Minnesota, how are other states faring in the race to be ethanol king?

Nationwide, there were only 54 ethanol plants in 2000; by 2008 that number had grown to 139, with an additional 62 refineries under construction. Stimulated by the Energy Policy Act of 2005, which requires that the U.S. gasoline supply include 7.5 billion gallons of ethanol by 2012, the existing and proposed plants will have the capacity to produce 12 billion gallons of ethanol. Yet refining that much ethanol will consume 48 *billion* gallons of water. And that's just for the production process. First, farmers must have water to grow the corn.

Environmental writer Ted Williams recently examined the adverse effects of the ethanol boom on rivers and fisheries. "No crop grown in the United States consumes and pollutes more water than corn," says Williams. "No method of agriculture uses more insecticides, more herbicides, more nitrogen fertilizer. Needed for the production of one gallon of ethanol are 1,700 gallons of water." Seventeen hundred gallons multiplied by 12 billion is a lot of water. And California's Water Education Foundation thinks the number should be higher: 2,500 gallons of water to grow enough corn to produce 1 gallon of ethanol. The state of California has a goal of producing a billion gallons of ethanol a year. To grow enough corn to refine that much ethanol would take 1.7–2.5 *trillion* gallons—more than all the water from the Sacramento–San Joaquin River Delta that now goes to Southern California cities and to Central Valley farmers. The Delta supplies water to two-thirds of the state's residents and irrigates more than 7 million acres of some of the nation's most productive agricultural land.

Because American farmers already grow lots of corn each year, using some of it to produce ethanol is not a new demand for water. The new water demand comes from the increased production of ethanol, the increased irrigation water for new acreage devoted to corn, and the additional water needed to change from growing, say, wheat to growing corn, which is a more water-intensive crop. And most of these increases are occurring in states that are already strapped for water.

In some states, the water to grow the corn comes from rainfall. But corn farmers in major ethanol-producing states—North and South Dakota, Nebraska, Kansas, Colorado, Texas, and California—irrigate their fields, either diverting water from rivers or pumping it from aquifers. As corn prices soar, ethanol producers are competing for corn that would otherwise be used for food, for water that would be used for other crops, and for water that, if left in rivers, would nurture plants and fish.

Some visionaries dream of developing "cellulosic ethanol," that is, ethanol produced from the stalks and husks of corn—the waste product—rather than from the kernels. In 2007, the Department of Energy awarded six companies $385 million in grants to work on developing an economically competitive cellulosic ethanol. But immense technological challenges ensure that the success of that venture is a long way off. Even if it is ultimately successful, cellulosic ethanol would not solve the water problem.

The rapid ramp-up of corn production in anticipation of an ethanol boom produced a glut of corn on the market in late 2007 that temporarily depressed the price of ethanol. Most industry experts suggest that ethanol refiners have paid insufficient attention to transportation and distribution concerns. Infrastructure problems pose challenges, especially transportation bottlenecks due to scarcity of the specialized railroad cars needed to ship a flammable liquid. Ethanol supporters describe the turmoil as temporary supply-chain problems. Thanks to the United States Congress, the industry faces a bright future. The Energy Independence and Security Act of 2007 requires a 500 percent increase in ethanol production, to 36 billion gallons a year, by 2022.

Producing petroleum, natural gas, coal, or methane also consumes a lot of water, but much less than ethanol. In a 2008 study, researchers at Virginia Polytechnic Institute and State University quantified the amount of water required by various energy-generating technologies to produce a standardized unit of power—1 million British thermal units (BTUs). Natural gas—an efficient energy source—takes as little as 3 gallons of water to produce 1 million BTUs, but ethanol may require as much as 29,100 gallons.

Our energy-intensive economy imposes demands on water in other ways, some of them surprising and rarely reported. Even as the demands of traditional agriculture—the actual growing of things—

threaten our water resources, a new kind of farm is emerging on the landscape: "server farms," which bear no resemblance to Old MacDonald's. Server farms are what Google, Microsoft, and Yahoo call their data centers. When most people think of Google, they imagine its Mountain View, California, campus, crawling with bearded, sandaled geeks writing code, and other such things. While the campus is indeed the brains of the operation, it's not what runs Google's most famous product, its search engine Web site, www.google.com. Server farms, which are clusters of thousands of unmanned computers, process every Internet search request, direct deposit, online money transfer, online video game, video conference, chat room, www.youtube.com clip, panda-cam, or blog site. The computers actually take up only one-quarter to one-half of the space on the server farm. Backup diesel generators, battery systems, hundreds of miles of wire, routers, and the network's operation offices occupy the bulk of the space.

The use of this many computers demands large quantities of electricity, perhaps 2 megawatts per 10,000 square feet. Between 2000 and 2005, the energy needed to power the Internet doubled, and it is expected to jump by another 75 percent by 2010. In 2006, server farms in the United States required 5,000 megawatts of power—1.5 percent of *all* the electricity generated in the United States. The explosion in online video, Internet telephone services, and digitization of university libraries ensures that Google will need more server farms in the future. A Google official told me off the record that Google's "data centers have a complete copy of every website on earth." That's a lot of stored data to search.

In 2007, Google announced the opening of three new server farms in Tulsa, Oklahoma; Goose Creek, South Carolina; and Lenoir, North Carolina. The company may locate two more server farms in Columbia, South Carolina, and Council Bluffs, Iowa. Google's vice president of technical operations, Lloyd Taylor, said that industrial parks with ready access to water and electricity are attractive sites for the company. "We're like a heavy manufacturing facility," Taylor said, "but what we manufacture is Internet services." Oklahoma's governor, Brad Henry, welcomed Google's decision to select Tulsa for a $600 million server farm. The farm sits on 800 acres and will employ 200 people at an average annual salary of $48,000. In Lenoir, Google will take over the site of a former lumberyard, which went out of business as a result of the overseas competition that has decimated North Carolina's furniture industry.

Powering the millions of computers in server farms across the country generates a considerable amount of heat. Imagine thousands of toaster ovens sitting inside a windowless concrete structure. Getting rid of the heat so that it does not adversely affect the computers poses a challenging engineering problem—one that requires great amounts of energy to solve. Indeed, for every dollar Google spends on powering its server farms, it spends another dollar on air-conditioning. But rather than using traditional air-conditioning, Google (and Microsoft and Yahoo) use large quantities of water to cool the computers. Water has extraordinary heat-absorbing capacity. Although Google's public relations representative refused to disclose the specific quantity of water used, for fear of disclosing trade secrets, he acknowledged that Google uses water for its cooling system because water-cooled technology uses less energy. None of the major Internet providers will discuss these processes or the amount of water they consume. Perhaps as a result, no one has even estimated how much water server farms consume. But it's a huge quantity. One server farm, AT&T's Ashburn, Virginia, facility, pumps 13.5 million gallons a day for four immense "chillers." Even the supposedly nonmaterial information economy imposes huge water demands.

As our economy becomes increasingly high-tech, it requires more and more energy, which in turn requires even more water. Simply put, it takes a lot of water to produce energy.

In March 2007, Reland Kane, Tucson Electric Power Company's environmental superintendent, took me on a tour of the company's Tucson power plant. It's a coal-fired plant that uses groundwater for its boiler, turbine-feed system, and cooling tower. In the United

States, most electricity is produced by alternating current (AC) generators driven by steam turbines. Burning coal in a power plant produces heat that boils water that creates steam that, under high pressure, spins a turbine. The turbine spins a shaft that turns the AC generator, which produces an electric current at high voltage for distribution to substations and step-down transformers. This process, the magical one by which I can flip a switch and a light goes on, converts moving energy—the energy of the spinning turbines—into electrical energy. When hydroelectric energy is generated by moving water through turbines at dams, the process is nonconsumptive: once the water has passed through the turbines, it continues downriver. But coal-fired and nuclear power plants consume lots of water.

Kane explained that the biggest demand for water in his plant is for the cooling process. I could hardly hear him speak because of the turbine's noise. The vibration of the whole building made me think that my back teeth would rattle loose. As I had these thoughts, Reland casually mentioned that the noise and vibration levels were low that day because . . . I couldn't hear his response, but it occurred to me that I lead a cloistered academic life. Immense cooling towers at the plant capture the steam and run it through pipes surrounded by water that cools and condenses the steam back into water.

Water is a critical input for the energy industry, not only for producing power but also for mining, refining, processing, and transporting oil, natural gas, coal, and other fuels. It takes 2 to 2.5 gallons of water to refine 1 gallon of petroleum. So how much water does it take to produce electricity? Between 0.5 and 0.7 gallon per kilowatt of electricity for a coal-fired power plant. A typical 1,000-megawatt plant consumes 10,000 gallons of water a minute through evaporation. The average American annually uses some 12,000 kilowatts. To put these numbers in perspective, consider the calculations of the Virginia Tech researchers. One 60-watt lightbulb that burns for twelve hours per day consumes 3,000 to 6,300 gallons of water per year. If you want to save water, turn off the lights.

After agriculture, the energy industry is the second-largest user of water in the United States. Nuclear and fossil fuel production of energy require approximately 140,000 *billion* gallons per day of fresh water, or 39 percent of the country's total use, though most of this is nonconsumptive. Even discounting those open-loop plants that return warm water to rivers or lakes, consumption of water in energy production exceeds one-quarter of all nonagricultural water use.

To support a growing population and an expanding economy will require 393,000 megawatts of new generating capacity by 2020, according to the 2001 report of the National Energy Policy Development Group. That amount of power will require 1,300 to 1,900 new power plants, or more than one plant built every week for twenty-five years.

Just as the energy industry uses lots of water, the water industry uses lots of energy. The 60,000 water systems and 15,000 wastewater systems in the United States use approximately 75 billion kilowatts per year of electricity—about 4 percent of the nation's energy consumption. It takes a lot of energy to pump, transport, treat, and distribute water and to collect and treat wastewater. Desalination of ocean or brackish water consumes huge amounts of power. And because water is heavy, about two pounds per quart, it takes considerable energy to move it, whether it's pumped from the ground or transported through canals and pipes. In Texas, farmers who pump groundwater from 500 feet below the surface receive monthly electric bills for thousands of dollars per well. In California, 19 percent of the state's electricity, 30 percent of its natural gas, and 88 billion gallons of diesel fuel are used to convey, treat, and distribute water and wastewater. Nationwide, the delivery of water and the treatment of wastewater consume substantial amounts of power.

In 2004, in the shadow of 9/11, U.S. House and Senate subcommittees asked the secretary of energy for a report on threats to energy production resulting from limited water supplies. The Department of Energy's 2006 report, "Energy Demands on Water Resources," projects that the country's population will grow by 70 million in the next twenty-five years and the demand for energy will grow by 53 percent. The water used for electric power generation needed to satisfy these increasing demands could consume as much additional water per day as would 50 million people. Most of this growth will occur in the Southeast, the Southwest, and the far West, where water supplies are already limited. Providing the additional electricity needed for the interior West will require 270 million gallons of water per day. But regulators in Idaho, Arizona, and Montana recently denied permits for new power plants because of water shortage concerns. And a 2008 Associated Press investigative report found that water shortages might require 24 of the country's 104 nuclear reactors to throttle back or temporarily shut down.

Energy use and water are connected on a grand atmospheric scale as well. Carbon dioxide (CO_2), an odorless, colorless gas, plays a critical function in maintaining the balance of life on this planet. As a by-product of the combustion of fossil fuels, it's also one of the greenhouse gases that can trap heat in the earth's atmosphere. The sun's light waves enter the atmosphere and strike the earth's surface, which absorbs the heat and then sends back some of the light waves. The light waves that enter the atmosphere are shorter in wavelength than the ones reflected. Carbon dioxide and other greenhouse gases effectively act as a mirror reflecting the long (infrared) waves back toward the earth rather than allowing them to escape into the outer atmosphere.

Credible scientists no longer doubt the reality of global warming. Lingering doubts evaporated with the release of three reports in 2007. In February, the Intergovernmental Panel on Climate Change (IPCC), a group of 2,500 of the most respected scientists worldwide, released its long-awaited Fourth Assessment Report. Operating under the auspices of the United Nations, the IPCC declared for the first time that global warming is "unequivocal" and that there is a more than 90 percent probability that the earth's warming is caused mainly by the release of CO_2 and the other greenhouse gases resulting from human activities. This assessment should settle the question of causation, especially after the group won (with Al Gore) the 2007 Nobel Peace Prize, but some skeptics, primarily Fox News pundits, still maintain that this is all some sort of a ruse. They allege that the IPCC allows political concerns to shape its scientific findings. Political agendas may drive some IPCC members, but, because the IPCC operates by consensus, it cannot release a report unless the entire membership agrees with the findings. Any findings that the IPCC does release represent the least common denominator of what the scientific community believes. Thus, the IPCC's findings almost assuredly understate, rather than overstate, the human contribution to global warming.

Even if one could shrug off the IPCC as political, these skeptics must also deal with groups that concur with the IPCC, including such prestigious scientific organizations as the National Academy of Sciences, the American Geophysical Union, the American Meteorological Society, and the American Association for the Advancement of Science. The only dissent comes from the American Association of Petroleum Geologists, which has close ties to the fossil fuel industry.

But there is a larger point. Putting aside the debate over the human causation factor, no one doubts that the earth *is* warming. Regardless of who is to blame and for what, the effects of global climate change will create major water supply problems. An IPCC working group released an April 2007 report that predicts more frequent climatic events, including heavier rainstorms and more extreme droughts.

In February 2007, the National Research Council, an arm of the National Academy of Sciences chartered by Congress as an independent organization that provides information on scientific issues, released a report assessing the consequences of global warming for the Colorado River Basin. It soberly warns of "a future in which warmer conditions across the Colorado River region are likely to contribute to reductions in snowpack, an earlier peak and spring snowmelt, higher rates of evapotranspiration, reduced late spring and summer flows, and a reduction in annual runoff and streamflow."

Earlier peaks and smaller snowpacks mean less water, and less water means drought across the entire Southwest. The dust bowl of the 1930s could become the norm in the Southwest, the report somberly concludes, requiring "costly, controversial and unavoidable trade-offs" in water use. Kevin Trenberth, a scientist at the National Center for Atmospheric Research in Boulder, Colorado, predicted, "This is a situation that is going to cause water wars. If there's not enough water to meet everybody's allocation, how do you divide it up?"

These dire warnings are reason enough to pay attention to predictions about global warming, but the question remains, just how bad can conditions get? Scientists have quantified the likely reductions in water flow in the Colorado River, and the news is, indeed, bad. The predictions range from an 18 percent reduction in Colorado River stream flow by 2050 to a 45 percent decline by 2060. The Colorado River Basin is not the only western river system likely to suffer lower flows. One study predicts that snowpack in the Cascade Range, relatively low-elevation mountains in Washington and Oregon, will decline by 59 percent by 2050. In the Sierra Nevada, snowmelt already begins at least a week earlier than it did before World War II. And more precipitation falls as rain than as snow. Here is the double whammy: only one state in the country depends on the Colorado River *and* the Sierra Nevada for its water supply—California, home to 38 million Americans.

Lester Snow worries about snow conditions in the Sierra Nevada not because he's a skier but because he's the director of the California Department of Water Resources. He's widely regarded as one of the most thoughtful and savvy water managers in the United States. Says Snow, "Climate change is the single biggest challenge we face when we're looking at water resource management 50 and 100 years out." With deeper droughts and higher flood peaks, it's the uncertainty that poses the greatest challenge for him. As temperatures change, he notes, environmental issues change. "More intense ecosystem management becomes part of your water resource strategy," he says. It's a scary time to be in charge of California's water supply; snowmelt from the Sierra supplies 40 percent of California's water. The potential impact of global warming on the Sierra is frightening. By 2050, Snow envisions "at the low end, a 25 percent reduction in the snow pack and, at the moderate end, a 40 percent reduction." As California's cities, farms, and environment compete for less and less water, Snow will have his hands full. Jeff Mount, director of the Center for Watershed Sciences at the University of California, Davis, predicts "a world-class hog wrassle."

In addition to causing decreasing water levels, global warming reinforces the water-energy nexus. The California Energy Commission has concluded that the 44 million acre-feet of water that California annually uses results indirectly in the emission of 44 million tons of CO_2 from the fossil fuel processes used to generate the electricity. To make matters worse, 14 percent of California's electricity comes from hydroelectric facilities—the most water-efficient form of power generation. A smaller snowpack will reduce the power generated by turbines at dams, which will require more electricity to be generated by other sources, which will require more water to generate that power. Bottom line: higher CO_2 emissions and higher water consumption.

Our thirst for energy has frightening implications for our economy and our environment. Energy to power our automobiles, run server farms, and light our homes takes a lot of water. And we seem not to have enough of either water or energy.

CHAPTER 4

Fouling Our Own Nests

"We're all downstream."
—Anonymous

THE REVEREND Bob Greene recently moved from Atlanta to become pastor at the First Southern Baptist Church in Lake Havasu City, Arizona. He received a rude welcome to this spring-break party city, which had achieved notoriety when an episode of MTV's *Spring Break*, cohosted by Jenny McCarthy and Chris Hardwick, highlighted its debauchery and drunkenness. Lake Havasu City's reputation was solidified with *Girls Gone Wild: Spring Break* and lots of clips on www.youtube.com. Greene was sitting in a fast-food restaurant with his wife and thirteen-year-old son when two women walked in wearing only bikini bottoms and two stickers instead of tops. "I was shocked," said Greene. "My son's eyes were about to pop right out of his head." Greene observed, "I guess the no-shirt, no-shoes thing doesn't apply."

Although the revelers on Lake Havasu could hardly care, Lake Havasu City has the dubious distinction of being the largest city in the United States without sewer service. Its 55,000 residents rely on septic systems to dispose of their waste, which has gradually leached into the Colorado River, causing large spikes in nitrate levels. A 2005 report by the U.S. Bureau of Reclamation documented disturbingly high nitrate concentrations in drinking water. Lake Havasu City's residents and tourists are literally soiling their own water supply. And potentially that of others. Just downriver from Lake Havasu City, Arizona and California withdraw Colorado River water for more than 20 million people.

Severe water quality issues in the Colorado River prompted Arizona's governor, Janet Napolitano, in 2005 to create a stakeholder group, the Clean Colorado River Alliance, to address water quality problems. The alliance's 2006 report identified the presence of nitrates and also of fecal coliform bacteria (*Escherichia coli*), which are present in human and animal feces, as significant problems for water quality in the Lower Colorado River. The human health risks from high nitrate levels and fecal coliform bacteria are substantial. In young children, high nitrate levels can cause a condition known as "blue baby syndrome," which occurs when oxygen levels in the blood drop to dangerously low levels. Ingestion of fecal coliform bacteria can cause severe intestinal problems that can be fatal. According to a UNESCO study, *E. coli* and related waterborne pathogens infect 940,000 Americans each year and cause approximately 900 deaths. Lake Havasu City is a microcosm of a national problem. Millions of Americans obtain drinking water from unregulated and untested private wells, making nitrate contamination from human and animal waste that has leached into aquifers a significant national concern.

A related problem occurs when cities combine sewage and storm water in a single collection system. During storm events, the combination overflows local wastewater treatment systems, and once again there is human waste in the water supply. In 2004, Milwaukee dumped more than 4 billion gallons of raw sewage into Lake Michigan after heavy rains inundated its treatment plants. Other cities, including Pittsburgh, Atlanta, St. Louis, Detroit, and Washington, DC, have the same problem. In 2004, approximately 1.4 billion gallons of raw sewage from the District of Columbia ended up in the Anacostia River, a tributary of the Potomac. The U.S. Environmental Protection Agency (EPA) estimates that more than 860 *billion* gallons of sewage are annually dumped into America's rivers and lakes.

The environmental group American Rivers estimates that 3.5 million Americans become ill each year from contact with contaminated waters. In a grisly example, in 2007, fourteen-year-old Aaron Evans of Lake Havasu City became the sixth victim of *Naegleria fowleri* after being infected while swimming in Lake Havasu. This disgusting amoeba enters the nasal cavity in water, makes its way up into the brain, and kills its victim within seven days by eating the brain. It is always fatal, though mercifully rare. Another vicious waterborne parasite, cryptosporidium, causes severe diarrhea, abdominal cramps, and nausea. It can be fatal, as in Milwaukee in 1993, when an outbreak killed more than 100 people, mostly the elderly and those with compromised immune systems. In 2007, almost 2,000 people in Idaho and Utah suffered from infection by cryptosporidium, believed to have been spread at "splash parks" where children play.

In *The Omnivore's Dilemma*, Michael Pollan observes that the turning point in the history of corn and the industrialization of our food occurred in 1947, when the Muscle Shoals, Alabama, munitions plant switched to making chemical fertilizer. "The discovery of synthetic nitrogen changed everything—not just for the corn plant and the farm, not just for the food system, but also for the way life on earth is conducted." Ammonium nitrate fertilizer revolutionized farming. But, in using it, farmers have poisoned their own water supply. Fertilizers, pesticides, and herbicides have made their way into aquifers around the United States, including the Ogallala Aquifer, which stretches from South Dakota to Texas. The U.S. Geological Survey has found that certain pesticides, used for only the past thirty years, have already leached through the ground and contaminated many aquifers.

The most serious groundwater pollution problem facing the United States, according to the U.S. Geological Survey's Robert Hirsch, comes from the skyrocketing use of nitrogen fertilizer, which is generally unregulated. In Iowa, where corn is king and nitrogen-based fertilizers are in wide use, nitrate levels in surface water have almost doubled over the past decade. Between 30 and 40 percent of public water supplies have nitrate levels that exceed five parts per million. More than one in five private wells contain nitrate levels that exceed the EPA's maximum contaminant level. Iowans are literally poisoning themselves.

No river is immune from nitrogen-laced runoff, especially the granddaddy of them all, the Mississippi. Upstream diversions; agricultural runoff of pesticides, herbicides, and especially fertilizers; and discharges from municipal wastewater plants present water quality problems for downstream cities, especially St. Louis, Memphis, and New Orleans. The freshwater flows entering the Gulf of Mexico contain high concentrations of nitrates and phosphorus, which stimulate the growth of immense algae blooms. As the algae decompose, the process depletes oxygen in the water, a condition called hypoxia. Each year, usually from April through the summer, a "dead zone" forms of 5,000 to 8,000 square miles—roughly the size of New Jersey—where fish, shrimp, and other marine organisms die from the lack of oxygen. A recent USGS study found that 20 percent of the nitrates in the Mississippi River comes from Iowa.

The most notorious example of a polluted river is the New River, which flows north into the United States from Mexico at Mexicali, California. It carries with it more than 20 million gallons per day of raw sewage, thirty known viruses, twenty-five agricultural pesticides, and unregulated industrial wastes. Undocumented immigrants also use the river as a crossing. Some people fear this poses a public health problem, given that undocumented immigrants often seek employment in grocery stores and the agricultural industry, thus exposing thousands of Americans to communicable diseases. One film clip shows U.S. Border Patrol agents standing by and watching as immigrants cross the New River; knowing how polluted the river is, the agents were reluctant to enter the water themselves.

In a 2005 study, an environmental research firm, the Environmental Working Group, found that tap water in forty-two states contains more than 140 unregulated chemicals—ones that lack EPA-set standards for health-based limits. Although the pollution comes from many sources, the group focused particularly on agricultural chemicals. Analyzing the results of water quality tests submitted by water suppliers as required by the federal Safe Drinking Water Act, the group found tap water "contaminated with 83 agricultural pollutants, including pesticides and fertilizer ingredients." This tap water serves more than 200 million people. According to data from the U.S. Department of Agriculture, the agricultural industry annually applies 110 *billion* pounds of fertilizer to one-eighth of the landmass of the continental United States.

In addition to nitrates and raw sewage, another serious water quality problem plagues the country. Perchlorate, a rocket-fuel chemical that disrupts thyroid function in adults and may damage babies' brains, has turned up in drinking water, fruit, lettuce, and even breast milk in places ranging from Salinas, California, to Cedarville, New Jersey. Recent studies by the American Water Works Association and the EPA detected perchlorate in hundreds of sites in thirty-six states. In 2006, researchers at the University of Alaska Anchorage discovered that perchlorate upsets the hormone balance in fish, making some female fish hermaphrodites and generating abnormal testes in male fish. In 2006, Massachusetts became the first state to set limits on perchlorate levels in potable water, at two parts per billion; California followed suit in 2007 with a maximum contaminant level of six parts per billion. So far, EPA officials have balked at setting a national cleanup standard. In May 2007, the city of Pasadena, California, shut down nine of its thirteen drinking water wells because of perchlorate contamination from NASA's Jet Propulsion Laboratory.

In the 1950s, Kerr-McGee Chemical Corporation produced perchlorate compounds at a Henderson, Nevada, manufacturing plant. For several decades, the company conveyed perchlorate wastes to unlined storage ponds. As the material leached into the ground, it moved toward and then entered the Colorado River. One scientist estimates that the contamination plume from the Kerr-McGee plant contains 20 million pounds of perchlorate dissolved in more than 9 billion gallons of water. The company has embarked on a very aggressive remediation program, but the treatment process may take decades to complete. Meanwhile, an estimated 50 pounds of perchlorate per day enter the Colorado River. A 2007 study by the U.S. Government Accountability Office found almost 400 sites nationwide where perchlorate has contaminated drinking water, groundwater, sediment, or soil. About half of these are in California and Texas.

Another alarming contaminant is MTBE, methyl tertiary butyl ether, which oil companies have added to gasoline since 1990 as an oxygen enhancer to reduce carbon monoxide emissions from cars. Marcia Lyford's home in Dover, New Hampshire, reeks of it. The odor in Marcia's home came from a gas leak at a filling station a few blocks away. MTBE-treated gasoline had leached into the aquifer and contaminated the air in her home. This kind of contamination has spawned hundreds of lawsuits around the United States, as well as an unsuccessful effort by the petroleum industry to persuade Congress to immunize oil companies from liability suits filed since 2003. The health consequences of exposure to MTBE are uncertain. Marcia's thirteen-year-old son, Sam, has had respiratory and ear infections that Marcia suspects are due to MTBE. The EPA has concluded that it is a "potential human carcinogen," but there is no conclusive evidence about its health risks at low doses. At least seventeen states have banned MTBE, and the EPA has taken steps to ban it nationally.

New Jersey officials have found MTBE in 430 water systems that serve 4.8 million people. A national survey by the Environmental Working Group concluded that MTBE has been detected in 1,861 water systems in twenty-nine states. In Santa Monica, California, half of the city's water supply was lost in 1996 as a result of the discovery of MTBE in its groundwater wells. Water suppliers in South Lake Tahoe, Sacramento, and the San Gabriel Valley have also shut down drinking water wells because of contamination by perchlorate or MTBE.

From an environmental perspective, MTBE spills are much worse than regular gasoline spills, which are bad enough. At least with gasoline, which adheres to the soil, the remedy is to bring in a backhoe and remove the contaminated soil. If gasoline percolates into an aquifer, it floats on top of the groundwater because it's lighter. A municipality can usually deepen its well to avoid pumping gasoline into its water system. But MTBE is extremely water soluble, does not biodegrade well, and doesn't readily adhere to anything. Hence, MTBE stays in solution and travels farther and faster in the ground than gasoline, making it very likely to seep into water supplies. Lyondell Chemical Company, the nation's largest MTBE manufacturer, acknowledges that even one part per billion gives water "a distasteful odor and taste," making it "unsuitable for consumption."

The Safe Drinking Water Act mandates that municipal water suppliers annually test their supply for an array of chemicals. Until recently, however, scant attention was paid to private domestic wells, which supply water to more than 45 million Americans. New Jersey requires private well testing upon the sale of a house, but most Americans who rely on domestic wells for their drinking water have no information about the quality of that water. These wells are more susceptible than municipal wells to groundwater contamination because they are not as deep or as well constructed. In a California agency's recent study of domestic wells in Tulare County, one-third tested positive for *E. coli* and more than 40 percent had nitrate levels that exceeded the national drinking water standards.

In 2006, the USGS completed the largest study of private wells ever undertaken, testing some 18,000 wells. The results were startling. Atrazine, a widely used herbicide, was detected in 24 percent of the wells. Levels of MTBE, chloroform, and arsenic exceeded the EPA's drinking water standards in 11 percent of the wells sampled, and nitrate exceeded standards in 8 percent. The good news is that the levels of contamination of these polysyllabic compounds are quite low; the bad news is that we have little knowledge about how various chemicals react in combination with one another. Some scientists think that combinations of chemicals may be especially damaging to vulnerable populations such as the very young, the very old, and those with compromised immune systems.

Animal-feeding operations also pose a major pollution problem for domestic and municipal wells. Cattle, turkeys, chickens, sheep, and lambs are fattened up before slaughter in tens of thousands of pens around the United States. These immense pens hold the animals for as long as twelve months, creating a major manure problem, as do dairies. Each dairy cow, for example, produces 120 pounds of manure a day, and California's Central Valley alone is home to 1.4 million cows. That's 84,000 tons of manure a day. The manure is transferred to storage lagoons and land applications, but storms and leaching inevitably release much of it to waterways and aquifers. California's Central Valley Regional Water Quality Control Board sampled wells near eighty-eight dairies and found high levels of nitrate in one-third of the wells, a result that is consistent with findings in other states. By one estimate, 95 percent of feedlot cattle receive hormones for growth promotion, much of which is excreted, and of the 25 million pounds of antibiotics used in animal production, approximately 75 percent is excreted in animal waste. Seventy percent of the arsanilic acid fed to chickens to encourage growth is excreted, introducing 2 million pounds of arsenic into the environment each year.

Chicken excrement poses another threat to America's water supplies. To accommodate a dramatic shift in American consumers' diets away from beef and toward chicken, thousands of chicken houses have sprung up across the South, especially in Georgia, Alabama, and Arkansas. A $29 billion market in 2004, the American poultry industry is led by the world's largest poultry producer, Tyson Foods, headquartered in Arkansas. Old-fashioned chicken coops bear little resemblance to the poultry factories of today. Chickens are crowded into pens and sit on mesh wire that allows for easy removal of the manure. Food moves on conveyor belts past the cages. The coops are as long as a football field and hold as many as 25,000 chickens.

Manure presents a vexing problem for Tyson because there is so much of it. Tyson developed "poultry litter," a combination of manure and wood chips that fertilizes fields where Tyson grows chicken feed. In Arkansas, many of these production farms are located along the Illinois River. Runoff from rainstorms sweeps across the fields and into the river, carrying enormous quantities of phosphorus from the poultry litter. Phosphorus in water encourages algae blooms, creates a foul smell that affects the taste of water, and threatens fish by reducing the oxygen.

In 2001, the downstream city of Tulsa, Oklahoma, sued Arkansas poultry companies for violating Oklahoma's water quality standards. Tulsa's ultimate victory forced the industry to ship tons of poultry litter out of the Illinois watershed. But the problem continued as new and expanded poultry farms generated increased amounts of chicken litter. In 2005, Oklahoma's attorney general, Drew Edmondson, sued Tyson Foods and other poultry companies in an effort to stop the pollution and to require the companies to pay for cleanup of the existing contaminants. Arkansas' attorney general, Mike Beebe, who was running for governor at the time, responded by filing his own suit in the United States Supreme Court. He won the election in 2006 but lost the case in 2007, when the Court allowed Oklahoma's suit to proceed in the lower courts. In 2008, Edmondson asked the federal judge to order a change to the companies' practices. This interstate fight pits downstream Oklahoma interests, which want to protect the $9 billion poured into the state's economy each year by 350,000 recreational users of the Illinois River, against upstream Arkansas' $5 billion poultry industry.

Volatile organic compounds, or VOCs, have given rise to some of the country's most notorious groundwater pollution crises, from Love Canal to the disputes documented in the books and movies *A Civil Action* and *Erin Brockovich*. VOCs pose potentially great health risks, and in 2007 scientists found low levels of VOCs in thirty-three of thirty-five wells tested, most commonly TCE, or trichloroethene. In 2004, the EPA estimated that two-thirds of the Superfund sites dedicated to VOC issues had controlled the risk of contaminated groundwater migrating away from the site. This does not mean that the groundwater was cleaned up, only that it does not threaten to contaminate other groundwater supplies.

Chlorinated solvents, especially TCE, are powerful degreasing chemicals used in military and industrial applications. TCE, which gets into groundwater through spills or leaks from storage tanks, is so toxic that the EPA has set the maximum drinking water standard at five parts per billion. In Arizona alone, some forty Superfund and other contaminated sites contain large amounts of chlorinated solvents in groundwater left by the operations of electronics manufacturing plants, military facilities, dry cleaners, and landfills.

Linked to numerous cancers, brain damage, skin diseases, and immune disorders, TCE poses particularly grave health risks because it's devilishly hard to remediate. TCE interacts with organic soil matter, forming a dense nonaqueous phase liquid (DNAPL). TCE is denser than water and therefore travels downward through groundwater, contaminating it along the way. Then it settles at the bottom as a DNAPL. Customary ways to cleanse TCE-contaminated water are either to pump the contaminated groundwater to the surface, where exposure to air converts the TCE to a harmless gas, or to pass the water through granulated activated carbon filters. Both options are very expensive and, worse, only partially eliminate the pollution. A large plume on the south side of Tucson, Arizona, has been the subject of litigation and cleanup efforts for almost twenty-five years. Despite enormous expenditures, TCE is still in the groundwater, thanks to DNAPLs. Pumping the contaminated groundwater insidiously causes a TCE rebound effect as the DNAPLs slowly dissolve into the groundwater, replacing the contaminants that have been pumped out. After the failure of repeated cleanup efforts, the EPA finally declared one pocket of groundwater in Tucson so severely contaminated that it is not worth cleaning.

Groundwater pollution is costly. The case made famous in the movie *Erin Brockovich*, starring Julia Roberts, ended with a $333 million settlement. In 2006, Pacific Gas and Electric Company agreed to pay an additional $295 million to settle another round of associated water contamination suits. And still other plaintiffs have yet to settle. A 2004 EPA report noted that financial pressures were impeding the agency's ability to clean up Superfund sites. It alarmingly predicted that as many as 355,000 hazardous waste sites would require cleanup over the next thirty years, costing $250 billion. In a sobering comment on how long we have to live with contamination once it has occurred, regulators in California found pesticides in 98 of 107 monitoring stations in the Central Valley in 2006, including DDT, which was banned in 1972.

Now, some of you may be halfway out the door to Home Depot to buy a home water filtration system. Go ahead, desert the rest of us. The country has a potentially larger problem if those with financial means choose to opt out of the public water supply system and leave those without means to drink water contaminated with perchlorate, MTBE, arsenic, or TCE. In third-world countries, this is what happens: the affluent do not suffer from waterborne illnesses. This ethical and moral issue poses a substantial challenge in our democratic society. And if my exegesis on morality failed to give you pause, consider that your home filtration system may cost a lot of money, use a lot of energy, and waste a lot of water. For every gallon of potable water produced by a home reverse osmosis filtering system, between two and four gallons of water go down the drain, adding to the demands placed on the municipal wastewater treatment system. Moreover, most of them will not remove these exotic chemicals.

Nor is bottled water a panacea. Elizabeth Royte, author of *Bottlemania*, points out that the Food and Drug Administration does not require bottled water companies to treat their products to remove a variety of pesticides, heavy metals, and bacteria. To be sure, most bottled water is safe to drink, as is most tap water. But not all of either one.

Water pollution is an acute national problem that threatens our existing water supply. And things are going to get worse because it may take years or decades for contaminants that have been dumped or spilled to percolate into the ground and eventually arrive at the aquifer, where they sit ready for you or your municipal water provider to pump out. Recent spills will compromise potable water years or decades from now.

We have the technological capacity to clean up most of these messes. But here's the rub: it will take scads of energy for the required pretreating, filtering, screening, backwashing, aerating, pumping, mixing, and disposing. Producing this energy, we now know, requires water.

CHAPTER 5

The Crisis Masked

"Isaac's servants dug in the valley and discovered a well of fresh water there. But the herdsman of Gerar quarreled with Isaac's herdsmen and said, 'The water is ours!'"

—Genesis 26:19–20

A COUNTRY FACES an urgent water crisis, bordering on catastrophe, when levies break, wells go dry, rivers peter out, taps sputter, pipes collapse, dams burst, power plants close, sewage overflows, pollution mushrooms, workers lose jobs, land subsides, aquatic species go extinct, wetlands dry up, factories are shuttered, fountains shut off, tap water discolors, reservoir levels drop, water tables plummet, cropland fallows, salt water intrudes, and drought persists. That country is the United States. But few Americans have suffered these consequences, thanks to Steve Robbins, Lester Snow, and other water managers around the country.

"We've done too good a job," explains Robbins, general manager of the Coachella Valley Water District in southeastern California. A 2007 report by the U.S. Geological Survey found that groundwater pumping had caused an alarming drop in the earth's surface in the Coachella Valley. Despite considerable publicity that accompanied the release of the report, no one paid much attention. "Every day," says Robbins, "the first thing I do is look at the letters to the editor, but there hasn't been a single one." He thinks that's because "people have always gotten up, turned the tap on, and there was water there, and their sprinklers came on." Despite warnings by Robbins and other water managers that there's a water crisis brewing, "we have been able to keep chewing the gum and sticking it in the holes in the dike," says Robbins. The biggest challenge facing water managers in the United States today, he believes, "is to convince the public that there really is a problem. Sometimes, the only way you can do that is to have a disaster." He's right. Long before the 2005 New Orleans floods and the 2007 San Diego fires, we knew that flooding could inundate New Orleans and that building subdivisions in the midst of bone-dry forests could produce a conflagration in Southern California. But it took terrible catastrophes to get our attention.

Lester Snow, director of the California Department of Water Resources, similarly finds it frustrating as he tries to persuade Californians that the state faces a water crisis. "Recent surveys," Snow observes, "have shown that people don't have a clue where their water comes from." In this survey, most Southern Californians replied that water "comes out of my tap." When pushed to explain, as many as 80 percent answered that it comes from a well down the street, even though Southern California imports more water than any other place in the world. When asked whether they'd ever heard of the Delta—meaning the Sacramento–San Joaquin River Delta, the source of much of Southern California's water—the number one response was, "Airline." An exasperated Snow wonders, "How do you compete in the media-intensive society we live in to get a water resources message out?"

The current drought besetting Metro Atlanta has raised awareness of water scarcity, but most proposals for reform involve quick fixes—short-term palliatives—such as bans on washing cars or watering lawns except on alternate days. Once the drought ends, Georgians expect to wash their cars and water their lawns whenever they wish. In other words, drought is an aberration. But the water crisis facing the United States involves more than a temporary and localized decline in precipitation.

With two-thirds of the earth covered by water, how can we be facing a water crisis? For starters, less than 1 percent of the planet's water is drinkable. About 96.5 percent of the earth's water is ocean water, which is too salty to drink. Another 1.7 percent is frozen in polar ice, and 1 percent of fresh water is too brackish to drink. That leaves only 0.8 percent in lakes, rivers, and wetlands, in the ground, and in the atmosphere. Still, 0.8 percent of the earth's water is a lot of water.

Because the earth's atmosphere inhibits water from entering or escaping, the quantity of water on the planet is a fixed amount. We're drinking the same water that the dinosaurs did. We don't destroy water when we use it; we change its character and its location. After we drink a glass of water, our bodies absorb some into our tissues and organs, release some through our pores as perspiration, emit some as water vapor when we exhale, and excrete some. Similarly, when farmers water crops, such as tomatoes, the fruit absorbs some water, some water evaporates into the atmosphere, the leaves emit some into the atmosphere via a process known as transpiration, and some percolates into the ground and eventually reaches an aquifer, where it may become available for reuse. After we use water, we return almost all of it, in one form or another, to the environment.

Water moves through the hydrologic cycle as various forms of energy act on it. The sun's heat causes water to evaporate off the surface of oceans and other bodies of water. Wind currents carry moisture-laden air over land, where the relative humidity sharply increases; as the water vapor condenses, gravity causes precipitation to fall to the earth. Evaporation returns half of the precipitation to the atmosphere. Three things happen to the other half. First, trees and plants absorb and transpire some of it. Second, gravity causes rain and snowmelt to flow over the surface of the earth in search of the lowest point. This surface runoff replenishes lakes, rivers, wetlands, and, as one river flows downstream into the next, eventually

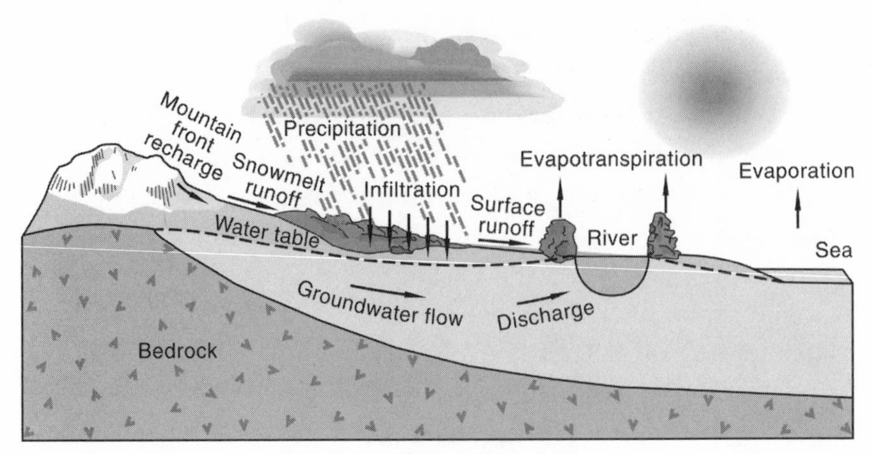

Figure 5.1. The hydrologic cycle.

the ocean. Third, water infiltrates the ground, percolating downward under the force of gravity to reach aquifers. Even then, groundwater continues to move through the aquifers, eventually augmenting nearby rivers and lakes.

This hydrologic cycle begs an important question: if our water supply is fixed and we can neither make nor destroy water, how can we run out of it? The answer is that some uses of water preclude future reuse. Water may not be *where* we want it *when* we need it in the *form* that we need. Many uses of water inadvertently preclude its reuse for years, centuries, or even millennia. Consider coastal areas around the country where municipalities pump groundwater, deliver it to residents, treat the resulting sewer water at regional wastewater treatment plants, and finally dump it into the ocean.

This once-potable water will not be available for reuse until it evaporates off the ocean, moves by wind currents over land, precipitates out, infiltrates the ground, and percolates into the aquifer, where it can be pumped from a well. Our water crisis is therefore a function of the timing and location of water use, compounded by current demographic shifts. Our national water crisis is not uniform across the country. It has acute local conditions, which are not confined to the arid West. In 2003, the U.S. Government Accountability

Office surveyed states about their water supplies. Only one state reported that it would not have supply problems in the event of drought. Even more ominously, thirty-six states expected to suffer water shortages in the next ten years under normal conditions. The looming crisis is sharply etched: water is an exhaustible resource.

"IT'S NOT A HAPPY TOWN," Herman Friedman says of Ramapo, New York, the largest town in Rockland County, a mere twelve miles northwest of New York City. Friedman, a candidate for Ramapo's town council, wants to preserve Ramapo's typical northeastern suburban character. Most residents live in very nice single-family homes. Rockland County, one of the most heavily Jewish American counties, is home to a number of Orthodox Hasidic villages. Because Hasidic Jews do not drive on the Sabbath, they prefer densely clustered housing, which is not allowed by Rockland County's zoning plan. They claim exemption from the zoning under a federal law, the Religious Land Use and Institutionalized Persons Act of 2000, which prevents local land use ordinances from substantially burdening the exercise of religion by groups, such as the Hasidim. Supporters of growth controls, such as Marlaine Paone, claim that the struggle is "about development, about the environment, it's not about religion. I'm Jewish. Two-thirds of my organization is Jewish." In 2005, *New York Times* reporter Peter Applebome noted that Rockland County "feels like suburban New York's answer to the Middle East, people of different cultures with different attachments to the land who seem permanently at odds."

One axis dividing Rockland County's residents is water. The Preserve Ramapo group fears that new Hasidic development, especially of Patrick Farm, a bucolic area of 200 acres, will overwhelm the region's water supply. A Hasidic spokesman counters, with some support from the U.S. Geological Survey, that the real stress on the water supply comes in summer with the watering of innumerable expansive lawns. Hasidic villages don't have large lawns. Regardless of who is right on this issue, the increased population at Patrick

Farm and other developments will place new demands on the water supply. But so what, you may be thinking: this is New York, not Nevada. There must be plenty of water.

Bordered by the Hudson River, dotted with lakes and ponds, and traversed by the Ramapo River, Rockland County seems unlikely to suffer water shortages. Yet it has. The county declared three drought emergencies between 1995 and 2002 and was on the edge of declaring a fourth in 2005. A 2005 Columbia University study revealed that the recent "droughts" were mild downturns in precipitation in comparison with severe droughts that have beset Rockland County in the past. Recent emergencies were due to a "significant mismatch between supply and demand." Over the past three decades, construction has placed new demands on the county's water system, but the supply of water has remained constant. "Even a few months of lower-than-normal precipitation," warned Nicholas Christie-Blick, one of the study's authors, "is sufficient to trigger an emergency." Rockland County declared "drought emergencies" because it had run out of water.

The situation is getting worse. Groundwater provides approximately three-quarters of Rockland County's public water supply. The Rockland County Department of Health reported that, beginning in 2000, groundwater withdrawals began to exceed "recharge," the term hydrologists use to describe water that replenishes the aquifer. Think of an aquifer as a checking account: if you consistently withdraw more than you deposit, you'll eventually run out of money. Your only options are to curb your spending, borrow from the bank, or get a second job. Rockland County faces a similar predicament.

Rockland County's solution seems like an easy one: it's sitting next to the Hudson River, with a daily flow of 12 billion gallons. But numerous sewage treatment plants dump effluent into the Hudson, mercury enters the river from several power plants, and past contamination by the General Electric Company has resulted in a massive Superfund site with toxic materials on the bottom of the Hudson. A cleanup is under way, but drinking Hudson River water is completely unacceptable to many Rockland County residents. At the very least, it would require construction of a desalination plant to filter the water because the Hudson is tidal near Rockland County. Other options, including aggressive water conservation and sharp increases in water rates, would reduce demand but would also generate fierce political resistance. Rockland County residents love their emerald green lawns. And because industrial users currently enjoy volume discounts, it's politically dicey to ask the voters to absorb higher water costs for domestic water use. They'll end up paying more anyway. Even though desalinating seawater is a very expensive process, United Water New York, the local water supplier, announced plans in 2007 to construct a desalination plant by 2015.

But Rockland County is not unique in the Northeast. In water-rich New England, "hundreds, if not thousands of little streams are getting hammered" from diversions and groundwater pumping, according to Kirt Mayland, director of Trout Unlimited's Eastern Water Project. "The flow is too low, and fish can't live there, but it's not always obvious to the public." In Connecticut, students at the University of Connecticut used so much water in 2005 that a quarter mile of the Fenton River dried up, killing 8,000 brown trout in one of the worst fish kills in Connecticut history. Connecticut has more than 60 rivers that suffer from "flow impairment"; Massachusetts has more than 160; Vermont, 50; and Rhode Island, 35. A 2007 Trout Unlimited report astonishingly concludes that "the majority of watersheds in New England suffer from unnaturally low stream flow levels." This is in New England, not in Nevada.

And New England is not unique in the East. Charlotte, North Carolina, the birthplace of Billy Graham, is one of the fastest-growing cities in the country, with a population of roughly 600,000; with the suburbs included, the metropolitan population exceeds 1.5 million. Nine Fortune 500 companies have headquarters in the Charlotte metropolitan area, including Lowe's, Nucor Corporation, Duke Energy Corporation, and Bank of America. With BofA's acquisition in 2008 of Merrill Lynch, Charlotte has become a financial services epicenter, second only to New York. Perhaps more important culturally, the National Association for Stock Car Auto Racing has offices and racecourses that make the city of Charlotte NASCAR's de facto capital. The NASCAR Hall of Fame will open there in 2009. The city has also secured National Football League and National Basketball Association franchises and is negotiating with the Florida Marlins to relocate that major league baseball team. No longer a southern backwater, Charlotte has become a handsome, eclectic, cosmopolitan hub.

A suburb of Charlotte, a town named Kannapolis, has ambitious plans stoked by a 2005 proposal for the North Carolina Research Campus. The plan is the brainchild of David H. Murdock, owner of Castle and Cooke and Dole Food Company, and Molly Corbett Broad, then-president of the University of North Carolina's sixteen-campus system. In 2000, Pillowtex Corporation went bankrupt and closed its Kannapolis manufacturing facility. The plan calls for that facility to become the anchor of the 350-acre research campus in downtown Kannapolis and to be part of North Carolina's transformation from an economy based on textiles and tobacco to one rooted in biotechnology.

But Kannapolis and an adjacent suburb, Concord, are strapped for water. The cities are growing at a rate of 4 percent a year, a robust pace for the East. Sandwiched between the Catawba and Yadkin river basins, the cities rely on three small lakes fed by the Rocky River, a tiny rivulet compared with the Catawba. In 1999 and 2000, Concord and Kannapolis almost ran out of water when the Rocky River came close to drying up and levels in the three lakes plummeted. Just to the north, the community of Statesville did run out. According to Annette Privette, public relations spokesperson for the city of Concord, the city targeted its top twenty water customers and told them it needed help. Companies such as S & D Coffee, which provides coffee to McDonald's and Dunkin' Donuts, undertook plans to reduce their water consumption. The city also slapped a large rate increase on those who used more than 7,000 gallons a

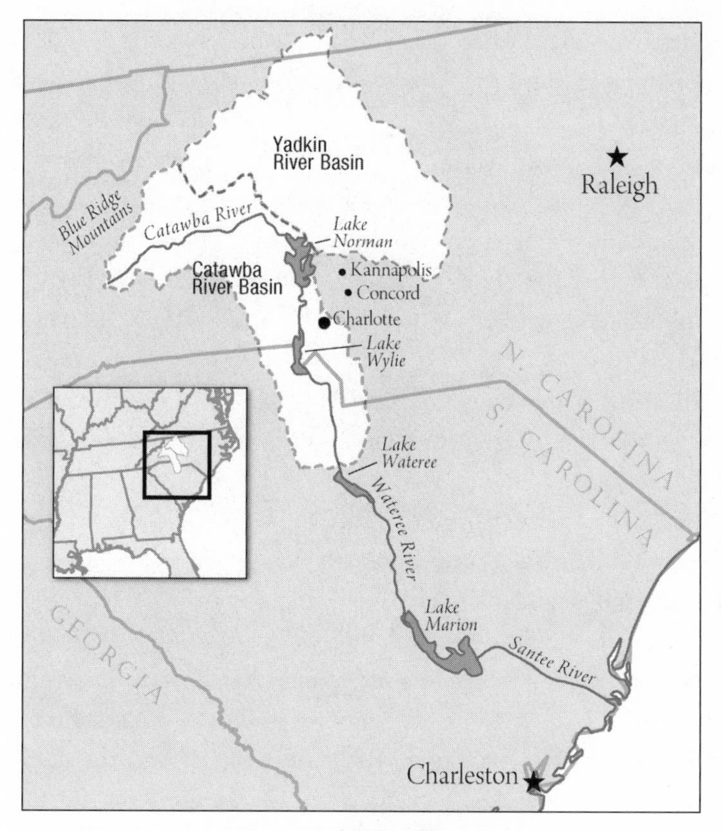

Figure 5.2. Catawba River Basin region.

month. Average household usage at the time was about 6,700 gallons per month—for an entire household. That's roughly 222 gallons per day, a relatively modest amount compared with Las Vegas' domestic use. Still, Concord stayed the course with its water reduction plan, and after eight years average household use is down to 4,600 gallons per month. But a 4 percent annual population growth rate has more than offset those improvements.

Rising in the Blue Ridge Mountains of North Carolina, the Catawba River flows east, turns south, and passes through a series of flood control and hydroelectric reservoirs, the most prominent being Lake Norman, and then passes west of Charlotte and becomes the border of North and South Carolina for ten miles or so before it empties into Lake Wateree. Eventually, it becomes the Santee River before emptying into the Atlantic Ocean northeast of Charleston. A major river whose flow averages 5,000 cubic feet per second, the Catawba fills eleven lakes, powers eighteen electrical plants, and attracts countless bass anglers and boaters. It's a magnet for industry and home buyers. The recreational opportunities are enticing thousands of baby boomers to relocate to the area. "The Catawba River is our lifeblood. It defines us," says Donna Lisenby, the Catawba Riverkeeper.

In 2004, Kannapolis and Concord applied to North Carolina's Environmental Management Commission for an Interbasin Transfer (IBT) certification, which would allow them to divert more than 10 million gallons per day from the Catawba River (and a lesser amount from the Yadkin River) and transport it into the Rocky River Basin for Kannapolis and Concord. The reaction from Catawba River Basin residents and towns, including Valdese, was hostile, reflecting fear that the diversion would compromise their future opportunities, increase the frequency of droughts, degrade the bass fishery, and increase costs to treat water. Reduced water flows, the Catawba Riverkeeper Foundation argued, would harm both the freshwater and estuarine ecology, deprive downstream communities of water needed for daily life, and concentrate water pollution.

The community reaction surprised Concord's mayor, J. Scott Padgett. Speaking at a public meeting in Valdese that attracted 700 people, the mayor said, "We never anticipated it would be controversial," drawing gales of laughter from most of the audience. A Baptist minister from Kannapolis received an even less civil reaction. When he told the audience that his solution, as a Christian, would be to ask, "What would Jesus do?" the audience responded with loud, sarcastic boos. Not all Baptists think alike. The slogan on the billboard in front of the First Baptist Church proclaimed: "It is God's World, Including Our Rivers. Vote 'No' on IBT's That Alter God's Plan." Another public meeting drew 1,300 people, a surprising number for a hearing on a water supply issue. In September 2006, North Carolina's governor, Mike Easley, received eighteen letters from sixth-graders at Concordia Christian Day School. Handwritten, simple, and unpolished, some with misspelled words and others with drawings of fish and ducks, all carried the same message: "Please, save our river."

In January 2007, despite public protest, North Carolina's Environmental Management Commission granted the cities' request for the IBT. The Southern Environmental Law Center, on behalf of the Catawba Riverkeeper Foundation, immediately appealed the state's decision, as did a coalition of eight cities and seven counties along the Catawba.

The state's decision also caught the attention of South Carolina. In Rock Hill, a small town just over the border, Mayor Doug Echols fears a loss of drinking water. In the 2007 drought, algae blooms made the drinking water smell rotten, boatyards closed, and river-dependent businesses struggled. Sometimes the solution to pollution is dilution, but if a river's flow is low, industrial discharges can exceed the river's "assimilative capacity" and thus violate the Clean Water Act. The low flows in the Catawba prevented a South Carolina paper company, Bowater, from discharging its wastewater into the river. The plant had to shut down, and employees lost their jobs.

In 2007, South Carolina's attorney general, Henry McMaster, announced he would file suit in the United States Supreme Court to block the transfer. "Rivers affect people's lives more than anything else except the air we breathe," said McMaster. Concord officials attribute McMaster's actions to political motivation, but other South Carolina officials, including Governor Mark Sanford and members of the state's legislature and congressional delegation, also weighed in against the transfer. McMaster asked the Supreme Court to prohibit North Carolina from depriving its downstream neighbor of its rightful share to the water. In 2008, the Court appointed a special master to hear the dispute. Meanwhile, the North Carolina Division of Water Resources disclosed that Union County in North Carolina has also filed a petition to transfer additional water from the Catawba. Stresses on the Catawba led the environmental organization

American Rivers to rank it number one on its 2008 list of America's most endangered rivers.

Leaving the East Coast and turning next to the American West, one might think that the South Platte River Basin in eastern Colorado is an odd place to farm. The Rocky Mountains to the west block moisture from reaching the eastern plains, which receive only ten to twenty inches of rain a year. But the river has provided farmers with ample irrigation water, until recently. In the 1870s, Colorado adopted the prior appropriation doctrine, a rule that rewarded the earliest diverters of water from rivers with senior rights to the water. If there is not enough water for all farmers, this rule of "first in time, first in right" can prevent more recent diverters from taking any water in order to protect the senior diverters. Under this system, enterprising farmers, miners, and cities, even the Coors Brewing Company, claimed most of Colorado's abundant surface water. Newer arrivals found themselves out of luck until the 1940s, when advances in technology, such as high-lift turbine pumps, gear-driven pump heads, and power-line electricity, allowed them to pump groundwater from the Ogallala Aquifer. A stunning expansion of irrigated agriculture in eastern Colorado occurred thanks to commercial wells that could extract 1,200 gallons per minute from 3,000 feet below the surface.

But these groundwater-dependent farmers were living on borrowed time and borrowed water. Groundwater is connected to surface water. In the South Platte River Valley, groundwater pumping intercepts subsurface water that is moving toward the river. Pumping eventually lowers the river's flow and deprives senior diverters of their water rights. Colorado's water users, collectively, were using too much water, creating a problem akin to the game of musical chairs: after each round with a chair removed, someone has no place to sit. Some Colorado water user had to stop using water, but the Colorado official charged with administering water rights, state engineer Hal Simpson, looked the other way. He didn't want to offend either group of powerful farmers—the diverters or the pumpers. So he did nothing. In 2002, a devastating drought deprived senior diverters of adequate water. As their crops withered in the fields, well owners kept irrigating. So the senior diverters, as well as the cities of Boulder, Highlands Ranch, and Sterling, sued, and the Colorado Supreme Court ordered the state engineer to protect the seniors' rights.

In May 2006, another extremely dry year, Simpson made "the toughest decision" of his career and turned off 440 pumps that irrigated 200 farms totaling 30,000 acres along the South Platte River. Now it was the pumpers' turn to watch their crops wither. Millions of dollars' worth of wheat, corn, sugar beets, and melons were left to die. The controversy has fostered ill will between the diverters and the pumpers. Farmers who rely on groundwater, such as Harry Strohauer, who invested $700,000 in seed and fertilizer in the spring of 2006 for his 1,100 acres of potatoes and onions, blame the state engineer for letting things get out of control. Farmers who depend on surface water, such as Jim Aranci, who farms 1,200 acres of alfalfa, have no sympathy for well owners who pumped when they knew they didn't have rights to do so. "They're parasites," says Aranci. "They're no different than a mosquito."

As spring turned to summer in 2006, Carol Ellinghouse, water resources coordinator for the city of Boulder, received a tip from a farmworker that some farmers were still pumping groundwater. So Boulder hired private investigators to spy on suspected violators. After the investigators discovered "puddles" around sprinklers and "lush corn" on fifty plots of land, Boulder turned the information over to the state engineer's office. News of the spying angered many farmers, including David Knieval, who threatened: "I wish I could have caught them on my property. We shoot trespassers."

The palpable anger comes from farmers who are struggling to grow a crop in the field, make a payment at the bank, provide clothes for the kids, and put a meal on the table. Without water, there is no hope. One farmer, seventy-nine-year-old Bill McCracken, faced with a turned-off pump, made a little money doing construction work in Wyoming and paid his property taxes with Social Security checks.

He's sold off most of his cattle. "I had a farm worth probably close to a million dollars," he notes sadly. "Now it's worth almost nothing."

Many struggles over water involve squabbles among states. Let's consider a three-state struggle over the Republican River, which begins in Colorado, flows east into Nebraska, and then turns south into Kansas. Colorado reduced flows into Nebraska, which in turn encouraged Nebraska farmers to sink countless large-capacity irrigation wells along the river just before it enters Kansas, while sanctimoniously lambasting Colorado for its actions. In theory, none of this should have happened because the three states agreed to the Republican River Compact in 1942. But unconstrained power and greed explain how states act when it comes to water.

In 1998, Kansas dragged Nebraska before the United States Supreme Court for violating the compact. Nebraska had the good sense to settle before the Court had a chance to administer a good scolding. But in Nebraska, agricultural interests that depend on groundwater contribute $4.5 billion to the state's economy. Few Nebraska politicians have the stomach to deliver bad news to their powerful constituents. And the news is gloomy: a 2005 study concluded that Nebraska may owe Kansas tens of millions of dollars for shorting Kansas on its rights to Republican River water under the compact.

In 2007, the Nebraska legislature took a small step in the right direction with appropriations to acquire water rights that will augment the flow of the Republican River into Kansas. What the legislature did not do, however, was impose restrictions on groundwater pumping. Perhaps the legislature learned a lesson by watching local voters oust two Natural Resources District board members who favored imposing restrictions on groundwater pumping. One of them, Robert Ambrosek, whose father was one of the first to drill a well in the area, wants to protect the river. "I want to see my grandchildren and great-grandchildren have an opportunity to do better than what we've done with it. What gives us the right," he ponders, "to use up all the water?"

Kansas had no intention to sit idly by as Nebraska's water policies left it high and dry. Kansas' attorney general, Paul Morrison, promised to act decisively to force Nebraska and Colorado to cut their water use in order to increase the flow in the Republican River. Neither Nebraska nor Colorado denied that it was in violation of the compact. Kansas' chief water resources engineer, David Barfield, calculated that Nebraska had exceeded its allocations in 2005 and 2006 by 82,000 acre-feet—enough water to supply a city of 100,000 for ten years. He demanded that Nebraska shut off all wells within two and a half miles of the Republican River, which would fallow 500,000 acres of Nebraska farmland and cause land values to plummet.

When Nebraska's Natural Resources Districts announced in 2007 that their solution would be to cut down water-thirsty vegetation along the river rather than control pumping, Barfield summarily dismissed the idea. After decades of Nebraska's foot-dragging, Barfield is fed up. "Cutting down trees is a grand idea," he said, "but it doesn't necessarily send water to Kansas. We want water. We don't want plans or possibilities or whatever. We want water."

Farther south, Texas, the second most populous state in the country, is also running low on water. By 2060, the Texas Water Development Board expects the state's population to increase demand by 27 percent and the water supply to decline by 18 percent. No one knows where the water will come from to close the gap, but some Texas water districts hope it will be from Oklahoma. In 2007, the Red River Rivalry—the annual football game between the University of Oklahoma and the University of Texas—took on new meaning as the two states found themselves embroiled in a controversy over Red River water. Texas' Tarrant Regional Water District serves 1.6 million people, including the fast-growing cities of Fort Worth and Arlington. It needs additional water, but its most convenient source, the Red River, which forms part of the border between Texas and Oklahoma, contains naturally occurring salts that make the water too salty to drink. The district applied to the Oklahoma Water Resources Board (OWRB) for a permit to divert

411 million gallons per day from Cache Creek, Beaver Creek, and the Kiamichi River, tributaries to the Red River in the Oklahoma panhandle that have better-quality water.

This Texas raid on Oklahoma's water got the same reception as a "Hook 'em, horns!" cheer would in Norman. The Oklahoma legislature enacted a moratorium on exporting water to other states, a dubious bit of protectionism that is probably unconstitutional. The OWRB's director, Duane Smith, announced that the state was in the midst of a long-term water planning process, and, until that was finished, it would be inappropriate to grant new permits to export water. "We're going to do our very best," said Smith, "to adhere to the moratorium." Texas' Tarrant Regional Water District promptly sued, claiming that the Oklahoma law unconstitutionally discriminates against Texas. This action made Oklahoma state representative Jerry Ellis, a cattle rancher from southeastern Oklahoma, apoplectic. Mixing his metaphors, Ellis raged: "To allow this issue to be decided by the courts would gut democracy and the result would be Communism without a firing squad." Whatever he may think the courts are for, it's obviously not to interfere with Oklahoma's sovereign right to tell Texas to buzz off. Go Sooners!

The Tarrant Regional Water District is not the only agency in Texas with its eye on Oklahoma's water. The Upper Trinity Regional Water District has applied for 104 million gallons a day. Other Texas districts, including the North Texas Municipal Water District, are closely following the Tarrant litigation. Oklahoma should be concerned. Its population of 3.6 million is dwarfed by north-central Texas' 6.4 million, with increases expected to add another 2.7 million by 2030. Oklahoma's Jerry Ellis has offered not exactly an olive branch but sort of a solution to those in northern Texas who need water: "You all come over here," says Ellis, "and you can have the water you want if you pay taxes, employ our people and build plants." Otherwise, Ellis says, "I'll sell it to you one bottle at a time."

As the arid southwestern states fight wars that have raged for years, new wars are emerging in wet states. For most of the year, weather forecasters in Portland, Oregon, predict some variation of "Morning fog will yield to mist, followed by scattered showers. Toward evening, expect rain, heavy at times." That's why it's so surprising that Portland suburbs should be scrambling to find water. The Clackamas River, which flows from Mount Hood to the Willamette River, contains runs of endangered chinook and coho salmon, as well as winter steelhead. A 2005 Oregon law requires water utilities to show that diversions from rivers will not harm endangered fish species before they can begin diverting water. To protect the fish, Oregon's Water Resources Department in 2007 recommended a 62 percent increase in Clackamas River flows in summer. That announcement sent a seismic tremor through the planning departments of the water utilities that serve Lake Oswego and the other fast-growing suburban communities in Clackamas County. Dan Bradley, manager of the North Clackamas County Water Commission, a junior rights-holder on the river, described the impending restrictions as "a catastrophe" that puts "fish recovery on the backs of municipalities." In contrast, John DeVoe, executive director of WaterWatch, an environmental group that sued to protect the fish, dismissed Bradley's rhetoric as hyperbole. The six-week summer window for water restrictions affects lawn watering, not drinking water, said DeVoe.

And in California, Los Angeles experienced its worst drought on record in 2007 when the city went 150 consecutive days without rain. As 2008 rolled around, the usual winter rains did not materialize, and the city suffered the driest spring on record. Yet Long Beach was the only major metropolitan area in Southern California to mandate conservation.

Farther north, in California's Bay-Delta ecosystem, the contentious interface between San Francisco Bay and the Sacramento and San Joaquin rivers, a dry winter in 2007 devastated fish populations and forced the California Department of Water Resources to reduce water deliveries to San Joaquin Valley farmers and to Southern California cities. Also in 2007, the Supreme Court of California held that the environmental analysis for Sunrise Douglas, an 18,000-home development under construction in Rancho Cordova, near Sacramento, was inadequate because it failed to identify sources of water for the project. In the San Francisco Bay Area, in May 2008 the East Bay Municipal Utility District imposed mandatory water rationing on its 1.3 million customers, including residents of Berkeley and Oakland. Residents of single-family homes were required immediately to reduce their water use by 19 percent; golf courses, by 30 percent; and refineries and manufacturers, by 5 percent. Yet across the bay, San Francisco's water utility took no such drastic action.

An exasperated Governor Arnold Schwarzenegger declared the state officially in a drought condition in June 2008, paving the way for more widespread rationing. Lester Snow observed, "The governor is ringing the bell. We're heading over a cliff." Some communities are tone-deaf. After an east Sacramento couple, Anne Hartridge and Matt George, decided to pitch in by no longer watering their lawn, the city threatened to fine them $748 for failing to "irrigate and maintain" their front yard.

The economic consequences of the water crisis for the nation's largest economy will be substantial. Throughout the state, local governments have denied or delayed permits for dozens of housing and commercial developments. In June 2008, federal officials told farmers in the Westlands Water District, which sprawls over 600,000 acres and generates $1 billion in farm revenue, that water allocations would be cut by 60 percent. At a meeting of 400 farmers, the district's general manager, Tom Birmingham, told a reporter, "Half the people in this room are going to go broke."

Around the country, people resort to extreme measures to forestall water shortage and the economic dislocation it brings. In Arizona, hauling water has become a routine part of life in rural areas and on the Navajo Reservation, where more than half the population hauls water to their hogans. In Ash Fork, a hamlet west of Williams, Arizona, Tracy and Eddie Hunter and their three children have a 5,000-gallon storage tank at their home, which holds enough water to last approximately three weeks. They drive their pickup, fitted with a smaller tank, to a water-filling station, where they pay to fill their tank. It's expensive and an annoyance. Sometimes Tracy must even select which clothes to wash. If she washes all the dirty clothes at once, "we might not have enough water to take showers." Horse rancher Danny Larm, who also lives in Ash Fork, has been hauling water for thirteen years. "A lot of people think hauling water is bad," he says. "But around northern Arizona, it's just what everybody does." The nearby town of Williams exhausted its water resources recently and had to drill 3,700 feet to find a new supply.

The need to haul water is not confined to Arizona. In the tiny town of Orme, Tennessee, forty miles west of Chattanooga, the creek that serves as the town's water supply went dry in 2007. Now, three days a week, the fire department sends a tanker to fetch water. In Lucerne Valley, California, a high desert community near San Bernardino, thirsty residents rely on water trucked in by local water haulers. When a 2007 sting operation by the state health department put unlicensed haulers out of business, the residents were not happy. They depend on that water, even if they are gouged $100 for 3,400 gallons. More than 100 angry residents turned out for a meeting at which the state highway patrol tried to explain what had happened over shouts of "We need the water!" One little girl pleaded into the microphone, "I don't want to die." Melodrama notwithstanding, the problem is acute.

Some westerners have their sights set on water-rich states as potential saviors, but they'll have to think again. Not even the Great Lakes, which journalist Peter Annin calls the Saudi Arabia of fresh water, have excess water. As the largest surface freshwater system in the world, the Great Lakes contain 95 percent of the fresh surface water in the United States. Roughly 40 million Americans and Canadians in eight states and two provinces rely on the Great Lakes Basin for their drinking water. This apparent abundance of fresh water has created fierce local pride in the Great Lakes and has generated fear over water demands of outsiders. A recent billboard in

Michigan shows four grisly men from Texas, Utah, New Mexico, and California sipping from straws in the Great Lakes. The caption reads "Back off suckers. Water diversion . . . the last straw." Whenever I give a talk in the Great Lakes region, I try to reassure the locals that we in the Southwest have no intention of diverting water from all the Great Lakes. We'd be content with merely one—Superior. Sometimes they laugh; usually they stare stonily, proving again that great comedians know their audiences.

Wild-eyed proposals to tap into the Great Lakes periodically crop up. In 1998, the Nova Group of Ontario proposed to fill tankers with water and ship 160 million gallons per year out the St. Lawrence River to Asia. In 2001, the city of Webster, New York, located on Lake Ontario, ran "Water for Sale" ads in the *New York Times* and *Wall Street Journal*, prompting inquiries from a Texas businessman who envisioned filling railroad cars with Great Lakes water and shipping it south. But these proposals had no financial viability: water is simply too heavy and too cheap to warrant such investment, never mind the political and environmental objections. In truth, the real threat to the Great Lakes comes not from states in the Southwest or from Asia but from neighboring states and cities located just outside the Great Lakes Basin.

Still, it was the distant threats that prompted the Great Lakes states and provinces to craft a new management scheme for controlling diversions from the Great Lakes. Four years of negotiations led, in 2005, to the Great Lakes–St. Lawrence River Basin Water Resources Compact. The proposed compact treats the Great Lakes and associated groundwater as a shared resource. With limited exceptions, it outlaws new or increased diversions. Existing withdrawals of Great Lakes water remain subject to regulation by the individual states, as are new or increased diversions for use within the Great Lakes Basin. The compact bans diversions outside the basin, with an exception for communities that straddle the perimeter of the basin. By 2008, all eight Great Lakes states (Minnesota, Wisconsin, Michigan, Indiana, Ohio, Pennsylvania, and New York) had ratified the compact, which Congress then approved and President Bush signed. (A parallel agreement requires approval by the two countries.)

In the late nineteenth century, Waukesha, Wisconsin, was known as the "Saratoga of the West" for its mineral springs and hotels. The twentieth century saw the city evolve into a mix of factories and farms as its population grew, while neighboring Milwaukee's declined. The city of Waukesha's Web site boasts that in 2006 *Money* magazine named it the "36th best small city to live [in]," a dubious honor, or so I thought until I contemplated just how many small cities exist across the country. As Waukesha grew, it increasingly depended on groundwater from a deep aquifer. The strain of new subdivisions caused the groundwater table to plummet by more than 600 feet and the concentration of radium, a naturally occurring contaminant, to rise to more than double the maximum level set by the Environmental Protection Agency. The aquifer is now unusable.

In response, the Waukesha Water Utility wants to divert 20 million gallons per day from Lake Michigan, fifteen miles away. But it faces a problem: Waukesha sits five miles west of the boundary of the Great Lakes Basin, on the other side of an imperceptible rise, the subcontinental divide, from which water flows west into the Mississippi River system, not east into the Great Lakes. The utility's proposed diversion runs afoul of the compact's ban on new diversions out of the Great Lakes Basin. To secure Great Lakes water, the utility will need the approval of all eight states, and it may not even get Wisconsin to agree. Elected officials in Milwaukee, the state's largest city, oppose sending Lake Michigan water to suburbs whose growth has come from siphoning off residents, businesses, and the tax base from Milwaukee.

To outsiders, it may seem paranoid for the Great Lakes states and provinces to worry about diversions from such enormous bodies of water, but only 1 percent of Great Lakes water is replenished each year. The Great Lakes are essentially a nonrenewable resource, not a giant reservoir available to quench the thirst of any and all customers. Diversions in excess of inflows could seriously damage the

Great Lakes Basin's surprisingly fragile ecosystem. Lest anyone think bodies of water that large can't be destroyed by humans, the Aral Sea, in the former Soviet Union, offers a somber lesson. Pursuing an ambitious agricultural irrigation program in the 1950s, Soviet planners diverted so much water from rivers that flow into the sea that the world's fourth-largest inland body of water lost 90 percent of its volume and 75 percent of its surface area in mere decades. Peter Annin, in his book *The Great Lakes Water Wars*, describes this ecological calamity: "The Aral has receded so far that it takes more than five hours of driving on the old seabed in a four-wheel-drive vehicle to get from Muynak, on the old south shore, to the edge of what's left of the shrunken Aral—a distance of more than sixty miles." To Annin, the Aral Sea is a vivid, if distant, reminder of how quickly humans can destroy an ecosystem.

Lake Superior, which is the size of South Carolina, is experiencing the longest period of below-average levels since the U.S. Army Corps of Engineers began taking measurements in 1918. Each one-inch drop represents a loss of 500 billion gallons of water. Lowered lake levels are even more dramatic in the Georgian Bay area, at the northeastern portion of Lake Huron. Here, large expanses of lake bed are exposed and dry, wetlands are desiccated, and what were once islands are now part of the mainland. Prime swimming holes have turned into mudflats. The Georgian Bay Association, a homeowners' group, suspected that the declining lake level was due to the Corps of Engineers' dredging of the Saint Clair River, which connects Lake Huron and Lake Saint Clair. The Corps has dredged the river, north of Detroit, several times to allow freighters access to the upper Great Lakes. As the outlet for Lake Huron, the river allows water to flow from the upper lakes, especially Michigan and Huron, into Lake Erie.

Mary Muter, who owns a cottage on Lake Huron, and other Georgian Bay Association members went door-to-door soliciting $100 or $200 donations to fund a blue-ribbon study of the causes of the declining lake levels. They collected $200,000 and promptly

Figure 5.3. Boat Dock on Georgian Bay.
Photograph courtesy of Georgian Bay Association.

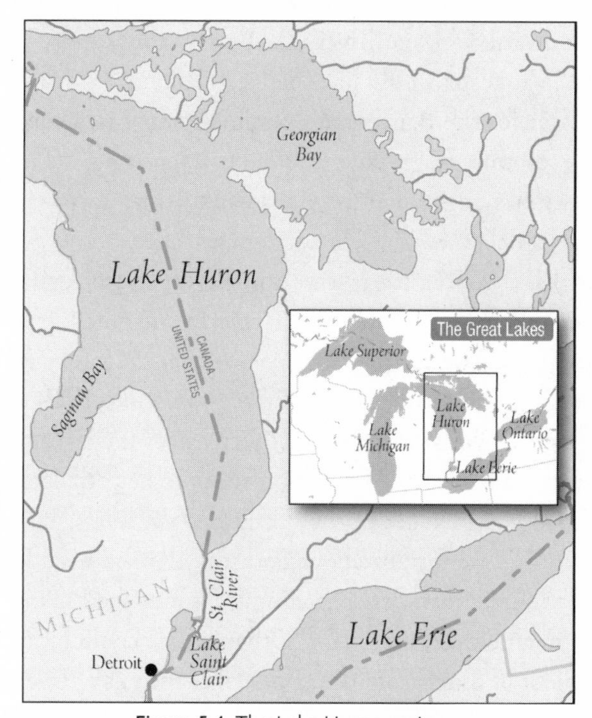

Figure 5.4. The Lake Huron region.

commissioned W. F. Baird & Associates, a prestigious Canadian engineering firm, to study the problem. In a 2005 report, Baird concluded that the navigation channel was deepening through erosion, allowing more water to flow from Lakes Michigan and Huron to Lake Erie. The Corps concedes that its dredging has lowered the levels in Lakes Michigan and Huron by sixteen inches, but the Baird report found that scouring of the channel lowered the upper lakes' levels by another twelve inches. And some people think the erosion is continuing.

In 2007, Bill Bialkowski, an engineer and member of the Georgian Bay Association, made fresh calculations with startling results. He figures that the "drain hole" created by the dredging has increased the amount of water flowing from Lakes Michigan and Huron by a staggering 2.5 billion gallons every day. That's more water than is used by the city of Chicago (2.1 billion gallons per day) to supply 7 million residents of the Chicago area and to help with navigation on the Chicago River. That's 100 times more water than the city of Waukesha would like to divert. Robert B. Nairn, the principal engineer for the Baird study, commented: "I was surprised that something of this magnitude could be happening." The bathtub drain theory has angered some Canadians: "The Yanks are stealing our water," proclaimed one Canadian newspaper.

Great Lakes water levels rise and fall in cycles usually lasting about fifteen years, in response to changes in snowmelt, precipitation, and evaporation. The Corps is not persuaded that the Baird report is correct, and the International Joint Commission (IJC), the U.S.-Canadian agency with regulatory authority over the Great Lakes, has commissioned a five-year, $15 million study to be completed by 2009 that will assess the causes of the declining lake levels. Homeowners in Georgian Bay are furious about both the claimed need for another study and the time it will take to conduct it, but the IJC is a cautious organization and much is at stake. Other causes could be accelerating the movement of water from the upper lakes into Lake Erie, including some geologic changes. When the glaciers crept down and gouged out the Great Lakes, receding roughly 10,000 years ago, their enormous weight pushed down the earth's crust, which is slowly, and unevenly, recovering. Scientists call this process isostatic rebound. One theory is that perhaps a slight uplifting has tilted water away from Georgian Bay.

But the Baird scientists don't buy it. As the water levels in Lakes Michigan and Huron dropped over the past three decades, the level in Lake Erie should have dropped as well. But it hasn't. In 2007, Lakes Michigan and Huron remained at the same level as in 2006, but Lake Erie rose six inches. "We're seeing drastic sustained decline in the Michigan–Huron system at the same time that Lake Erie is rising," notes Bill Bialkowski. "This is indicative of water loss independent of naturally occurring fluctuations."

Ironically, the good news would be for the IJC to find that the drain hole is the explanation. In that case, engineers could install physical structures to slow the movement from Lake Huron to Lake Ontario. Such a technical fix would be politically explosive, enormously expensive, and a regulatory nightmare. But it would probably work. However, scientists have identified global warming as another important variable, especially for Lake Superior. Warmer temperatures have decreased the lake's winter ice cover and exposed more surface area to loss through evaporation. If Great Lakes levels are declining as a result of global warming, the Corps is helpless to fix it.

The tinderbox of the Great Lakes water fight exploded again in 2007 when Democratic presidential candidate Bill Richardson, governor of New Mexico, called for a national water policy with a dialogue between states to deal with water problems. Campaigning in Nevada, he explained that water-rich northern states could help alleviate shortages in southwestern states. In an interview with a Las Vegas newspaper, Richardson announced, "States like Wisconsin are awash in water." His popularity took a quick nosedive in Wisconsin and other Great Lakes states. Richardson quickly backpedaled after

Michigan's governor, Jennifer Granholm, attacked his idea, saying simply, "Hell, no!" But the seed of fear had been planted, once again, that distant states are plotting to get their hands on Great Lakes water.

A CRISIS IS A decisive or crucial time, a moment when the ultimate ending remains in doubt, when action might avert catastrophe—a disastrous ending. Whether the water crisis will turn out benignly or will result in economic collapse, mass migrations, environmental devastation, and more bitter interstate battles depends on how we respond—or fail to.

Part Two

Real and Surreal Solutions ~

CHAPTER 6

Business as Usual

"When you put your hand in a flowing stream, you touch the last that has gone before and the first of what is still to come."
—Leonardo da Vinci

IN THE PAST when we needed more water, we engineered our way out of the problem by diverting rivers, building dams, or drilling wells. Today, with few exceptions, those options are not viable solutions.

Between 1852 and 1916, steamboats regularly traveled 400 miles up the Colorado River from its terminus at the Delta in the Sea of Cortés in Mexico to Fort Yuma and other outposts just south of present-day Las Vegas, servicing the needs of gold miners and other free spirits on the frontier. Steamboat traffic eventually petered out, eclipsed first by the intercontinental railroad and then by the automobile. Today, that steamboat journey would be impossible, as the mighty Colorado River no longer reaches the ocean, its flows having been depleted or reduced to a trickle by more than 1,600 dams and diversions on the Colorado and its tributaries. As hard as it is to believe, the Colorado River usually runs dry before it reaches the ocean.

A dry river that is not an oxymoron should frighten all of us. But the Colorado has lots of company. In most years, the Rio Grande doesn't reach the Gulf of Mexico. A similar pathetic fate besets the

Hillsborough River in Florida; the Walla Walla River in Washington; the Santa Fe River in New Mexico; the Gila, Salt, and Santa Cruz rivers in Arizona; the Big Wood River in Idaho; the Ipswich River in Massachusetts; the San Joaquin and Kern rivers in California; the Frenchman River and Pumpkin, Lodgepole, Prairie Dog, Beaver, and Sappa creeks in Nebraska; the Beaver and North Canadian rivers in Oklahoma; the Big Hole and Sun rivers and Bear, Middle, Indian, and Big creeks in Montana; and the Cache la Poudre, North Fork of the Gunnison, San Miguel, Fraser, and La Plata rivers in Colorado. A section of each river goes bone-dry from our diversions, dams, and wells.

Some rivers, such as the Santa Cruz, are dry year-round except after rainstorms. Others, such as the Ipswich and Hillsborough, suffer seasonally, usually in summer. For still others, such as the Gila and Rio Grande, certain sections dry up. Two hundred miles of the Gila, west of Phoenix, carry no water. Cities and irrigation districts regularly dewater the Rio Grande for fifty to sixty miles south of Albuquerque. And the Los Angeles River, once the primary water source for the city, has become a fifty-one-mile concrete-lined perversion of a once-vibrant river. Despite hopes for its restoration, it flows only intermittently, carrying storm water mixed with urban detritus.

Many other rivers, and countless springs, creeks, and wetlands, have similarly suffered, though researchers and government agencies have not systematically documented either the names or the number of dried-up watercourses. It's surely a frighteningly large number, yet it is only a tiny percentage of the total harmed by diminished flows. The severity of the impact of diversions on rivers depends on the size of the river and the amount of water diverted; withdrawals of water typically cause a spike in the temperature of the water left in the river. Temperature increases lead to oxygen depletion, algae blooms, and fish kills. Levels of industrial and agricultural pollutants jump, as there is less dilution of these noxious chemicals.

It was once thought that water left in a river was wasted, but we now know—thanks to the development of the fields of biohydrology and ecohydrology—that freshwater flows into oceans are critical to the survival of aquatic species and to the maintenance of spawning habitats. In estuaries, the intertidal zone receives nutrients from upstream and wards off saltwater predators. The larval and juvenile stages of fish, oysters, crabs, and shrimp thrive in the nursery conditions of the intertidal zone. A 2004 report of the National Wildlife Federation disturbingly concludes that increased diversion of water by Texas cities and farmers endangers most of the Texas bays in the Gulf of Mexico, including Galveston, Matagorda, San Antonio, and Corpus Christi. Equally essential to the health of our rivers, intermittent floods scour sediment, nourish habitat, and impede the encroachment of invasive species. Additional diversions of surface water would diminish these peak flows. The bottom line is this: in most parts of the country, increasing withdrawals of surface water from rivers as a solution to water shortages would threaten public health and cause environmental degradation. Many of our rivers are already dead or on life support.

Nonetheless, to some people, the solution to our water woes is to build still more dams. We're good at that—so good that only sixty rivers in the United States remain free-flowing. Early settlers in New England dammed rivers to provide power to run small grist mills, and later, during the industrial revolution of the nineteenth century, large dams powered textile factories, processing cotton grown on plantations in the South. In the twentieth century, the Tennessee Valley Authority became a political force to be reckoned with, as its tentacles spread over seven states, by providing dams for flood control, navigation, and electric power generation. The U.S. Army Corps of Engineers, a Department of Defense agency created in 1775, has assumed major responsibility for damming rivers to control flooding and enhance navigation around the country, but it is best known, or most ignominious, for its work on the Mississippi River, the Missouri River, and the Columbia River.

Nowhere is the effort to tame Mother Nature to serve human needs more evident than in the American West. The U.S. Bureau of Reclamation, an agency of the Department of the Interior, embarked on a dam-building frenzy in the 1930s, beginning with the construction of Hoover Dam in 1931. The remarkable construction of Hoover Dam symbolized the ability of the United States to do almost anything, even harness the mighty Colorado River. To a nation in the throes of the Great Depression, it was a welcome boost, exemplifying the kind of ambition that promised a bright economic future. And the power for that future would come from dams, lots of them. The water aided western farmers, and the power enabled American companies such as Boeing, Lockheed, and Martin Marietta to become aerospace giants. By the end of the dam-building era in the 1960s, most major western rivers had been dammed, some repeatedly. Indeed, in the United States as a whole, there are 75,000 dams at least six feet high and as many as 2.5 million smaller dams. Former secretary of the interior Bruce Babbitt has poignantly observed that we have built "on average, one large dam a day, every single day, since the Declaration of Independence."

Thanks to dams, we move huge quantities of water out of watersheds, often to cities hundreds of miles away, pumping water over mountains and drilling tunnels through mountains. In Arizona, a concrete-lined canal called the Central Arizona Project moves water from the Colorado River east to Phoenix and eventually south to Tucson, a distance of 330 miles and almost 2,900 feet up in elevation. Dams in the Colorado River Basin also supply water to major cities outside the basin, including Denver, Albuquerque, Salt Lake City, San Diego, and Los Angeles, thanks to an elaborate system of canals, aqueducts, and tunnels. Denver gets Colorado River water pumped up and through the Rocky Mountains; Albuquerque, from the San Juan River through diversion structures and tunnels; and Los Angeles and San Diego, through the Colorado River Aqueduct, which transports the water 242 miles across the Mojave Desert and over the San Jacinto Mountains. We literally move water uphill to wealth and power. Thanks to dams, enormous growth has occurred in places without adequate water resources, such as Los Angeles. The growth of these urban areas depends on an elaborate infrastructure that alters Mother Nature's hydrologic boundaries and profoundly harms the environment.

Advocates of new dams frequently portray them as a way to augment water supplies. But dams don't add water to a river. They help farmers and cities by providing water at a time when Mother Nature doesn't. In the West, most water comes from winter snowfall in the mountains. During the spring thaw, snowmelt creates cascading rivers that provide more water than farmers or cities need for that season. But by mid- to late summer, when farmers and cities actually need water, river flows may be minimal. By creating storage reservoirs, dams provide year-round water for municipal, industrial, and agricultural use.

In 2007, California's governor, Arnold Schwarzenegger, began what he dubbed a "holy war," a fight to build two new dams and expand an existing reservoir to increase the capacity of the state's water storage system. A new dam on the San Joaquin River would create the Temperance Flat Reservoir in northern California near Fresno; the Sites Reservoir, to be located sixty miles north of Auburn, would store water from the Sacramento River off-stream in a barren valley; and the third project would expand Los Vaqueros Reservoir, north of Livermore. The combined dams would hold 3.2 million acre-feet, enough water to satisfy the needs of more than 6 million families for a year. The state would collect water during wet years for use during dry years. At least that's the theory.

The governor has a battle on his hands. Supporters think that the political equivalent of "a perfect storm" makes conditions right for building the first major dams in California in more than thirty years. The continuing drought has diminished the supply from the Sierra Nevada. The Colorado River, an important water source, has had below-normal flows for eight straight years. Attention to global warming has heightened the sense that California is vulnerable. And, finally, the decision to turn off pumps in the Sacramento–San

Joaquin River Delta to protect endangered delta smelt crimped the supply system.

Lester Snow, director of the California Department of Water Resources, defended his boss' decision as "one way to manage the effects of climate change." Snow links new dams to controlling floods, which are expected to worsen as a result of global warming. The state currently releases water from existing dams on the San Joaquin and Sacramento rivers before snowmelt or floods can overwhelm them. This water is not needed by farmers or cities when it's released. New dams, Snow argues, would enable the state to capture this water and store it off-stream for future use. In a state such as California, with an elaborate water infrastructure, new dams would give Snow's agency greater operational flexibility to respond to changing conditions.

The proposal for new dams is popular with Republican legislators, powerful Central Valley agricultural interests, urban water system managers, and, naturally, the construction industry. The governor picked up new allies, including comedian Paul Rodriguez, who chairs the California Latino Water Coalition. The coalition emphasized the reality that many Latinos depend on the state's $37 billion agriculture industry. Quipped Rodriguez, who was born in Sinaloa, Mexico, "We came to this country to make sure Caucasians have salad." However, opposition surfaced quickly. California assembly speaker Fabián Núñez, a Democrat from Los Angeles, proclaimed simply: "Water storage is not going to happen." His party advocates conserving, recycling, developing underground storage, and adding to the height of existing dams. Similarly, fifteen major environmental and fisheries groups, including California Trout and the Natural Resources Defense Council, favor investing "millions in efficient water-use technologies and programs that we know will reduce demand" rather than "billions in costly and environmentally destructive dams."

If the dams are built, the cost will certainly be in the billions; just how many billions is the subject of debate. In early 2008, the best estimate was $10.3 billion, but the costs are likely to rise, a lesson learned from another controversial dam proposal on the American River—the Auburn Dam. The dam was first authorized by Congress in 1965, but construction was halted a decade later when seismic studies, after a nearby magnitude 5.7 earthquake, revealed a fault line directly under the dam site. One would have thought this revelation would permanently end discussion of resuming construction, but, because most good dam sites in California already have dams on them, the Auburn Dam proposal has limped along for decades. A revised feasibility study, released by the Bureau of Reclamation in 2007, projected the total cost for the Auburn Dam at $9.6 billion—ten times the original estimate and three times an estimate made by the Army Corps of Engineers only two years previously. In 2008, the California Water Resources Control Board hammered the final nail in the coffin of the Auburn Dam when it revoked the water rights for the project because the Bureau of Reclamation had not completed the dam "with due diligence."

The components of these Auburn Dam estimates demonstrate why so few dams are being built today. The "field cost" of the dam penciled out at $5.4 billion for the actual dam, a power plant, an electronic substation and transmission lines, highway and road relocation, site preparation, and "contingencies." Here's where things get interesting. It would also cost $2.3 billion for land purchases (beyond what has already been purchased) and $1.5 billion for environmental mitigation. (The American River has popular recreational activities and endangered or threatened species.) The combined $3.8 billion in land and mitigation costs actually exceeds the costs for constructing the dam itself. No wonder the private sector has shown little interest in underwriting the cost of dams.

The logistics of planning and permitting a dam—with environmental impact statements, drafts, reviews, reports, and decisions—ensure it will take years or even decades to build a dam. Planning for the Sites Reservoir began in the early 1990s. And major issues remain unresolved, such as finding the electric energy needed to pump the water out of the river, over a hill, and into the reservoir. A power plant will generate electricity when water is released from the reservoir. Nonetheless, Peter Gleick, a MacArthur Fellow and founder of the Pacific Institute, an environmental think tank, concludes: "This reservoir will use more energy than it produces—that's a fact. That means that it's going to be a producer of greenhouse gases because we will be using fossil fuels to run the pumps." Add to these objections that some of the stored water will evaporate off the surface of the new and expanded reservoirs.

Because most rivers in California are already tapped, new dams would compete with existing ones for the same water. Barry Nelson of the Natural Resources Defense Council notes that the existing dam on the San Joaquin River already diverts the entire flow of the river in seven out of ten years. In those years, there is no surplus water for the Temperance Flat Reservoir to store. Critics think it's folly to build more dams when the existing ones don't fill up regularly.

Also left unanswered are the questions of who gets the water and who pays for the dams. The Metropolitan Water District of Southern California, which supplies 18 million Californians, has refused to support Governor Schwarzenegger's proposed dams unless there is a commitment to fix the conveyance system. In the 1990s, the district purchased water from northern California farmers only to find out that the canal system for transporting the water to Los Angeles was already full. From the perspective of the district's general manager, Jeffrey Kightlinger, the new dams will provide no benefit to Southern California without a new plumbing system.

The idea of a new transportation system has revived bad memories of one of the most notorious water feuds in California history. In the early 1980s, Southern California interests wanted the state to build the Peripheral Canal, a conveyance system that would move water around the Sacramento–San Joaquin River Delta. Northern Californians considered the canal a scheme to grab water and, after an epic struggle, defeated it at the polls. But it's back on the table, with initial cost estimates ranging from $2 billion to $3 billion. Dam supporters argue that the new dams would also benefit Central Valley growers, but Bob Alvernay, a seventy-six-year-old cattleman and rice farmer, is dubious. "The water won't be for us. It will be too expensive. It's for the cities down south."

Finally, who's going to pony up $10.3 billion for the dams and the extra $2 billion or $3 billion for the canal? In the dam-building era from the 1930s to the 1960s, the federal government paid the lion's share. Under a cozy relationship with the Bureau of Reclamation, state interests got the federal government to front the money on the theory that the beneficiaries of these large water projects would eventually pay back the government. It seldom turned out this way. Massive subsidies, such as zero-interest loans to farmers for fifty years and Reclamation's discretion to describe benefits of a dam, such as flood control, as a federal nonreimbursable purpose, drove down repayment obligations. In Arizona, we think that's precisely the role of the federal government: to arrive with wheelbarrows full of cash, build our water projects, and then go away and leave us unregulated. After all, we're independent westerners. But the federal budget is in bad shape these days with massive deficits as the result of waging two wars, providing tax cuts, and bailing out private companies. In the days when Floyd Dominy was the commissioner, Reclamation was the best friend of large federally financed water projects. California water interests probably didn't expect the current commissioner, Robert Johnson, to waffle on federal financial support. "How much money we can bring to the table is a tough question," said Johnson. "That will be a real struggle for us."

The California governor wants the legislature to enact a $9 billion bond package; otherwise he'll launch a ballot initiative. But even water utility lobbyists, such as Randy Kanouse of the East Bay Municipal Utility District, are reluctant to ask state taxpayers to bear the costs of a project that will benefit only some ratepayers. Assuming that the governor succeeds with his bond funding, that leaves several billion dollars to be paid by water users. Governor Schwarzenegger has shown great staying power and remarkable comeback ability in the past. For this "holy war," he'll need all the political savvy he can muster to get the new dams.

Arnold Schwarzenegger is not the only governor who'd like to build new dams. Idaho's governor, C. L. "Butch" Otter, favors building more dams and expanding existing ones on the Snake River in order to help move water in Idaho. His 2007 proposal has as little chance of success as the proverbial snowball in hell. Dams on the Columbia-Snake river system have cut off access to salmon spawning streams, extirpating several species of salmon and threatening others.

Critical attention has been focused on four dams in particular—Ice Harbor, Lower Monument, Little Goose, and Lower Granite—all of them on the Snake River above its confluence with the Columbia River. After these dams were completed in the 1970s, salmon runs plummeted. Since then, the Bonneville Power Administration has spent tens of millions of dollars on fish ladders to bypass dams, and even barges and trucks to transport salmon fry downstream through the locks on out to the Pacific Ocean. I love the image of one-inch fish riding on barges and trucks. It's a Rube Goldberg solution to the problem of dams blocking fish passage, but the massive amounts of hydroelectric power generated by these dams give industries and residents of the Northwest some of the cheapest electric rates in the country and have encouraged a search for creative ways to save the fish while operating the dams to maximize power production. Despite these efforts, four runs of wild salmon remain in peril, with one species, sockeye, in danger of extinction.

This dire situation has led to strident calls for removing or "breaching" the four Lower Snake dams. The calls gained traction when critics pointed out that the dams add only small amounts of hydropower to the grid. These dams principally provide an artery of transportation for Idaho wheat farmers to barge their wheat down the Snake and out the Columbia on the way to Asia. Other farmers use railroads.

In 2002, the Corps of Engineers concluded a seven-year study of methods for improving salmon passage through the four dams.

The Corps rejected breaching the dams in favor of additional tinkering with the operation of the dams, expensive tinkering to be sure. The Corps' proposals, which would cost $390 million over ten years, did not satisfy its critics. Rob Masonis of American Rivers said, "There is apparently no end to the Corps' appetite for retrofitting the Snake River dams with expensive devices that provide little, if any, benefit to salmon." U.S. District Court judge James A. Redden agreed with Masonis. The judge's 2005 ruling acerbically dismissed the Corps' contention that the dams were an immutable part of the landscape that could not be removed. In 2007, a federal appeals court upheld Redden's decision that the federal government was violating the Endangered Species Act by not taking more aggressive steps to protect the salmon.

Governor Otter's pitch for new and expanded dams may play well with his constituents, but the governor is walking into a hornet's nest of environmental opposition, regulatory dead ends, and legal constraints. The governor claims there are two or three good dam sites, but he didn't happen to mention where they are.

Proposals for new dams rarely pass the straight-face test. Consider Black Rock Reservoir, in eastern Washington adjacent to the Columbia River. In 2003, Congress authorized the Bureau of Reclamation to initiate a feasibility study for diverting water from the Columbia and pumping it several miles to the reservoir. But there's a catch. It's mostly uphill to the reservoir site, a 1,400-foot elevation gain. The proposal, styled a "water exchange," would secure for farmers in the Yakima River Valley a storage facility in exchange for reductions in their direct diversions from the Yakima River, thus improving instream flows and benefiting salmon.

A seasoned environmental lawyer, Rachael Paschal Osborn, and her organization, the Columbia Institute for Water Policy, are having a field day with this one. The Black Rock Reservoir dam would be a very large dam, larger even than Grand Coulee Dam on the Columbia River, yet would contain only 1 percent of the water held behind Grand Coulee. Initial cost estimates came in at $4.2 billion in construction expenses and $78 million in annual operating costs; by 2007, the figure exceeded $6 billion. Can the local apple, grape, and hops farmers possibly afford to repay these costs to the federal government? Not a chance. The Columbia Institute concludes that this project, costing in excess of $4 billion, "would yield only $4 million per year in new agricultural receipts." That's a 0.1 percent return on investment. That's why the Bureau of Reclamation is casting the project as having fisheries benefits—a federal purpose backed by federal dollars. But the Columbia Institute's dissection of Reclamation's cost-benefit analysis showed that the approximately $29 million in fishery benefits involved $900,000 for the actual fishery and $28 million for "nonuse" benefits. This nonuse category is how economists try to quantify the public's desire to protect a resource they'll never see, fish, hunt, or use, such as caribou herds in Alaska. As the Columbia Institute sees it, if the goal is to protect salmon, put the $4-billion-plus into concrete fish-recovery measures.

Then there's the energy cost. The Bureau of Reclamation estimates that the Black Rock project could generate $8.8 million per year in hydropower as water is released from the reservoir into irrigation canals with power plants. But that's offset by an estimated loss of roughly $4 million in power revenues on the Columbia as a result of the water diverted. The Columbia Institute ridicules Reclamation for not explaining how two irrigation canal power plants would generate twice the electricity of five dams on the Columbia. And "it would cost $62 million each year to pump water from the Columbia River, 1,400 feet up to the Black Rock reservoir." If that is true, the federal government would annually spend $62 million to net $4 million. That's good news for someone but surely not for the federal taxpayer.

But the death knell for the Black Rock Reservoir may have come in September 2007, when a Reclamation study disclosed that water seeping from the reservoir might spread radioactive contaminants. Five miles west of the reservoir site is the Hanford Site, once part of the Manhattan Project, which produced plutonium for nuclear weapons, and now the largest Superfund site in the country—some 586 square miles, or half the size of Rhode Island. If the Black Rock Reservoir site fills with water, seepage would percolate into the ground and begin migrating toward the Hanford Site. Jane Hedges, the Hanford program manager for the Washington Department of Ecology, is concerned that the seepage "would raise the water table and rewet, remobilize contaminants." So is the U.S. Department of Energy, which manages the Hanford cleanup. A toxic soup of tritium, iodine-129, technetium-99, and uranium-238 could eventually leach into the Columbia River, creating significant health risks. Before the Black Rock Reservoir is built, it must overcome a daunting price tag, speculative benefits, public safety concerns, and environmental objections.

In the 1950s, it seemed inconceivable that there could be a downside to building dams. But there was. Decades later, we came to realize how profoundly dams alter watersheds. Dam construction inundated some of the most beautiful canyons in the West, such as Hetch Hetchy and Glen Canyon, and transformed the rivers below the dams. Water flowing from a dam has a constant temperature, as opposed to the temperature of river water, which fluctuates with the seasons. Most dams increase water temperatures in downstream rivers, but some decrease temperatures as they release cold water from the bottom of the reservoir. The flow itself depends on the decisions of engineers, not on a natural rhythm. Native fish and other aquatic species suffer from these changes when the nutrients that formerly sustained the downstream aquatic habitat become trapped in the reservoirs upstream. Anadromous fish—Pacific salmon and steelhead on one coast and Atlantic salmon and shad on the other—have found the paths to their spawning grounds blocked by impassable edifices. Some species have become extinct; others have merely suffered. Dams also decrease the level of oxygen in reservoir water. The release of this oxygen-deprived water can kill fish downstream. Today, opposition to the building of new dams is quite substantial. Having witnessed the profound alteration of the hydrologic regime of dammed rivers, the environmental community is adamant that new dams are not an environmentally acceptable solution to problems of water shortage.

Rather than building dams, we have begun to decommission them. We are actually removing dams faster than we're building them. This movement began with the push of a button detonating explosives that destroyed Edwards Dam in Maine in 1999. Since then, almost 300 other dams have been removed, restoring thousands of miles of free-flowing rivers. There is traction to this movement as communities realize the considerable economic and environmental value to be gained from eliminating aging dams that generate little hydropower.

Thus far, most of the dismantled dams have been small, insignificant, and uncontroversial. But the removal of Edwards Dam immediately produced stunning results. Within a year, Atlantic salmon, alewives, sturgeon, and shad surged back into the river. Now attention is shifting to larger, more important dams, some of which generate substantial hydropower. Support for this movement came from a surprising source—the George W. Bush administration. In 2005, the U.S. Department of Commerce's National Oceanic and Atmospheric Administration spearheaded NOAA's Open Rivers Initiative to provide funding for dam removal. Partnering with local communities, NOAA has removed more than eighty dams, opening 700 miles of river to migrating fish.

The most expensive dam removal project in history will begin in 2009 in Olympic National Park in Washington. The 105-foot-tall Elwha Dam and the 210-foot-tall Glines Canyon Dam in the Elwha River Valley will be torn down to restore access to seventy miles of spawning habitat for steelhead and all five species of Pacific salmon. Even though Congress was prepared to provide $180 million for the deconstruction, most local citizens in the town of Port Angeles opposed the dams' removal until solid economic studies demonstrated that revenues from tourism would dramatically increase. Townsfolk then realized that the dams had outlived their usefulness. The

Bush administration's 2008 budget for NOAA included $10 million to help remove two more dams, the Veazie and Great Works, on Maine's Penobscot River. Their removal will open up almost 1,000 miles of rivers and streams for the endangered Atlantic salmon and other sea-run fish.

The relicensing process for privately owned dams is driving the decision to dismantle many dams. The Department of Energy's Federal Energy Regulatory Commission (FERC) regulates and licenses hydroelectric dams for periods of thirty or fifty years. Five years before expiration of the license, the dam owner must apply for a new license, a process that considers the strictures of the Clean Water, Endangered Species, and National Environmental Policy acts, three laws enacted *after* the dams were built. The owner must also demonstrate that relicensing is in the public interest, a test that looks at not just the dam's power-generating capacity but also its impact on environmental values and recreational uses. These factors led FERC to deny the owners of Edwards Dam a license renewal and prompted the owners to demolish it.

Bringing an old dam up to environmental and safety standards often greatly exceeds the current revenues generated by hydropower. In the end, it may be cheaper for the owner to tear down the dam. Left to its own devices, FERC would normally rubber-stamp most renewal permits, but the Federal Power Act and the Endangered Species Act require FERC to accept the judgment of the U.S. Fish and Wildlife Service and the National Marine Fisheries Service about the consequences of a dam's operation on threatened and endangered species. In addition, environmental organizations and other stakeholder groups get to participate in the relicensing process, which can stretch out over many years—in one case sixteen years.

This relicensing process is profoundly affecting the decisions about dams owned by PacifiCorp, a Berkshire Hathaway/Warren Buffett–owned company, on the Klamath River. Emerging from snowmelt in the Cascade Range of Oregon, the Klamath provides irrigation water to Oregon farmers before it crosses the border into

California; then it flows through a 200-mile reach designated a national wild and scenic river before it empties into the Pacific Ocean. Upstream diversions in Oregon have decimated steelhead and salmon runs on the lower river, where 1.2 million fish once spawned, prompting the federal government in 2001 to withhold water from 1,000 farmers in order to protect the downriver salmon. Armed with pitchforks and shotguns, outraged farmers protested this threat to their livelihood. Their protest was heard. The Bureau of Reclamation restored water to farmers in 2002, leaving thousands of salmon to flop pathetically in a few inches of water. Approximately 70,000 chinook salmon died from disease caused by the shallow, warm water—the largest fish die-off in American history. Indignation around the country rose as photographs and news clips documented the killing of these magnificent fish. Pitted against the farmers are the Yurok, Hoopa, Karok, and Klamath tribes, who have lived along the river for more than 10,000 years and whose diet and culture center on salmon. Also affected are commercial fishermen, whose 2008 season was canceled because of the decline in numbers of returning fish.

In 2007, the U.S. Fish and Wildlife Service and the National Marine Fisheries Service made a dramatic decision that may initiate what the *Washington Post* calls "the largest dam-removal project in world history." The agencies rejected the proposal of PacifiCorp to trap salmon and haul them around the dams in order to restore depleted fish runs. To obtain a renewal license, PacifiCorp would have to install fish ladders and other devices to allow the fish to bypass the dams. Okay, that sounds easy enough. Not quite. Modifying the aging dams could cost approximately $470 million, or $285 million more than the cost of removing them. In the face of the agencies' decisions, PacifiCorp's public relations spokesman, Dave Kvamme, bravely announced that the company would carry on, but added, "We are going to have to look at costs and risks."

In late 2008, PacifiCorp decided to endorse a dam removal agreement with Oregon, California, and the Bush administration. The company agreed to contribute $200 million toward the costs of dam removal. California will kick in as much as $250 million in additional funds, and a new federal agency will relieve PacifiCorp of liability during the dam removal process. Dam removal on the Klamath—and four dams are at stake—represents a watershed event that will open up more than 300 miles of habitat for the salmon and steelhead.

The water stored behind the dams on the Klamath mostly irrigates potato, sugar beet, and alfalfa fields, and the hydropower produced, while substantial, serves only 70,000 homes, 1.7 percent of PacifiCorp's output. That's a considerable amount of power but still minuscule compared with one of the most notorious dams ever built in the United States, the O'Shaughnessy Dam. Never heard of it? Well, it floods Hetch Hetchy, Yosemite National Park's sister canyon to the north, which was dammed in the 1910s over the vehement protests of John Muir. No one alive today knows firsthand the extraordinary beauty of the canyon it flooded, but historical photographs, from an era before Ansel Adams captured the rapture of Yosemite, hint at its majesty. To environmentalists, the restoration of Hetch Hetchy embodies the ultimate restitution. But the water stored in Hetch Hetchy Valley behind O'Shaughnessy Dam supplies 2.6 million people and 80,000 businesses in the San Francisco Bay Area. The hydropower from the dam lights San Francisco's schools and runs Bay Area Rapid Transit (BART). The idea of removing the dam strikes San Francisco officials as "just plain goofy." Yet in 2007 President Bush included $7 million in the National Park Service budget for "Hetch Hetchy restoration studies." To be sure, it was a trivial amount of money amid the red ink in a $2.9 trillion budget, and the appropriation committees in both houses of Congress omitted Hetch Hetchy funding, but sometimes symbolism sheds light on reality.

The reality is that the era of dam building in the United States is over. A few smaller dams will still be built, but anyone who thinks that we'll solve our water crisis by building lots of new large dams is willfully ignoring the absence of viable dam sites and the enormous costs and substantial environmental objections.

AFTER THE FEDERAL GOVERNMENT cut off their water in 2001, Klamath Basin farmers drilled more than 100 new wells and began pumping groundwater. Within three years, the groundwater table had dropped as much as twenty feet, and residential wells had begun to go dry. "We're seeing the result of the pumping stress," observed Ken Lite, a hydrologist with the Oregon Water Resources Department. Between 2001 and 2004, the department issued permits for approximately 130 wells, some capable of pumping 10 million gallons per day. Oregon law authorizes the department to issue new wells only "within the capacity of available sources" and requires the department to determine and maintain "reasonably stable ground water levels." Wells quickly affect the aquifer because the dense volcanic ground beneath the Klamath Basin does not hold a large volume of water, according to Michael Zwart, a hydrologist for the state. In 2002, environmental groups petitioned the department to impose a moratorium on new wells in the Klamath Basin, but the department refused. Even though a 2001 study found that wells will cause a lasting decline in groundwater levels, the department's Barry Norris reasoned that the department didn't know enough about the basin's hydrology to justify denying new permits. To WaterWatch of Oregon's Steve Pedery, what the farmers are doing "is like writing a series of huge checks when you don't know how much money you have in your account."

Groundwater was once thought to be as inexhaustible as the air we breathe. We now know better. In the nineteenth century, when the science of hydrology was in its infancy, American judges were perplexed as to how to divvy up rights to pump groundwater. Lacking guidance from scientists, judges threw up their hands and proclaimed, "If you can get it out of the ground, it's yours." And the free-for-all began. Since then, the science of hydrology has matured, but the legal rules have not kept pace. In most American states, groundwater law is governed by the "reasonable use" doctrine, which allows individuals to pump as much groundwater as they desire so long as it is used beneficially. Other states use an even more bizarre legal rule—the right of capture—which treats groundwater like a deer in the forest belonging to the hunter who shoots it. In the 1940s, technological breakthroughs vastly expanded the capacity to pump huge quantities of water from extraordinarily deep areas beneath the earth's surface. Commercial irrigation wells can now pump thousands of gallons per minute. Because the legal rules are so permissive, farmers have installed millions of wells and feverishly begun pumping.

Groundwater is a renewable resource unless we pump more than is recharged naturally, which is what we are now doing. Think of an aquifer as a giant milkshake glass, and think of each well as a straw in the glass. The water in the glass is limited, but access to it is not. The reasonable use doctrine epitomizes what Garrett Hardin called "the tragedy of the commons"—limitless access to a finite resource. Anyone can pump an unlimited amount of water. There is no incentive to husband groundwater because pumpers do not own the resource. Instead, the system encourages willy-nilly development. As a result, groundwater tables are plummeting in many sections of the country. Excessive groundwater pumping has caused the ground to collapse; rivers, lakes, and springs to dry up; and riparian habitat to die. If we continue to exploit our groundwater resources in this way, we will eventually run out.

The Ogallala Aquifer underlies parts of eight states, from South Dakota to the Texas Panhandle. This aquifer was left over when a huge inland sea receded about 80 million years ago. In the 1940s and 1950s, large-capacity wells increased groundwater pumping by more than 1,000 percent. By 1957, wells irrigated more than 3.5 million acres and, by 1990, 16 million acres to grow corn, milo, wheat, and alfalfa, turning the High Plains into "the breadbasket of the world." Water from the Ogallala supplied almost two-thirds of the total irrigated acreage in the United States. All this pumping eventually exacted a toll as the groundwater table began to drop. By 1980, these declines exceeded 150 feet in some places, and, despite farmers' attempts to conserve water, they continued for another 40 feet in portions of Texas and Oklahoma by 1995. Some wells have gone dry; countless others have been deepened. Still the water table continues to decline. Since 2000, groundwater levels in parts of Nebraska have dropped an additional 30 feet. During the same time, the Arbuckle-Simpson Aquifer in south-central Oklahoma has fallen more than 21 feet.

Excessive groundwater pumping is not confined to what was once the dust bowl. On Long Island, New York, excessive pumping has lowered the water level in parts of Nassau and Suffolk counties, home to 2.7 million people, to more than 20 feet below *sea level*. It doesn't take a rocket scientist to grasp Long Island's problem. Farther down the eastern seaboard, in Calvert County, Maryland, the water table, 30 feet below the land's surface in 1960, dropped to 150 feet by 2004. In Arkansas, farmers used almost 8 billion gallons of water per day in 2000, half of that to grow rice. The state of Arkansas is, surprisingly, the fourth-largest user of groundwater in the United States, even though the state gets an average forty-nine inches of rain a year. The Mississippi River Valley Alluvial Aquifer supplies water to Arkansas farmers and to water users in Missouri, Kentucky, Tennessee, Mississippi, and Louisiana. The U.S. Geological Survey (USGS) predicts that portions of the aquifer may go dry as early as 2009. The U.S. Army Corps of Engineers thinks the water may last until 2015.

Some areas of the country have experienced even more severe declines in their water tables. A tenfold increase in pumping in Baton Rouge, Louisiana, between the 1930s and the 1970s, resulted in a water table decline of 200 feet. In eastern Washington, the water table has declined by 200 feet in some locations in the past twenty years. In Houston, it's down 400 feet. Groundwater supplies drinking water to approximately 8.2 million people in the Great Lakes watershed, yet pumping has lowered the water table by 900 feet in portions of the aquifer beneath Chicago and eastern Wisconsin.

Groundwater pumping has profound consequences for nearby surface water. Think about the following riddle: Where does water

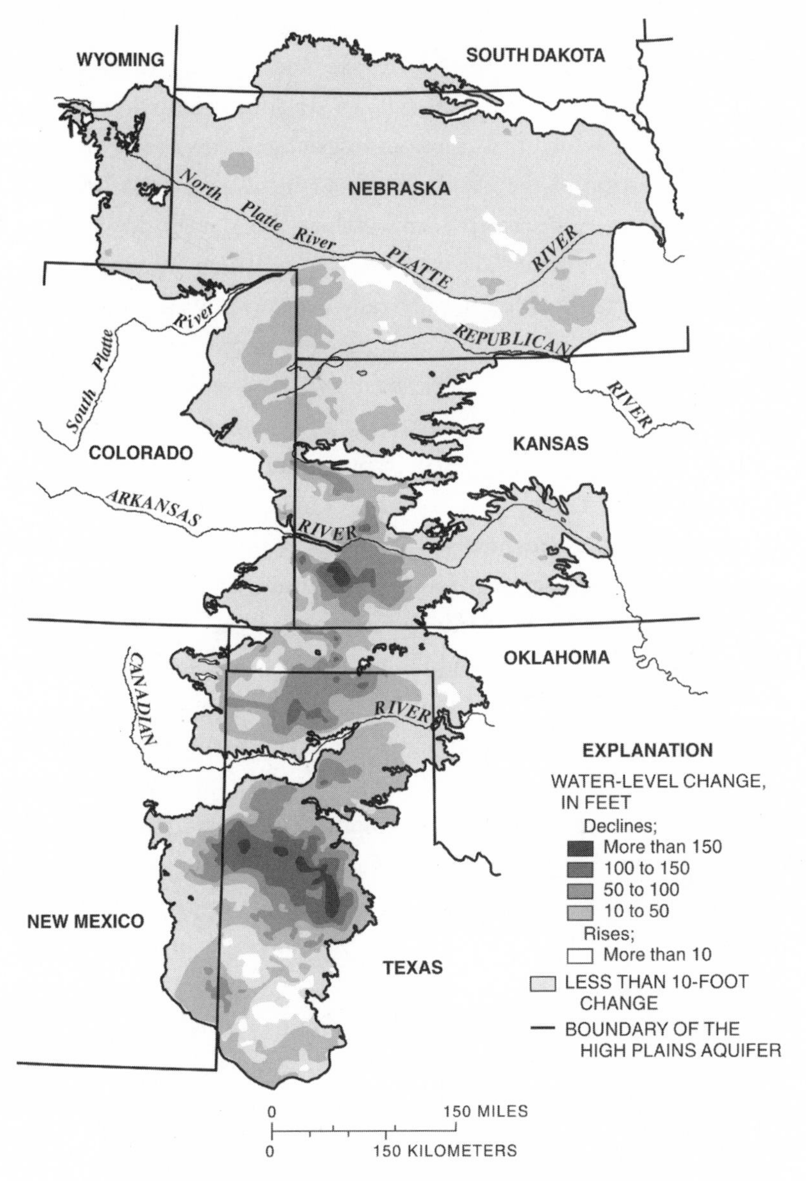

Figure 6.1. Ogallala Aquifer Groundwater Declines, 1950–2005.
Courtesy of U.S. Geological Survey.

in a river come from if it hasn't rained recently? It comes from groundwater that percolates downward and laterally to replenish rivers, creeks, springs, and wetlands. In Arizona, pumping dried up the Santa Cruz River in Tucson and desiccated the river's riparian habitat. A section of the San Pedro River in southeastern Arizona, one of the country's premier bird-watching sites, went dry in 2005 for the first time in recorded history. And in northern Arizona's Yavapai County, parts of which have already run out of water, the city of Prescott wants to build a thirty-mile pipeline to transport groundwater from the Big Chino Aquifer, near the headwaters of the Verde River—one of the last free-flowing streams in the desert Southwest. The proposed pipeline led the environmental organization American Rivers to put the Verde on its 2006 list of most endangered rivers.

Laurie Wirt, a geochemist with the USGS, loved to kayak the Verde River. Before her death in 2006, at the age of forty-eight, in a kayaking accident in Colorado, Wirt studied the relationship between groundwater in the Big Chino Aquifer and surface water in the Verde. Using isotope analysis, stream flow and precipitation records, documented well pumping, and geologic analysis, she concluded that at least 80 percent of the flow in the Upper Verde comes from the Big Chino Aquifer. Development interests tried to rebut her conclusions with a hydrologic study of their own, but, despite quibbles with some of the data, they were forced to acknowledge that her conclusion is accurate. Drawing down the aquifer will devastate the river.

And the water table is declining partly as a result of private domestic wells. As of 2007, Yavapai County had more than 27,000 private wells, which were immune from any state control. At least 2,500 wells were located close to springs that flow into the Upper Verde.

It's not only rivers in the arid West that have dried up from pumping. Just outside Boston, the Ipswich River has dried up in five of the past eight summers. In Florida, Tampa Bay Water has pumped so much water from rural areas of Pasco County, just north

Figure 6.2. Two photographs of the same section of the Santa Cruz River south of Tucson, Arizona. Figure 6.2A, taken in 1942, courtesy of Arizona Game and Fish Department. Figure 6.2B, taken in 2002 by Raymond M. Turner, courtesy of the U.S. Geological Survey, Desert Laboratory Repeat Photography Collection.

Figure 6.3. Ipswich River. Photograph courtesy of Ipswich River Watershed Association.

of Tampa and St. Petersburg, that *scores* of lakes have dried up. Imagine being in the shoes of Steve and Kathy Monsees, who built their dream retirement home on Prairie Lake, forty miles north of Tampa Bay, only to watch, horrified, as the lake dried up and their beautiful lakefront property became a mudflat.

Property-rights advocates often argue that property owners have an inherent right to drill wells on their property. Restrictions on this right, it is claimed, would violate the takings clause of the U.S. Constitution and require government compensation. But groundwater is not a private resource owned by the overlying landowner. It's a public resource owned by the state. Citizens can use it, but use rights profoundly differ from ownership rights. Moreover, what these property-rights advocates endorse is actually the antithesis of property rights: it's a circular firing squad. When it comes to unconstrained groundwater use, your "right" is simply to pump water until your well dries up because a neighbor, a large-scale commercial farming operation, or a municipality has caused the water table to plummet. The legal rules give you no "right" to exclude the new pumpers. That's not much of a right.

To grasp this chaotic system, consider the state of Texas, which uses the right of capture to govern (or, more accurately, not govern) groundwater withdrawals. Befittingly for a Wild West state, each Texan is free to pump, or capture, unlimited amounts of groundwater, regardless of the consequences to neighbors. The San Antonio Water System (SAWS), the city's water utility, has spent $20 million over the past ten years building the Carrizo Aquifer Storage and Recovery Project, a program designed to take currently unneeded Edwards Aquifer water and store it in the sandy Carrizo Aquifer for future use. Since 2004, SAWS has annually stored as much as 13.4 billion gallons in the aquifer. This recharge-and-recovery system makes sense except that across Hardy Road from the project is land owned by the Bexar Metropolitan Water District, a utility that supplies water to residents on the outskirts of San Antonio.

BexarMet's service area is booming and the district needs additional water supplies, so it decided to drill a well field adjacent to SAWS' project. San Antonio officials fumed at this blatant attempt to steal the stored water. BexarMet officials couldn't care less. A spokesman for BexarMet, T. J. Connolly, explained, "We're going to win this fight because we've got the law on our side and we've got the wells on our side and we've got the demand, both residential and commercial, that we've got to meet." In 2005, BexarMet told regional water planners that the district would pump 5 million gallons a day from the Hardy Road site; in 2007, the district decided to triple that amount. SAWS is helpless to do anything about it.

Nationally, we already pump 83 billion gallons of groundwater per day—which accounts for one-quarter of our water supply. More than half of us in the United States rely on groundwater for drinking water. Consider this statistic: in 2000, we pumped an astounding 30 *trillion* gallons of groundwater—that's 274 gallons per day for every man, woman, and child in the United States. And here's the scary part. These numbers represent the most recent statistics released by the USGS, but they are based on the year 2000. Since then, historic droughts have plagued many parts of the country, prompting cities, miners, farmers, and homeowners to scramble in search of new water supplies. Across the country, permissive legal rules have allowed them to turn to groundwater, and they have. In the United States, more than 15 million households get water from private domestic wells. And that number is increasing because each year approximately 8,000 well-drilling businesses across the country bore 800,000 new wells. It's a great time to be a well driller.

Figure 6.4. Earth fissure in Queen Creek, Arizona, 2005. Photograph courtesy of Arizona Geological Survey.

Unsustainable groundwater pumping causes other problems. Imagine the horrified look on the face of Joan Etzenhouser of Queen Creek, south of Phoenix, when heavy rains opened a fissure in the earth next to her home in August 2005. Within days, it had become a crevasse five to ten feet wide, as much as twenty-five feet deep, and more than a mile long. She was lucky. If the fissure had run under her home instead of beside it, the house could have collapsed.

Excessive groundwater pumping causes earth fissures by lowering the water table; in some places in Arizona the water table has dropped by several hundred feet. Once groundwater is pumped out, the sheer weight of the sediments eventually causes compaction and the ground surface sinks. To grasp the phenomenon of subsidence, remember the last time you purchased a box of Kellogg's Corn Flakes at the supermarket. When you opened the box, your first thought was "Kellogg has ripped me off!" The box was a quarter empty, yet you hadn't had a single bowl of cereal. If Kellogg had added a couple of cups of milk to the box, the flakes would still fill the box. But without the milk for support, the flakes settled as the box was jostled around during shipment. Without groundwater filling porous spaces in earth sediments, the earth settles.

Groundwater pumping has caused land subsidence in California's San Joaquin Valley, opened up enormous sinkholes in Florida, and increased the risk of flooding in Houston and Galveston. It's also a serious problem in the Coachella Valley, a two-hour drive east of Los Angeles and San Diego, where residents and tourists alike have long depended on groundwater for crops and fairways. With 125 golf courses in Palm Springs, Rancho Mirage, Palm Desert, Indian Wells, La Quinta, and Indio, the area pumps 32 billion gallons per year but gets only three inches of rain. Not surprisingly, the groundwater table has dropped more than 100 feet in some places. A 2007 study by the USGS found that parts of the Coachella Valley have dropped more than a foot in nine years. Even worse, the pace is quickening. The USGS's Michelle Sneed, the lead author of the study, notes that "the subsidence rates have more than doubled since 2000." One can only speculate as to what will happen as the Coachella Valley's population, one of the fastest growing in California, increases by hundreds of thousands in the next ten years. Steve Robbins, manager of the local water district, predicts dramatic damage to the infrastructure as roads crack and sewer lines break.

Beyond such obvious examples, the severity of the problem of land subsidence is difficult to measure because most compaction occurs beneath the surface, out of sight. Earth fissures and sinkholes are merely the surface manifestations of a deeper geologic transformation. In Arizona, state geologist Lee Allison estimates that there are as many as 300 earth fissures. Stan Leake, a hydrologist with the

Figure 6.5. Land subsidence in the San Joaquin Valley of California. Signs on pole show approximate altitudes of land surface in 1925, 1955, and 1977. Photograph courtesy of the U.S. Geological Survey.

USGS, reports that "more than 17,000 square miles in 45 states, an area roughly the size of New Hampshire and Vermont combined, have been directly affected by subsidence." Land subsidence, in addition to decreasing our water supply, also further compromises it. Subsidence compacts the soil, making it more difficult for new water to percolate into the ground and ultimately reducing the storage capacity of aquifers.

A parallel phenomenon to subsidence is saltwater intrusion. Gloria Schleicher, an eighty-four-year-old widow, lives on a saltwater river in Tenants Harbor, Maine. For water, she relies on her own well, drilled to a depth of 180 feet. The well adequately suits her needs during winter, but each summer the water acquires a salty taste, so she's begun taking samples to Aqua Maine, a bacteriological laboratory in West Rockport, where test results showed "some infiltration of seawater." Summer residents on her peninsula, which juts out into the Atlantic Ocean, pump so much groundwater that ocean water migrates laterally and contaminates the potable supply.

Because fresh water is lighter (less dense) than salt water, groundwater and salt water generally don't mix. Fresh water floats above seawater. But groundwater pumping creates a pressure gradient that tends to move all water inland, toward the wells. In Maine each summer, this process gradually increases salinity levels in Gloria Schleicher's well water until it's no longer suitable for human consumption. Salinity spikes have forced her to buy plastic gallon water jugs at the grocery store, which she refills from a faucet outside the town hall, about three miles from her home. Saltwater intrusion is rendering water supplies unpotable from Down East, Maine, to Cape Cod, Massachusetts, to Cape May County, New Jersey, to Hilton Head, South Carolina, on down to Miami, Florida, west across the Gulf of Mexico, and then up the Pacific Coast from Los Angeles to the San Juan Islands, near Seattle.

Seawater has intruded into aquifers along coastal areas of Los Angeles since the early twentieth century. To combat this problem in the 1950s, the Los Angeles County Flood Control District injected potable water into an abandoned well in Manhattan Beach to test whether it could create enough pressure to halt further intrusion. It did. Today, the Water Replenishment District of Southern California runs the barrier well program. Using approximately 300 injection wells and double that number of observation wells, the district annually injects 30,000 acre-feet of potable and reclaimed water into wells spaced every 500 feet over the sixteen-mile barrier project. It's quite a financial commitment for the district, costing some $20 million annually, according to Ted Johnson, the district's chief hydrologist. But the project has protected Los Angeles' groundwater aquifers from further saltwater intrusion.

As demand for water increases, our existing supplies face threats from global warming, pollution, subsidence, saltwater intrusion, and development. Unfortunately, these factors reinforce one another. If global warming causes sea levels to rise, saltwater intrusion will worsen—in every coastal area. Business as usual won't solve the crisis. No matter how many new dams we build or wells we drill, neither dams nor wells can outstrip the water cycle that fills them.

CHAPTER 7

Water Alchemists

"But what a fool believes he sees
No wise man has the power to reason away."
—The Doobie Brothers

IF DIVERTING RIVERS, building dams, and drilling wells won't solve our water crisis, perhaps water-short areas can import water from water-rich areas. To some dreamers, no scheme is too far-fetched to be considered seriously; no place is too far away to be the source. These water alchemists, as I call them, covetously gaze at the Mississippi River, the Columbia River, icebergs in Alaska, and rivers in British Columbia and wistfully imagine the problem is solved. In Winnipeg, Manitoba, a Canadian think tank, the Frontier Centre for Public Policy, suggests selling water to the southwestern United States. The Centre's 2007 proposal advocates solving Manitoba's financial woes with a thirty-foot-diameter pipe at the mouth of the Nelson River, near Hudson Bay. The $34 *billion* pipeline would carry an annual volume of 1.3 trillion gallons. This is a lot of water, but it represents only a few days' worth of the freshwater flow into Hudson Bay. An analyst for the Centre gushed about it in a *Wall Street Journal* column, but when President George W. Bush broached the idea to Prime Minister Jean Chrétien in July 2007, it went nowhere. Canada chose to keep its water.

Sometimes the more things change, the more they stay the same.

In the 1960s, the U.S. Bureau of Reclamation cast its eyes on the Columbia River Basin. As told by Marc Reisner in his classic, *Cadillac Desert*, the Bureau of Reclamation's legendary commissioner, Floyd Dominy, envisioned a set of dams on the Klamath, Trinity, and Salmon rivers to help move water south to Los Angeles. To generate the hydroelectric power needed to pump water under mountains, over mountains, and through mountains, Dominy proposed two additional dams—located in the Grand Canyon. This bizarre plan collapsed from its own weight, the opposition of Washington senator Henry "Scoop" Jackson, and an inspired public relations campaign orchestrated by the Sierra Club. Flooding the Grand Canyon, ran the Sierra Club's advertisement in the *New York Times*, would be like flooding the Sistine Chapel "so tourists can get nearer the ceiling."

A similar idea popped up in the early 1990s, when Alaska's governor, Walter J. Hickel, promoted the Alaska–California Sub-Oceanic Fresh Water Transport System, an undersea pipeline between the Copper River (home to world-famous wild salmon) and Lake Shasta, where the water would enter California's distribution system, a distance of 2,100 miles. The pipeline would deliver 4 million acre-feet per year to Southern California. That's almost as much water as California gets from the Colorado River. In its review of the proposal, the congressional Office of Technology Assessment concluded it would be one of the most complex and expensive engineering projects in history, rivaling the Panama Canal, the Trans-Alaska Pipeline, or the Channel Tunnel. A consultant's estimate came in at an eye-popping $150 billion, which did not include the costs of project financing, operation and maintenance, and taxes. The congressional assessment, which devoted not one word to environmental considerations, criticized the proposal as too expensive, given other water options available in California.

One such option materialized in 2002 when Alaska Water Exports applied for a permit from California's State Water Resources Control Board to divert water from the Gualala and Albion rivers in northern California's Mendocino County for ultimate delivery to

Southern California. The company proposed to capture the water and, using pipelines buried beneath the rivers, transport it to offshore loading facilities in the Pacific Ocean. Now, here's where things get interesting. Alaska Water Exports would load the water into gigantic polyfiber bags, each holding roughly 13 million gallons, and use tugboats to pull the bags to San Diego. This time, environmental factors doomed the project, prompting Alaska Water Exports to withdraw its application when it realized it would need to spend millions of dollars to conduct the necessary environmental impact studies before it could begin construction.

The common thread linking these ideas is an engineering mentality that assumes there must be a technological fix to water shortages. There's no need to live within our means, to tailor land use plans to the available water supply, because an engineering solution can solve the problem and allow growth and unsustainable water practices to continue for a while longer. This is how we've always conceived of water problems: there must be someplace to go for more water.

There will always be visionaries who dream of building a new pipeline to import water. The most recent incantation comes from a Colorado entrepreneur, Aaron Million, who thinks he's navigated the political, legal, financial, environmental, institutional, and engineering shoals that have sunk other dreamers. "Sometimes dreamers can pull things off," notes Trout Unlimited's Melinda Kassen. And if he does pull it off, says Robert C. Norris, a prominent Colorado rancher, "it's going to be one of the biggest deals of the century."

The inspiration for Million's idea came when he was a graduate student at Colorado State University. While studying a map of northwestern Colorado, he noticed that after the Green River flows out of Flaming Gorge Reservoir, straddling the border between Wyoming and Utah, it veers east into Colorado through Dinosaur National Monument before it hooks back west into Utah and then turns south to join the Colorado River. His background in ranch management had introduced him to water law. Colorado has rights

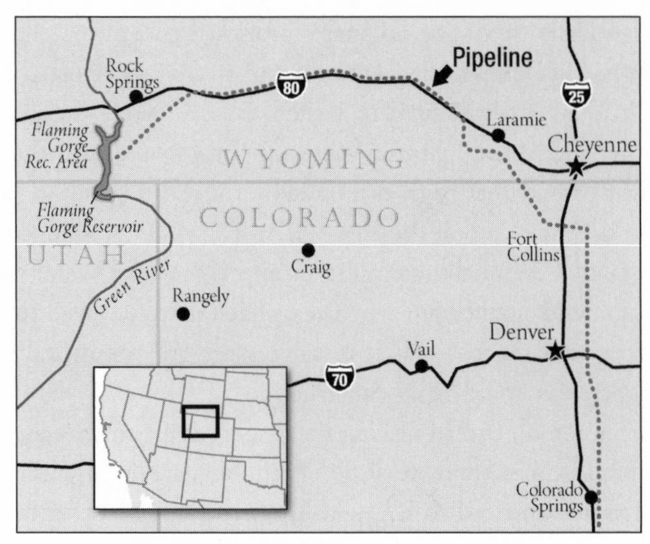

Figure 7.1. Aaron Million's proposed pipeline.

to 51 percent of the Upper Basin's share of the Colorado River, including the Green River. The forty-one-mile leg of the Green River that loops through Colorado gives Million, he believes, an opportunity to claim some of that water.

Million wants to move Green River water to Colorado's Front Range cities through a $4 billion, 400-mile pipeline. As Million currently envisions it, he'll divert water from Flaming Gorge Reservoir, at an elevation of approximately 6,000 feet, and put it in twin side-by-side forty-two-inch-diameter pipes. The pipes would run adjacent to Interstate 80; cross the Continental Divide at about 7,000 feet; drop down past Laramie; head south, skirting Fort Collins; and eventually terminate sixty miles south of Denver at Colorado Springs. The West has seen enormous water projects in the past, such as Arizona's Central Arizona Project, a federally sponsored boondoggle of the first order. But Million's proposal, quite remarkably, would be privately financed.

Although he initially expected the pipeline to carry as much as 450,000 acre-feet, he's scaled back the project to 165,000 acre-feet. The energy costs to pump even that much water over the Rocky

Mountains could reach $95 million a year. Still, Million is very optimistic that he'll build the pipeline. Even though he hasn't announced details about the diversion point, his own financial stake in the venture, or his sources of capital, he's effusive. When asked about his biggest hurdles, Million replied, "Our biggest hurdles? We don't have any. This thing is over." It'd be completed, he predicted in 2007, in three to five years.

Others aren't so sure. When Rick Gold, then-head of the U.S. Bureau of Reclamation office that operates Flaming Gorge Reservoir, was asked whether Million faced obstacles in securing a right to the water, he replied, "Sure, he has lots of them." Two of them are Utah and Wyoming. In order to take advantage of the easiest way over the Rockies, Million wants to divert water from the reservoir, not from the section of the Green River that actually loops through Colorado. But getting the water from Flaming Gorge Reservoir, located in Utah and Wyoming, will require the approval of those states, which may decide that it's not in their best interest for Million to take so much water. At the very least, they'll insist on their pound of flesh. Although the I-80 corridor, with an established highway, railroad tracks, and natural-gas pipelines, would make it easier for Million to get right-of-way permits for portions of his proposed route, he'll also need to secure right-of-way concessions from property owners along hundreds of miles of other sections of the route. Million can't simply exercise the power of eminent domain, as local governments can, to require property owners to allow the pipeline.

Environmental compliance poses another hurdle. Million will need a contract from the Bureau of Reclamation to divert water from Flaming Gorge and a right-of-way permit from the U.S. Bureau of Land Management to run his pipeline over federal land. Before either agency can act, they will need to prepare an environmental impact statement because the Colorado pikeminnow, an endangered species, makes its home in the Upper Colorado River below Flaming Gorge. A multiyear fish recovery program, carefully hammered out between water users and the environmental community, requires releases from the dam to benefit the fish. Environmentalists are opposed to Million withdrawing 165,000 acre-feet just upstream from the pikeminnow's habitat. The head of Utah's Water Conservation Board, Eric Millis, and the executive director of the Wyoming Water Development Commission, Mike Besson, are concerned about the impact on the fish recovery project.

The downstream states of California, Nevada, and Arizona will surely resist Million's proposal. As Scott Balcomb, a Glenwood Springs water lawyer, conceded, "any new water comes out of the hides of industry in California and Arizona." Even water users in Colorado are opposed, as Million's proposal has rekindled historical tension between users on the Western Slope and those on the Front Range. Eric Kuhn, director of the Colorado River Water Conservation District in Glenwood Springs, sees it as a matter of equity. "It's unacceptable that the unused portion of the river would all go to the Front Range," he says. Also weighing in against Million's proposal is a consortium of Colorado's largest municipal water interests who are competing with Million for the same water. To Million, his opponents are fighting turf battles and representing parochial interests. "These guys need to stand up like men and stop going around behind the scenes trying to stop this project. They need to look beyond their salaries and do what's best for the state." Until he explains the financing for his proposal, Million's idealistic pleas will fall on deaf ears in Colorado.

Finally, Million's proposal will never get off the ground unless he can secure rights to the water under Colorado law, a daunting task. Colorado's prior appropriation system prohibits speculating in water rights. To obtain a prior appropriation right, the diverter must demonstrate that the water will be used for a "beneficial purpose," such as by a farmer for irrigation or by a city for its residents. A speculative claim that Front Range cities are going to need the water to support future growth will not suffice. Russell George, former head of the Colorado Department of Natural Resources, explained that this legal rule prevents "a private company from having control over something as important as water, particularly in a time of shortage." What the state of Colorado does not want, explained George, is for "the public to get caught in a highest-bidder method of distribution of an essential commodity."

Million recognizes this problem. That's why he's trying to sign up cities to take the water. But he's caught in a catch-22. The cities won't commit to him without knowing his price, and he can't determine the price until he surmounts all the other obstacles. Without the cities under contract, he runs afoul of the anti-speculation doctrine. Cities such as Parker in Douglas County are keenly interested in Million's water, but only when the project is further along.

The hurdles facing Aaron Million beset all large-scale transbasin importation schemes. Immense costs, complicated environmental compliance, daunting engineering challenges, uncertain legal rights, political opposition, and institutional barriers dampen the spirits of the even the most intrepid entrepreneurs. Some large-scale deals are still being done, but, as we shall see, it helps to have an existing canal to transport the water from one place to another. Well, then, if it's too cumbersome to get additional water from so far away, how about someplace closer, such as the sky?

> "Everybody complains about the weather, but nobody ever does anything about it." —Mark Twain

DOWSERS HAVE USED divining rods to search for water for centuries despite the lack of even a shred of scientific evidence to support the method. To believers, dowsing works; how it works remains a mystery even to the American Society of Dowsers, which concedes that "the reasons the procedures work remain entirely unknown."

Rain dances, rooted in the customs of Native American tribes, particularly in the Southwest, attempt to induce precipitation and protect the harvest. These ceremonial dances follow prescribed rituals handed down orally over countless generations. For cultures whose very existence depends on water, rain dances are deadly serious business.

Other attempts to induce rainfall involve cloud seeding, which involves dispersing chemicals, usually silver iodide or dry ice (frozen carbon dioxide), from airplanes into clouds that radar analysis deems are moist enough to produce rain with a little human intervention. The use of silver iodide is like a Fourth of July fireworks event: silver iodide flares are ejected as the plane flies through the cloud, and an electrical switch inside the airplane cabin detonates the flares. In the 1940s, two General Electric chemists, Vincent Schaefer and Irving Langmuir, a Nobel Prize winner, pioneered the dry ice alternative when Schaefer literally shoved dry ice out the cargo door of a plane and into a cloud. While this sounds silly, they proved that it's possible to modify a cloud; where they dropped the dry ice, the cloud disappeared. People on the ground can try to seed clouds by spraying silver iodide across a propane flame, causing the particles to rise into the clouds above. In theory, silver iodide causes cloud moisture to freeze and create ice crystals, which, when they grow big enough, fall as snow or rain.

In the 1960s, the Unites States military, in Project Stormfury, tried to diffuse hurricanes over the Atlantic Ocean through cloud seeding. They may have succeeded in changing a hurricane's shape slightly, but only briefly, and the project ended when scientists realized that cloud seeding might perversely increase a hurricane's ferocity. During the Vietnam War, the military's Operation Popeye spent $21.6 million on cloud seeding to try to slow the movement of North Vietnamese trucks by turning the Ho Chi Minh Trail into mud. We still lost the war.

The cloud-seeding field has had more than its share of charlatans, most notably Charles M. Hatfield, who called himself a "moisture accelerator." In the early twentieth century, he traveled the country, coaxing large fees from farmers for shooting his secret formula of chemicals and gases into the air. In the 1970s, the federal government spent $20 million a year on weather modification. But the scientific community ultimately decided that the field was full of snake-oil salesmen, and federal funding petered out.

New efforts to secure federal appropriations have sometimes used fearmongering as a tactic. In 2006, John N. Leedom, a lobbyist for the Weather Modification Association, tried to persuade the Office of Science and Technology Policy to support a bill before Congress to establish the Weather Modification Research and Technology Transfer Authorization Act, sponsored by Senator Kay Bailey Hutchison (R-TX) and Representative Mark Udall (D-CO). Leedom argued that weather modification can mitigate damage from hurricanes, hail, droughts, floods, and tornadoes. Only a callous fool, implied Leedom, would not support the bill. The Office of Science and Technology Policy was not deterred from withholding support for the bill, citing unresolved issues of liability, foreign policy, and national security. Hutchison introduced the bill in 2007, but it never got out of committee.

Although the federal government does not currently support cloud seeding, some states are more sympathetic. By a 2001 estimate of the National Oceanic and Atmospheric Administration, sixty-six weather modification programs to suppress hail or enhance snow and rain are operating in ten states. In Colorado, the state no longer appropriates money for cloud seeding, but some ski areas, such as the Vail and Beaver Creek resorts, contract with private companies to seed clouds. Wyoming has the greatest commitment to weather modification; in 2005, the legislature approved a $9 million five-year pilot program.

A 2005 report by the Metropolitan Water District of Southern California estimates that cloud seeding in six areas of the Colorado River Basin could produce 1.7 million acre-feet of *new* water. Cloud seeding in Utah, the report claims, increased precipitation between 7 percent and 20 percent at a cost of less than $20 per acre-foot. In 2006, a report on augmenting the flow of the Colorado River through cloud seeding prepared by North American Weather Consultants, a cloud-seeding organization, reached similar results: a 5 percent to 15 percent increase in precipitation, or almost 1.4 million acre-feet at $5 per acre-foot. If weather modification can produce such a huge increase in precipitation at such a modest cost, it has a bright future.

One overarching question clouds the future of cloud seeding. Does it work? The Metropolitan Water District's report concedes that "there is no clear cause-and effect relationship between seeding a specific cloud and resultant precipitation in a specific location as proven by scientific method." The cause-and-effect issue is the Achilles' heel of weather modification. Here's the problem. Let's say a plane spreads silver iodide into clouds above Fresno, California, and a few hours later, or even the next day, it rains. Did the seeding cause the precipitation? Perhaps it would have rained anyway. Did the seeding *increase* the amount of precipitation? Perhaps it would have rained just as much without the seeding.

Before the United States government, state and local governments, and even ski resorts spend a lot of money on cloud seeding, it would be nice to have firm answers to these questions, but we don't have such answers and never will. Eric Betterton, a professor of atmospheric sciences at the University of Arizona, explains why: "You can never replicate the experiment and you can never do a controlled experiment. But that's what science is all about: the control and then the replication."

Betterton has conducted weather modification experiments that prove we can modify clouds. But to go further and "prove that you can increase precipitation on the ground is just a nightmare. It's really hard," he says. Absent such scientific proof, Betterton describes himself as an agnostic as to whether cloud seeding increases precipitation. Continuing his religious analogy, Betterton describes the practitioners of weather modification as "the believers and they all fall under a church . . . the Weather Modification Association. There are enough people out there, communities or local government, who also believe it and they hire them on a regular basis and they believe in the product that's delivered. What can I say? I'm an agnostic."

Organized in 1950, the Weather Modification Association promotes belief in the efficacy of cloud seeding, or intentional weather modification. It's an association primarily composed of people in the business of weather modification, such as vendors, scientists, engineers, economists, and consultants. The group argues that the test of scientific proof used by Betterton and others is so "stringent that few atmospheric problems could satisfy it," including global warming.

The National Research Council (NRC) of the National Academies is widely recognized as the last word on resolving scientific disputes. In 2003, it released a report, "Critical Issues in Weather Modification Research," which concluded that the decline in support for cloud seeding came about as a result of "extravagant claims, unrealistic expectations, and failure to provide scientifically demonstrable success." This report, an updated assessment of earlier reports of the National Academies, does not help weather modification advocates.

Advocates of weather modification must surmount a basic problem. We know that human activities can affect the weather and that seeding can cause some changes to a cloud, but we are unable to translate these changes into verifiable increases in precipitation. Past efforts to document precipitation increases have uniformly failed. The NRC report reaches the same result: "The Committee concludes that there still is no convincing scientific proof of the efficacy of intentional weather modification efforts." For the NRC to conclude that there is "no convincing scientific proof" is as negative a report as it can issue. "In evaluating the success or benefits of cloud seeding operations," the NRC report concludes, "the experience of six decades of experiments and applications that failed to produce clear evidence that cloud seeding can reliably enhance water supplies on large scale should be kept in mind."

Even if cloud-seeding advocates could persuade the scientific community that seeding is effective, weather modification must confront other challenges before it can become an effective supply augmentation tool. Who gets the water? Although there is little law on the subject, the legal issues are vexing. Downwind neighbors may claim, quite plausibly, that the increased water came at the expense of rain that would have fallen on them. This "rob Peter to pay Paul" argument creates one level of uncertainty about the water rights to augmented rainfall. And further questions arise after this augmented rainfall flows into a river. For example, states such as Colorado would protect the senior prior appropriator, not the junior cloud seeder, unless the cloud seeder could take advantage of the developed water doctrine that rewards rights to an importer of water notwithstanding senior appropriators. If all this sounds confusing, it is. And the lack of clear property rights to the water will discourage private sector actors from even initiating cloud-seeding projects. Another deterrent comes in the form of potential liability for causing flooding or avalanches should they succeed *too* well and for increasing air and water pollution by the intentional release of a toxic chemical, silver iodide, into the atmosphere.

Despite scientific skepticism, funding challenges, legal uncertainty, and liability exposure, state and local governments, as well as irrigation districts, remain interested in weather modification. In 2006, the seven Colorado River Basin states agreed to fund proposals in Utah, Wyoming, and Colorado in order to augment Colorado River flows through weather modification.

In Fresno, California, in January 2007, an extraordinarily dry month, Steve Johnson, general manager of Atmospherics, looked at the weather forecasts and prayed that his planes could fly. With the right forecast, his company uses a radar station to track storm paths and conditions and to direct his pilots to promising clouds. It had been a rugged winter for Johnson. Winter is California's rainy season and the time for most cloud seeding. But his planes remained on the tarmac. As the 2007 drought lingered on into summer, Johnson's company was hired by Florn Core, water resources manager for the city of Bakersfield, to fly cloud-seeding missions over the Kern River Basin. "[It's] a little bit of a gamble on our part," conceded Core, but "it's worth it to give it a try."

CHAPTER 8

The Ancient Mariner's Lament

"Water, water every where,
Nor any drop to drink."
—Samuel Taylor Coleridge

ON MARCH 20, 2007, Jim Cherry grinned and his eyes sparkled. As the area manager of the U.S. Bureau of Reclamation's Yuma, Arizona, office, Cherry oversees the Yuma Desalting Plant (YDP). From one angle, it's a tough job because the plant is so controversial. From a different angle, his job is a piece of cake. The plant hasn't operated for fourteen years. Since 1993, this $250 million plant has sat mothballed, a white elephant glistening under the Yuma sun.

Cherry took on the challenge of running the largest desalination facility in the Western Hemisphere in 1999. The plant was designed to reduce salt content in the Colorado River water that the United States delivers to Mexico under a 1944 treaty. After the nearly twenty years it took to build the plant, it finally started up in 1992, but floods on the nearby Gila River and engineering design flaws caused the plant's closure after only eight months of operation. It has sat idle, in "ready reserve" capacity, since then. Meanwhile, the saline water that the plant would have treated has flowed into a concrete-lined bypass canal that empties into a barren section of the Sonoran Desert, thirty-seven miles south in Mexico, at the northern end of the Santa Clara Slough, which drains into the Sea of Cortés. The inflow of 108,000 acre-feet of water per year has transformed the desert and created an artificial wetland, the Ciénega de Santa Clara.

In early 2003, as the ongoing drought worsened, water users in Los Angeles and Phoenix expressed interest in capturing some of this saline water. This prospect alarmed the environmental community because the Ciénega had become the largest remaining wetland in the Colorado River Delta, a critical stopping point for migrating waterfowl on the Pacific flyway and home to endangered species including the desert pupfish and Yuma clapper rail. In 2003, Secretary of the Interior Gale Norton sent a report to Congress that

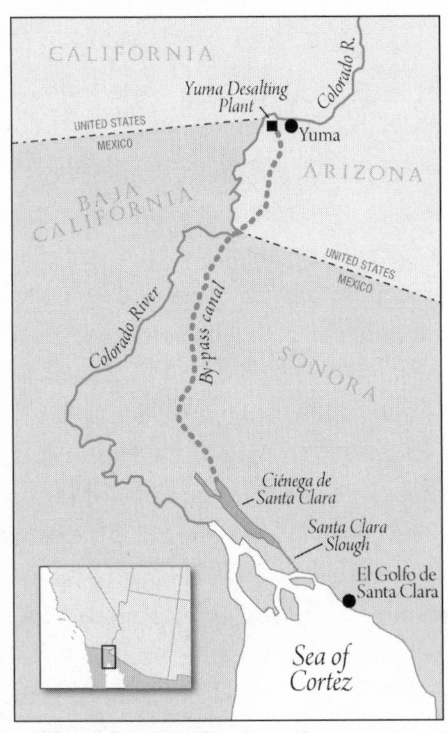

Figure 8.1. Yuma Desalting Plant region

Figure 8.2. Ciénega de Santa Clara.
Photograph courtesy of Karl W. Flessa.

explored the idea of restarting the Yuma Desalting Plant and compared options for securing other supplies of water for the cities. It turned out that it would be much cheaper—millions of dollars cheaper—to pay willing farmers to fallow some unproductive fields than to operate the plant. Given this huge disparity and given that operating the YDP would have profound environmental consequences for the Ciénega, one might have thought that Congress faced an easy choice. But in the world of western water, the most logical thing rarely occurs. Congress wasn't interested in saving money or the environment and directed the Bureau of Reclamation to prepare the plant to operate.

At this point, both municipal interests and environmentalists dug in their heels. The cities adamantly objected to the "waste" of water flowing to the Ciénega in the middle of a drought; environmental organizations resisted restarting the plant because it would cut off flow to the Ciénega and wipe out critical riparian habitat. The impasse ended thanks to the leadership of Sid Wilson, general manager of the Central Arizona Project, Arizona's canal that brings

Colorado River water to Phoenix and Tucson. Wilson assembled a stakeholder's group that included leading environmentalists, especially Jennifer Pitt of the Environmental Defense Fund and Peter Culp, a Phoenix attorney who was working for the Sonoran Institute. After nine months of meetings, they found common ground in 2005 with recommendations that would both preserve the Ciénega and allow operation of the YDP.

Agreeing to save the Ciénega represented a sea change for Wilson, who until a short time earlier had placed little value on it. "The Bureau of Reclamation *created* that wetland. It's not even supposed to be there," Wilson says. "But when you go down there, you recognize that artificial or not, it has some real value." Wilson, whose district supplies water to 1.5 million people, has come to realize that California, Arizona, and Mexico are confronting the same issues. "How," Wilson asks, "do you deal with growing demands for water by people, declining natural supply and trying to sustain environmental values?" The YDP provides an important piece of the puzzle.

That's why Jim Cherry was smiling on March 20, 2007, the first day of the Bureau of Reclamation's ninety-day demonstration run of the YDP. With dignitaries assembled and the media snapping photographs, Reclamation commissioner Robert Johnson pushed a button to start the flow of desalinated water into an aquarium. It had taken a year of work and an estimated $30 million to get the plant ready to run. Reclamation set up the demonstration to test a simple question: does the plant work? Some people doubted that the plant, having lain idle for fourteen years, could run again. The demonstration run would also provide an opportunity to verify the cost estimates of producing desalinated water and to utilize technologies developed since 1993.

Over the years, friends had teased Cherry about being manager of a plant that doesn't run: "What exactly *is* your job?" or "Is your paycheck an example of my tax dollars at work?" But Cherry and his staff, albeit reduced in size, had been working on cutting-edge issues of saltwater desalination. On the grounds of the plant, Reclamation operates the Water Quality Improvement Center, a site for field-testing new and improved desalination technologies. It's the only reverse osmosis testing facility in the country. Cooperative partnerships with private companies and local governments aim to reduce the time between the research and development phase and commercialization.

During the demonstration run, Cherry hosted visitors from around the country and the world who were interested in the center's state-of-the-art research. Does desalination offer the magical solution to the world's water problems? During the demonstration run, the plant (operating at 10 percent capacity) successfully treated more than 1.3 billion gallons, or 4,200 acre-feet, of water. Were it to operate at full capacity, it could process 78,000 acre-feet per year. But there are no plans to do so. "The plant has run well," said Cherry. "The challenge now is how the plant fits into the water supply picture for the Lower Colorado River." The Bureau of Reclamation has proposed another pilot run. Beginning in summer 2009, the YDP would run at up to one-third capacity, but only for a year. That an existing, successful $250 million plant is not operating at full capacity offers insight into the challenges facing proponents of numerous large-scale desalination plants.

Removing salt from ocean water offers a tantalizing possibility of a limitless new source of water. Desalination could answer the lament of Samuel Taylor Coleridge's ancient mariner: "Water, water, every where, / Nor any drop to drink." Following my talks about water shortages over the past few years, the first question I'm almost invariably asked is "What about desalination?" With the earth mostly covered in water, it seems painfully obvious that the ultimate solution to our water crisis is to remove the salts. In 1961, President John F. Kennedy suggested that achieving the capacity to desalt ocean water at a reasonable price would solve the world's freshwater shortage and "dwarf any other scientific accomplishments."

Surprisingly, the original impetus for trying to separate salt from water came from an effort to obtain the salt, a precious commodity, rather than the water, an abundant commodity. My, how times change.

Today, two principal technologies can desalinate ocean water. The first, distillation, mimics what the sun does to ocean water every day: it evaporates water off an ocean's surface and leaves behind the salts. Thermal evaporation, as its name suggests, heats water to the boiling point to produce water vapor and capture the salt in collectors. If done in a partial vacuum, water will boil at temperatures lower than 212 degrees Fahrenheit, which saves a lot of energy. Still, distillation requires immense quantities of energy, which is why most distillation plants are located in the Middle East, where oil is plentiful and water scarce. Indeed, Middle Eastern countries have so much oil that they use desalinated water to irrigate crops. Alas, scale eventually forms on the surfaces of a distillation system, restricting flows, reducing heat transfer, and driving up costs for maintenance and replacement of parts. But if money and energy are limitless, distillation is an option.

Distillation is also viable in small-scale operations with focused needs, such as on islands or aboard ships. Readers of *Life of Pi* may remember that as Pi floated across the ocean with a 450-pound Bengal tiger for 227 days, a homemade distillation system saved his life. A simple pot of water, covered with a metal umbrella whose diameter was greater than the pot, induced condensation of water vapor on the underside of the umbrella, which ran off as desalted water into a separate container. Along similar lines, cruise ships and yachts have begun using solar panels to generate the heat, hasten the process, and produce greater volumes of potable water.

The second desalination technology uses filters that allow the water to pass through but that block the passage of salt. The most popular technology, reverse osmosis, uses pressure to force water through semipermeable membranes, leaving the larger salt ions and some salt-laden water molecules behind in concentrated brine. In the Bureau of Reclamation's reverse osmosis system at the YDP, salt water enters the outer rings of the filter, fresh water eventually reaches the core tube, and the salty brine exits via a discharge pipe. Sounds simple enough. But these high-tech membranes are very expensive *and* prone to fouling. The performance of reverse osmosis systems depends on many variables, including water pressure, water temperature, pH, bacteria, and the level of total dissolved solids in the untreated source water.

That's why recent research at the Water Quality Improvement Center has concentrated on pretreating processes that prepare higher-quality salty water, if that's not an oxymoron, for the reverse osmosis process. The YDP first sends the salty water through a grit sedimentation basin, where the water moves slowly, allowing larger particles to settle out as sediment. Lime slurry coagulates particles, which a sludge thickener then removes. The clarity improves as the water percolates through sand and charcoal filters. Then high-pressure pumps push the water against the reverse osmosis membranes with an average pressure of 362 pounds per square inch. Here, another devilishly challenging reality rears its head: the membranes are fragile. At the YDP, maintaining the integrity of fragile structures under conditions of high pressure poses difficult engineering problems. They've taken great pains to protect the membranes from tearing, but this further drives up the cost.

As of 2005, more than 2,000 desalination plants operated in the United States, but about half of those desalinated brackish water and one-quarter desalinated river water. Brackish water and river water are easier and much cheaper to desalinate than ocean water, with its more concentrated salts. Still, interest in desalination is booming. Twenty-eight major desalination plants are under construction or in discussion. The Pacific Institute's Peter Gleick estimates that, even if all the plants get built, which is quite unlikely, desalinated water will constitute only three one-thousandths of the United States' water supply.

Yet some entrepreneurs think the future is bright. Aqua Genesis, a 2004 startup out of Las Vegas, thinks that its technology, powered by geothermal heat, will solve the energy consumption problem. Energy Recovery of California is pushing its PX Pressure Exchanger as a technology to recover and reuse some of the energy used in the reverse osmosis process. And Georgia's Aquasis wants to tap into the desalination market, which the American Water Works Association predicts will increase by $70 billion in the next twenty years. In 2008, however, Aquasis shelved a pilot project on the Turtle River near Brunswick, Georgia, when the company failed to attract private financing. Attempts to solve Georgia's water woes through desalination have failed in part because coastal waters have high sediment concentrations stirred up by rivers and tidal flows, which clog the expensive membranes.

But it's not just startups that are kicking the tires. In 2006, General Electric purchased a Canadian water filtration company, and in 2007 GE Water & Process Technologies agreed to invest in a proposed plant in Carlsbad, California, to be built by Poseidon Resources. GE's clever 2007 "ecomagination" advertising campaign includes a fishing trawler retrieving a net full of bottles of water as a voice-over notes that GE has desalinated 3 billion gallons of water. If two-thirds of the earth is covered with water, another GE ad asks, shouldn't three-thirds of the world's population be able to drink it?

Two states, Texas and California, actively support desalination. The Texas Water Development Board and the city of Corpus Christi are involved in proposed brackish water plants. In 2007, El Paso brought online the world's largest inland desalination plant, the Kay Bailey Hutchison Desalination Plant, with the capacity to treat 27.5 million gallons per day. It took fifteen years to design and build the plant, aided by $29 million in federal money, secured by Senator Hutchison. But the real action is in California, where in 2002 voters passed Proposition 50, which established a grant program for desalination projects. In 2005, the California Department of Water Resources awarded $25 million to support desalting plants; nineteen are proposed along the coast.

Governor Arnold Schwarzenegger's administration enthusiastically embraces desalination, but the California Coastal Commission is more cautious because of staff concerns about environmental consequences. A 2004 report of the commission notes: "Seawater is not just water, but habitat. It provides the matrix within which innumerable organisms live, and serves a critical role in everything from the food web to the climate." The discharge from a desalination plant may have double the normal salinity level of local seawater. Such a sharp spike in salinity may kill or gravely harm marine species and other organisms, particularly in the egg, larva, or juvenile life stages. Finally, the commission report raises a basic question about desalination projects proposed by private, for-profit corporations: "Should seawater, a public resource held in common for the benefit of current and future generations, be allowed to be expropriated by private business for profit?"

Desalination by reverse osmosis incurs large costs as a result of the high-tech membranes and the immense amount of energy required to run the plants. Reverse osmosis also generates a troublesome waste stream: every 100 gallons of seawater will yield 15 to 50 gallons of potable water, leaving 50 to 85 gallons as a supersaline by-product. Disposing of this brine poses significant engineering and environmental challenges. In Florida, fishery ecologists fear that the release of this supersaline water into sensitive estuaries off the West Coast will adversely affect the reproduction of clams and other marine organisms. Brine disposal for inland desalination facilities presents even greater challenges. A 2008 National Research Council study found "few, if any, options."

Before desalination becomes a widely available, cost-effective, environmentally friendly source of potable water, it must surmount a number of hurdles. Two desalination projects, one in Florida and the other in California, illustrate the problem. In 1999, Tampa Bay Water entered into a design-build-operate contract with Poseidon Resources, a major player in membrane technology. Poseidon then entered into partnership with a respected engineering firm, Stone & Webster, for overall construction of the largest desalination plant in the Western Hemisphere. The $110 million plant was expected to produce 25 million gallons per day of potable water at a surprisingly low price. Before construction even began, Stone & Webster filed for bankruptcy and Poseidon Resources replaced the firm with Covanta Energy, which also went bankrupt. A new contractor, Covanta Tampa Construction, began construction in 2001. In 2002, it became clear that Poseidon and its contractor had failed to secure long-term financing. Tampa Bay Water bought out Poseidon's interest, and Covanta Tampa Construction completed the project in 2003.

Unfortunately, the plant operated for only two weeks before problems cropped up. The intake pipes sucked in Asian green mussels—an invasive species that is creating headaches up and down the East Coast. Sediments quickly clogged the expensive membranes, requiring their immediate replacement. After another default on the contract, Covanta Tampa Construction also declared bankruptcy. In 2004, amid a flurry of lawsuits, Tampa Bay Water agreed to pay American Water/Poseidon, a joint venture between engineering companies in California and Spain, an additional $29 million to fix the pretreatment system. These troubles prompted Bob Stewart, board chairman of Tampa Bay Water, to note wryly, "Being on the cutting edge is not a very comfortable position." Water costs, initially estimated at $1.71 to $2.08 per thousand gallons, soared to $3.38 by 2008. The plant finally became fully operational in 2007, six years behind schedule and $48 million over budget.

Of the proposed desalination plants in California, the most controversial one is Poseidon Resources' $300 million plant in Carlsbad, which began as a potential partnership with the San Diego County Water Authority. It would eclipse Tampa Bay Water's plant as the largest one in the Western Hemisphere, pumping 100 million gallons per day and producing 50 million gallons of drinking water. Two environmental organizations, Surfrider Foundation and San Diego Coastkeeper, tried to derail the project because of potential harm to marine life near the plant from the discharge of 50 million gallons per day of salty brine, but California's State Water Resources Control Board rebuffed them in 2007. The groups then turned their attention to contesting a permit Poseidon must get from the California Coastal Commission.

The Carlsbad plant would benefit from its location next to an existing power plant, the Encina Power Station, because Poseidon could tap into the plant's once-through cooling system for its supply of ocean water and use its return pipe to discharge the brine stream into the Pacific Ocean. This location next to an existing power plant solves Poseidon's problem of acquiring oceanfront property in the most populous state in the country and eliminates the need to build the pipes and other infrastructure necessary to obtain ocean water and discharge the brine.

The Carlsbad project suffered two major setbacks in 2006. First, the San Diego County Water Authority's board of directors unanimously voted not to accept Poseidon's final environmental impact review because of changes in "the fundamental assumptions about the project." James H. Bond, chairman of the Water Authority, explained that the agency had failed, after years of discussion with Poseidon, to reach mutually agreeable terms for a partnership. The Water Authority's Bob Yamada said that buying water from Poseidon would be too expensive. "We quickly determined," said Yamada, "a purchase agreement would cost the agency $200 million more over 30 years." So the Water Authority walked away. Poseidon's senior vice president, Peter MacLaggan, bravely asserted that they would move ahead and find new partners. But without the San Diego County Water Authority (a government agency) as a partner, Poseidon faces significantly higher financing costs. And it also lost access to the agency's extensive network of pipes to distribute desalinated water, making it more difficult for Poseidon to find prospective buyers. Nonetheless, the city of Carlsbad and six other water districts have signed up to acquire water. The ongoing drought in California has made all water suppliers take a second look at projects that previously seemed too costly.

The second setback came from changes in "the fundamental assumptions about the project." The owner of Encina Power Station, NRG Energy, plans to stop using seawater for once-through cooling. Encina's intake and discharge system was central to Poseidon's proposal. Worse yet, NRG has commenced the permitting process to locate a new power plant on the site of what was supposed to have been the desalination plant. Poseidon may find itself without a site for its plant or the pipes to receive and discharge water.

In 2008, the California Coastal Commission, after tense debate, opposition from commission staff, and accusations that Poseidon had withheld critical environmental information, gave final approval to the Poseidon project, but the commission placed almost two dozen conditions on the permit. Poseidon will have to minimize the killing of marine organisms and offset greenhouse gases the plant would produce. For the moment, NRG has told Poseidon that it can use NRG's cooling system, but Poseidon may need to withdraw three times as much water as it needs in order to comply with its permit from the San Diego Regional Water Quality Control Board. Any lower amount would result in concentrated brine that could harm ocean life. Still, Poseidon's MacLaggan is upbeat. "We came here to get a coastal development permit; we're leaving with a coastal development permit." Yet years remain before Poseidon will be able to produce a drop of desalinated water at the Carlsbad facility. Meanwhile, environmental organizations filed a lawsuit challenging the commission's issuance of the permit.

Recent research, at the Water Quality Improvement Center and elsewhere, is driving down the costs of desalination, reducing its energy consumption, improving the yield of potable water (one WQIC test achieved 85 percent efficiency), and mitigating adverse environmental effects. Still, desalination is no panacea for our water crisis. Desalinated water remains considerably more expensive than other supplies, except in a few select locations. Residents of Santa Cruz, California, for example, face a 40 percent hike in water rates to pay for desalinated water. The energy demands vastly exceed those for other supplies. Even Poseidon's senior vice president for technical services, Nikolay Voutchkov, concedes that seawater desalination is "several times more energy intensive than conventional treatment of fresh water resources." Most of this energy for expanded desalination facilities will come from fossil fuel plants, thus increasing the emission of greenhouse gases and worsening climate change. The debate over desalination illuminates, again, the critical connection between energy and water. It takes a large quantity of water to produce a single kilowatt of electricity. And it takes megawatts of electric power to run a desalination plant.

As cloud seeding and desalination illustrate, many proposed "solutions" to our water crisis involve quick fixes that place a religious faith in uncertain technologies. To believers, desalination constitutes the silver bullet. To skeptics, it's a technology that, to paraphrase Winston Churchill, has great potential and always will. To Jim Cherry and other credible professionals in the desalination industry, the choice of such an advanced water treatment technology depends on the quality of the source water, the intended use of the finished product, and environmental considerations. To me, desalinated water deserves a place in the water portfolios of some American municipalities. But it is an emerging technology—not a magical solution to the problem of water scarcity. Its high cost, finicky technology, energy demands, and environmental complications make desalination most suitable for high-value commercial and industrial applications. Don't expect to see homeowners watering their lawns or farmers irrigating alfalfa fields with desalinated water any time soon.

CHAPTER 9

Shall We Drink Pee?

"Water is like our gold, and we have to treat it like that."
—California governor Arnold Schwarzenegger, June 2008

TALL AND LANKY, James O. Doyle bounds like a gazelle up the stairs of the Pima County, Arizona, wastewater reclamation plant. He's the plant superintendent and has palpable enthusiasm for his job. Passionate about the process of taking raw sewage and turning it into effluent, Jim wants others to share his fascination with the process. "Jim loves to give tours," said a staffer. It shows. The October 2006 tour was a field trip for my water law class, and Jim was visibly disappointed that we had only an hour to visit the plant. Speaking at a record-breaking number of words per minute, Jim tried to squeeze in as much information as possible. With no disrespect to Jim, his wastewater plant will never rival Disneyland as a tourist attraction. The plant has, shall we say, a certain perfumed odor.

We began our tour at the preliminary treatment area, where raw sewage first enters the plant. The odor was indeed ripe. As I glanced around, I saw Maya Kashak, a third-year law student, with the neck of her jersey pulled up over her mouth and nose, her eyes, wide as saucers, conveying an unmistakable look of horror. I'd taken field trips to treatment plants before, but I hadn't remembered the smell being quite so pungent. Perhaps this time when Jim asked me where I'd like to begin the tour, I shouldn't have said at the beginning.

The darnedest things end up in sewers. Roto-Rooter once asked its service technicians about strange things they had found while cleaning sewer lines. The technicians mostly encountered wandering tree roots, but they had also found several live bats; one baseball bat; one hockey stick; one sixty-pound pig; twenty-two pairs of dentures; a package containing $50,000 in cash, part of a murder payoff; a pair of women's silk panties, size sixty-two; and a plastic bag of marijuana, complete with pipe. Doyle's staff have found "anything anyone can stuff down a manhole," including a dead body, aborted fetuses, and automobile tires.

On our tour of Doyle's plant, our first stop is a large conveyor belt, much like the tread on a tank, that separates baseball bats, dentures, and other solids from the sewage. This debris is funneled into bins that empty into dump trucks for disposal at a landfill. The sewage next goes to primary clarifiers—million-gallon tanks where the solids of human waste and other heavy particles settle to the bottom. Thickeners and digesters turn this sludge into fertilizer for nonedible crops. In the next phase of treatment, microorganisms, stimulated by the introduction of high-purity oxygen, eat the remaining organic matter. By the time the sewage passes through secondary clarifiers, the water is clear and little odor remains. Finally, disinfection using chlorine kills any remaining pathogens.

Until recently, municipalities around the country disposed of effluent from treatment plants as cheaply and quickly as possible. After all, these facilities are called *wastewater* plants, signifying a valueless nuisance. Cities prefer to dump effluent into rivers, where it's diluted and flows downstream. Indeed, until the 1970s, cities often didn't even treat the raw sewage before dumping it into waterways. In 1969, Cleveland's oily, contaminated Cuyahoga River caught fire, with flames reaching as high as five stories. Fireboats extinguished the blaze, but the utter absurdity of a river on fire catapulted water pollution into the nation's consciousness. In 1972, Congress passed the Clean Water Act to halt the dumping of industrial and municipal waste into the nation's rivers. Aided by a large dollop of federal funds, cities erected wastewater treatment plants to clean up the sewage. The treated effluent still ends up in rivers. And while no one likes to think about being downstream of a city's wastewater plant, most of us live downstream of someone's wastewater.

As horribly as we've treated our rivers, our oceans have fared even worse. Eastern Massachusetts cities and towns routinely dumped raw sewage into Boston Harbor and bays until the 1980s, when a federal judge put an end to that practice. Massachusetts solved its problem by constructing a massive water treatment facility on Deer Island in Boston Harbor. Metropolitan Boston cities and towns now send their sewage to the treatment plant, where, after treatment, it is unceremoniously dumped into the Atlantic Ocean. That's par for the course in coastal areas of the United States. Although the narrow coastal fringe makes up only 17 percent of the nation's land surface, more than half of us, some 160 million people, live along coastal areas. Most wastewater generated along our coasts ends up in the ocean.

This disposal process profoundly affects the quantity and timing of our water supply. Wastewater discharged into the ocean will not be available for reuse until the hydrologic cycle brings it back around again—a process that may take scores or hundreds of years. Every time someone in eastern Massachusetts flushes a toilet, as much as six gallons end up in the ocean. Water is too scarce a resource to be disposed of so cavalierly. And wastewater turns out to have so many valuable uses that we've given it a new name, reclaimed or recycled water. Unlike new diversions, dams, wells, or cloud seeding, wastewater reclamation offers a viable way to expand our supply. And best of all, reclaimed water represents a renewable supply that literally increases as the population increases.

Tucson Water has delivered reclaimed water for more than twenty years. It takes treated water from Jim Doyle's plant, further cleanses it with anthracite coal and sand filters, and disinfects it a final time with chlorine. The finished product is not quite drinking water quality, but the water is not currently meant for human consumption. The reclamation system, with more than 160 miles of pipeline, serves 900 sites that include golf courses, parks, cemeteries, roadway medians, schoolyards, and some individual homes. The system's capacity exceeds 33 million gallons per day; in 2005, the system delivered more than 4.4 *billion* gallons, or 12,000 acre-feet. Tucson Water delivered 110,000 acre-feet of potable water that year, so reclaimed water currently constitutes about 9 percent of total water use, but it could be substantially higher because treatment plants process 70,000 acre-feet per year.

Golf courses use two-thirds of Tucson Water's recycled water because Arizona law requires them to convert from groundwater use when recycled water becomes available. Reclaimed water turns out to be a perfectly fine supply because the nitrogen and phosphorus in it provide excellent fertilizer for turf grass, ornamental plants, and some crops. Tucson Water has persuaded some local cotton farmers to turn off their groundwater pumps and use reclaimed water instead. Tucson's mayor and city council made a policy decision to encourage the use of recycled water by pricing it competitively with potable water.

Reclaimed water could also be used in Tucson Electric Power Company's generating plant as cooling water, in industrial applications, and in copper mining south of Tucson, but these large-volume users have existing rights to pump groundwater, and, notwithstanding favorable water rates, it's a lot cheaper for them to use groundwater. And to get water to the mines, Tucson Water would need to expand dramatically its delivery system for reclaimed water, at considerable expense. Meanwhile, the Pima County Regional Flood Control District, in partnership with Tucson Water, the Tohono O'odham Nation, and the U.S. Army Corps of Engineers, has embarked on a program to use reclaimed water to restore flows and riparian habitat along a thirty-mile section of the Santa Cruz River that has gone dry as a result of groundwater pumping.

For cash-strapped American cities, the cost of a reclaimed water system is an obstacle to reuse, as a major financial commitment is required to install a completely separate set of pipes and valves (in Tucson these are painted purple to avoid confusion with those of the potable supply). Another obstacle is the bad press that came in 2006 when hundreds of people in California became sick and five people died after eating spinach contaminated with *E. coli* bacteria. In 2007, wine growers in California's famous Sonoma County wanted no part of a $385 million recycled water project devised by the county to provide water for agriculture. Thanks, but no thanks, was their message.

Nevertheless, the use of recycled water is spreading quickly. Cities such as San Diego, Las Vegas, San Antonio, Boca Raton, Long Beach, St. Petersburg, and Los Angeles have begun to reuse water, as have locations as diverse as the village of Cloudcroft, New Mexico, population 749, and the borough of Manhattan. In 2007, the New York City Department of Health certified Tribeca Green, a twenty-four-story apartment building in Battery Park, to collect and treat about one-half the building's wastewater and use it in toilets, gardens, and the air-conditioning system.

Wastewater treatment technology can take sewer water and clean it up to drinking water quality, though most Americans would rather not dwell on this prospect. In 1998, San Diego floated a trial balloon along these lines, but the program—dubbed by a smart-aleck reporter the "Toilet-to-Tap" proposal—was dead on arrival. By the time Mississippi River water arrives in New Orleans, dozens of upriver communities have used it, treated it, and returned it to the river for downstream folks to drink. But public perception—the "yuck" factor—is difficult to overcome. In 2006, San Diego tried again with an advertising campaign that called for "reservoir augmentation," a nifty way to introduce highly treated sewage water into drinking water reservoirs. But this effort also was short-lived. The *San Diego Union-Tribune* editorialized, "Your golden retriever may drink out of the toilet with no ill effects. But that doesn't mean humans should do the same." The paper accused "zealots" of backing the plan, a charge that prompted San Diego mayor Jerry Sanders to repudiate the proposal.

Other communities are nevertheless moving forward with introducing recycled water into the potable water system. The tiny hamlet of Cloudcroft found itself with no other options. Situated at an elevation of 8,500 feet in southern New Mexico's Sacramento Mountains, Cloudcroft had simply run out of water. Sitting at the top of the watershed, Cloudcroft has no place else to turn. In 2007, the village began taking water from the conventional treatment plant, subjecting it to additional filtration methods, including reverse osmosis, mixing it with groundwater from the village's wells, and introducing the blend into its potable system. Other cities, including Tucson, have accomplished the same thing indirectly. Treated effluent is dumped into recharge basins, where it percolates into the ground, mixes with groundwater, and is pumped out again later.

Elsewhere in 2007, the Orange County Water District christened a $480 million plant to recycle water from treated sewage by using it to recharge aquifers that serve Anaheim, Huntington Beach, and other cities. Concerned about public perception and acceptance of the project, Orange County water officials held hundreds of meetings with community groups, hospitals, and religious leaders to foster an understanding of the additional filtration processes the district would use before the water would be reused. According to the district's general manager, Michael R. Markus, the "indirect potable water reuse" system uses traditional microfiltration and reverse osmosis and then further cleanses the water with peroxide and ultraviolet light. The resulting water, says Marcus, "is as pure as distilled water."

An insidious challenge faces reclaimed water. Despite the United States' war on drugs, Americans love drugs, especially painkillers, antidepressants, caffeine, and alcohol. We're addicted to so many prescription drugs that millions of us take one drug simply to counteract the effects of another. In 2006, pharmacists filled approximately 3.4 *billion* prescriptions, up by 59 percent since 1995. Just imagine the numbers when baby boomers start buying bigger pill bottles. Maybe it's not too late to invest in Pfizer. The connection to recycled water comes because hospitals and consumers routinely flush unused or expired pills down the toilet. Even with the pills we ingest, our bodies do not entirely absorb the chemicals in them. Instead, the chemicals in birth control pills, hormone supplements, antibiotics, and erectile dysfunction medicines pass through our bodies and end up at the local wastewater treatment facility, as do detergents, antibacterial soaps, deodorants, perfumes, and colognes. By one account, as much as 90 percent of drugs leave the body unchanged and still active.

Now here's the scary part. The treatment process does not remove them. In 2007, Oregon State University researchers tested wastewater in ten American cities for remnants of drugs, both legal and illegal. The researchers found large differences in methamphetamine use from city to city. This community-wide urinalysis found that cocaine and ecstasy use peaked on weekends in one fairly affluent community.

In 2008, an Associated Press national investigative team completed a five-month inquiry that found pharmaceuticals "in the drinking water supplies of at least 41 million Americans," including those living in Southern California, northern New Jersey, Philadelphia, Detroit, and Louisville. The situation is undoubtedly far more widespread, given that many cities, including New York, do not test for the presence of pharmaceuticals; others screen for only one or two. Users of bottled water or home filtration systems also risk exposure. Some bottlers merely repackage municipal water and do not remove pharmaceuticals. Nor do home treatment systems.

So if San Diego adopts its Toilet-to-Tap proposal, would men in San Diego get erections from drinking water with the residue of someone else's Viagra? No wonder Jerry Sanders backed away. In reality, it's an absurd idea because the concentrations of these chemicals are so small, measured in parts per *trillion*, that the human health implications would be minuscule. But the consequence of ingesting a veritable cocktail of tiny amounts of multiple drugs is unknown. The nation's premier water scientist, Robert Hirsch of the U.S. Geological Survey (USGS), notes that we have lots of good studies on threshold levels below which individual chemicals pose no human health concern, but no studies about the effects of small doses of multiple chemicals.

These pharmaceuticals and personal care products are part of a group of chemicals the U.S. Environmental Protection Agency (EPA) calls "emerging contaminants," signifying that the agency knows little about them but fears they may have great consequences for humans and the environment. The EPA and the USGS have begun to collect data on them. And the U.S. Food and Drug Administration has begun to study whether these compounds may encourage the development of drug-resistant bacteria. The most Orwellian threat derives from the power of these drugs to alter profoundly our biological makeup. Certain chemicals called endocrine-disrupting compounds interfere with the body's hormone-producing glands and organs, such as the thyroid, pituitary, adrenals, pancreas, ovaries, and testes. I know this sounds dry, but it basically boils down to who is a boy and who is a girl and why. Who is from Mars and who is from Venus?

While the consequences for us humans remain unclear, we do know that fish don't like our chemically altered rivers. In a University of Arizona study, fish immersed for three months in wastewater from Jim Doyle's treatment plant had altered genes and five times more hormones of the opposite sex than of their own. Elsewhere, scientists found that Potomac River smallmouth bass were "intersex" fish, with males producing immature eggs in their testes. Fishing organizations, including Trout Unlimited, fear that these compounds have led to decreased smallmouth spawning. I'm all for sexual freedom and cross-dressing, but I think it should be a matter of personal choice rather than of having drunk the wrong water. To be sure, no one has yet proven dire consequences for humans, but it's a profound warning against using reclaimed water for drinking water.

The good news is that a recent San Diego study found that with the "advanced treatment"—reverse osmosis, ultraviolet light, and peroxide disinfection—being used by the Orange County Water District, the water produced had no detectable levels of most emerging contaminants. The bad news is that this process adds additional cost and energy demand to an already expensive and energy-consuming process.

Reclaimed water *is* a viable way to address water shortages. It grows as our population grows, and it has confirmed uses that would save potable water for human consumption. Its multiple limitations include high costs for a dual system of pipes and valves, consumer acceptance problems, and implications involving endocrine-disrupting compounds. Still, it's a terrific, renewable source that expands our supply. As things stand now, we treat sewage to almost potable water quality and then discard it. Does this make any sense? The city of Los Angeles is desperate to find more water for its growing population, yet its Hyperion Sewage Treatment Plant produces a volume of treated water equivalent to the seventh-largest freshwater river in California, all of which the city discharges into the Pacific Ocean. That may soon change. In May 2008, persistent drought prompted Mayor Antonio R. Villaraigosa to support a proposal to use treated sewage to increase drinking water supplies.

Other communities have discovered that reclaimed water is a very valuable commodity. In 2007, the town of Prescott Valley, Arizona, became the first community to put effluent from its treatment plant up for sale. The town used an innovative two-day auction format with a price-floor bid process to sell rights and options to 2,724 acre-feet per year of reclaimed water for $24,650 per acre-foot. In contrast, it costs Tucson Water $80 an acre-foot to pump groundwater and farmers in California's Palo Verde Irrigation District $15 an acre-foot for Colorado River water. If the winning bidder, Water Property Investors of Port Washington, New York, exercises its options, the town will receive more than $67 million. Water Property Investors expects to peddle the rights to reclaimed water to prospective developers, who under Arizona law must demonstrate a water supply before they can build. The town's economic consultant claims that the town's auction uniquely "involved capital markets in a sophisticated way." Jaws dropped all over the West as water managers heard of the sale of reclaimed water for $67 million.

CHAPTER 10

Creative Conservation

"I believe it is our God-given right as Californians to be able to water gardens and lawns."

—U.S. senator Dianne Feinstein (D-CA), November 21, 2004

MILLIONS OF US apparently agree with Senator Feinstein, given that we spend $40 billion each year, and consume 270 billion gallons of water each week, in order to maintain more than 23 million acres of lawn. That's an area larger than the combined landmass of Connecticut, Massachusetts, Rhode Island, Vermont, and New Hampshire; no wonder the Scotts Company is happy. The actress Rene Russo is not a lawn lover. She's lending her high profile to a campaign to replace lawns with native, low-water-use trees and shrubs. Her Brentwood, California, garden, lush with St. Catherine's lace, octopus agave, California wild grapes, and California live oaks, contrasts sharply with yards filled with azaleas, camellias, roses, and, of course, immaculately manicured grass.

In 2007, the Los Angeles County board of supervisors signed on to Russo's campaign when it voted to require new housing developments to plant drought-resistant or native vegetation. For those Los Angelenos who love lush landscapes and for those visitors who marvel at the manicured gardens in Beverly Hills, it's worth noting that Los Angeles receives, on average, fifteen inches of rain a year—that's only three inches more than Tucson, in the middle of the Sonoran Desert. Los Angeles only acts as if it has abundant water resources. The city reclaims a meager 1 percent of all city water.

If governments expect citizens to conserve, they must set an example, as the city of Long Beach found out the hard way. When the city launched tough new watering restrictions in 2007, city officials asked residents to report broken sprinklers and other wastes of water. In the first month, callers to the hotline identified more than 400 wasteful practices, but 83 of them involved city parks and medians on streets and freeways. Matthew Lyons, director of planning and conservation for the Long Beach Water Department, conceded, "When people hear a message from the government, telling them to conserve, and they see the government wasting huge amounts of water, it makes people very cynical."

Nor are people blind to the incongruity of governments asking residents to conserve water while simultaneously issuing new building permits. In the midst of the drought in California in 2007, a common response to a plea for voluntary conservation, says Gary Arant, general manager of the Valley Center Municipal Water District, was "Why should we conserve so developers can build new homes?" Local governments must link land use and water supply, or opponents of growth may conclude that the best strategy is to use lots of water and provoke a crisis in order to halt relentless growth. In 2007, Pima County became the first county in Arizona to insist that developers provide information on the source of water for new development, and the environmental consequences of such water use, at the time of the request to rezone the property.

Can we solve the water crisis through municipal water conservation? (More on agricultural conservation later.) In many places, yes. In the 1980s and 1990s, developers in Phoenix had a passion for golf courses and artificial lakes filled with groundwater. Travelers who fly in or out of Phoenix's Sky Harbor Airport see golf courses and lakes scattered throughout the Valley of the Sun, a place that averages seven inches of rain a year. Upscale shopping malls lure patrons by cranking up their air-conditioning and leaving the front doors wide open even during the dog days of summer. Restaurants expand their seating areas with patios with misting systems, high-pressure nozzles that spray a fine mist that "flash evaporates," thus lowering the temperature by increasing the relative humidity. The latest craze is home misting systems, especially those that create a "fog effect" by the pool, reminiscent, I suppose, of fog on the Pacific Coast. To be sure, Native Americans and early Arizona settlers used evaporative cooling in their homes, but these new systems are outdoor features. Home patio systems typically use sixty gallons an hour; running a restaurant's misting system over a patio with a seating capacity of fifty when no one is sitting there bespeaks an attitude about water use.

Another example of wretched excess can be seen in Fountain Hills, Arizona, thirty miles northeast of Phoenix. The town brags about having one of the world's highest fountains, which "sends a snow-white jet stream of water 560 feet into the blue desert sky." Its height exceeds New York City's Delacorte Fountain, at 435 feet. High-pressure 600-horsepower pumps deliver water to a 2,000-pound nozzle. Water pressure of 375 pounds per square inch thrusts 7,000 gallons per minute through the nozzle at a velocity of 68 feet per second. Approximately 2,000 gallons, weighing eight tons, are in the air above the nozzle. Built in 1970, the fountain first used groundwater and ran more or less continuously. Criticism of this bizarre water use led the town to scale back operations. It now uses reclaimed water from the twenty-eight-acre lake at the base of the fountain and operates fifteen times a day at ten-minute intervals. The town boasts that evaporative loss is less than the water required to irrigate an acre of cotton. That may be true, but it's still more than 1.6 million gallons annually.

Areas of the East, including Florida, have yet to catch on to water conservation. Many cities and homeowners' associations actually require lawns and prohibit drought-resistant plants or artificial turf. On Marco Island, Ed Ehlen built a multimillion-dollar waterfront home. In 2005, he surrounded the house with a $19,000 artificial lawn, like those used on football fields. City officials told Ehlen he was violating the landscape code and ordered him to remove the plastic lawn. He refused, and he was miffed. The city has no restrictions on house paint colors, so Ehlen protested the city's artificial turf ban by painting an entire side of his house pink with purple and green polka dots. In Florida, 75 percent of residential water use is outdoors. As the fourth most populous state, Florida must eventually reconcile growth and water use.

Reducing the demand by using less water may require a cultural sea change when it comes to our love affair with grass, but some cities, such as Long Beach, have made remarkable progress in water conservation. Using aggressive conservation and recycling programs, the Long Beach Water Department has made the city one of the nation's most frugal water users. Per capita consumption is down to 121 gallons per day, compared with 168 for Los Angeles County, 250 for Riverside County, and 253 for San Bernardino County. In Texas, Karen Guz, director of conservation programs for the San Antonio Water System, boasts that San Antonio's water use, once 225 gallons per capita per day, declined to 136 gallons in 2006—a year of extreme drought conditions. With more than 1.1 million people, the country's seventh-largest city aggressively encourages *and* mandates water conservation. It has to.

San Antonio relies on groundwater pumped from the Edwards Aquifer. The aquifer was once believed to hold an inexhaustible supply, but overpumping has led to declining water tables and reduced outflows to springs that are home to endangered species. Several rounds of litigation led the Texas legislature to create the Edwards Aquifer Authority, which, after more litigation, eventually imposed restrictions on pumping. To comply, the San Antonio Water System sponsors comprehensive education programs and, as a stick, its conservation ordinance requires drought-tolerant grass for new developments and rain sensors on sprinkler systems. Drought restrictions prohibit fountains and waterfalls and limit the hours and days for watering lawns and gardens and for washing cars.

A popular and successful part of the system's strategy involves its "Kick the Can" program. The San Antonio Water System gives away high-efficiency toilets to homeowners whose residences were built prior to 1992. This restriction targets dwellings more likely to have older, high-water-use toilets. Low-flow toilets—1.6 gallons per flush (gpf) rather than 3.5 to 6 gallons—have not always worked well in the past, requiring two or three flushes simply to wet the toilet paper, never mind purge the waste. But the latest generation of high-efficiency toilets (HETs) use even less water (1.3 gpf) and dispose of waste effectively, and each toilet saves more than 11,000 gallons per year.

To overcome residents' perception that these toilets might not work, the San Antonio Water System built a portable toilet stand with a clear rectangular Plexiglas base beneath the toilet bowl. Then the question became, says Karen Guz, what they might flush down the toilet to prove convincingly that it worked. They settled on a baking potato, not a baby redskin but a full-size Idaho russet. The model toilet, equipped with a dual-flush system designed to dispose of liquid waste with 0.8 gallon per flush and solid waste with 1.3 gallons, handled the potato with either flush. The video of this contraption became a hit on www.youtube.com; the first night it aired, residents bombarded Guz's office with requests for the toilet.

San Antonio has other water-conserving programs, including rebates for efficient washing machines, on-demand water-heating systems, and water-saving landscapes. It's pushing for reform of plumbing standards industry-wide to take advantage of aerated faucets and showerheads. On San Antonio's famous River Walk, the Hilton Palacio del Rio, a four-star luxury hotel, retrofitted every room with aerated shower heads that convey a sense of high water pressure but use little water. Hotel managers made this decision only after a test run and guest survey demonstrated that guests gave a high rating to their "shower experience." The hotel's upscale guests apparently don't want the water trickling on them in some wimpy spray.

In 1995, the city of Albuquerque faced a declining groundwater table and legal limits on diverting more water from the Rio Grande.

So the city embarked on an ambitious conservation program that aimed to cut usage per person per day from 250 gallons to 175 gallons within ten years, a 30 percent reduction. The city achieved the target, helped in part by a rebate program that underwrote much of the costs for homeowners and businesses to convert to xeriscaping, using drought-tolerant plants. In 2007, the water utility chairman, Martin Heinrich, proudly announced that average daily use had dropped to 164 gallons, aided by an unusually wet spring. The city's revised target is now 150 gallons.

Utah has great opportunities to conserve water because its municipal use in 1995 topped out at a whopping 267 gallons per person per day, the second highest in the country. Utah's growth rate is the fifth fastest in the nation, thanks to immigration and high birthrates in the Mormon community. Eric Klotz, who manages water conservation for the Utah Division of Water Resources, notes that Utah's challenge is not indoor use, which is approximately 70 gallons per person per day, roughly comparable to that in the rest of the United States, but outdoor use. Using regulatory strategies, educational programs, and innovative technologies, such as "smart controller" irrigation clocks with evaporation-transpiration sensors, the Utah water resources agency achieved a 17 percent reduction by 2005. According to Klotz, a typical homeowner's programmable irrigation system uses 50 percent more water than the lawn and garden needs. Installing "smart controller" timers can save substantial water. Once available only to golf courses and commercial irrigators, these controllers now come in a simple model for residential use that employs wireless communication, similar to that used by your cell phone, to retrieve wind, heat, and rainfall measurements from local weather stations. The downside is their cost—$500 to $700. Achieving further reductions in Utah may be hard because Utah water rates are some of the lowest in the country and homeowners lack an incentive to invest in expensive but effective water conservation technology.

An extensive rebate program anchors the Metropolitan Water

District of Southern California's conservation strategy. Its $235 million commitment has replaced 2.5 million toilets with ultra-low-flow models. The Pacific Institute's Peter Gleick would like California to go further. Although "exciting developments in the high-efficiency toilet market may sound like an oxymoron," he estimates that replacing all existing toilets with the low-flow versions could save California 130 *billion* gallons per year. Replacing washing machines with more efficient models, says Gleick, could save an additional 33 billion gallons. Together, that's enough water for the annual needs of roughly 3 million Californians. Even seemingly trivial changes can save large quantities of water. A dripping faucet can waste as much as 2,000 gallons a year, while a leaky toilet can waste as much as 200 gallons a day. To Gleick, conservation makes much more sense than pricey dams and desalination plants.

Water conservation comes in many sizes. In San Ramon, California, the Food Service Technology Center tests equipment used in commercial kitchens. Dishwashing uses approximately two-thirds of all the water used in a restaurant, and half of that goes to prerinse dishes before they go in the dishwasher. After the technology center favorably evaluated a highly efficient spray valve, the California Urban Water Conservation Council replaced, free of charge, almost 17,000 valves in small and medium-sized restaurants. With each valve saving 156 gallons per day, or 57,000 gallons per year, the total water savings was almost 1 *billion* gallons per year. The program also significantly reduced the amount of hot water used, saving energy—more than 32 million kilowatts of electric energy and 5 million therms of natural gas per year.

Not all Californians have seen the light. While staying at San Diego's Westin Horton Plaza Hotel for a 2005 conference, I discovered the shower had *two* showerheads, each with enough pressure to take the paint off the walls. In July 2007, in the midst of the worst drought in memory, bathrooms in San Francisco's upscale Westin St. Francis Hotel had dual-head showers.

At the other end of the luxury spectrum, prisoners at California

State Prison, Corcoran, have found creative ways to use water and annoy prison officials. The *Sacramento Bee* reports: "By flushing the toilets in their cells, prisoners communicate with one another, relieve boredom, protest prison conditions, dispose of contraband and even create in-cell swimming pools." Prisoners use toilets to warn of the coming of guards and to get released from lockdown by flooding their cells. Prison officials claim that some inmates flush their toilets more than 100 times a day. Given that California has thirty-four prisons housing 173,000 people, that's a lot of water down the drain. In response, corrections officials have begun to retrofit California's prisons with the Sloan Valve Company's Flushometer, an electronic flush-control system that briefly delays all flushes and that limits flushes to not more than two in five minutes and three per hour.

As municipal water providers search for the most cost-effective conservation alternatives, multiple factors and variables preclude easy answers. However, the Water Conservation Alliance of Southern Arizona (Water CASA) recently analyzed eighty-eight separate conservation programs from around the American West to determine which programs work, how well, and at what cost. Hands down, toilet giveaways save the most water and cost the least per acre-foot of water saved.

In some communities, residents respond well to government requests to reduce their water use. In Tucson, where water conservation is a badge of honor, folks pitch in. On this score, San Diego has a long way to go. As the drought worsened in 2007, Mayor Jerry Sanders pleaded for his constituents to conserve. With great fanfare, the San Diego County Water Authority launched its "20-gallon challenge," a voluntary program that asks each person to reduce consumption by that much each day. But the agency's July 2007 water sales were the highest in history, and in October 2007 demand was 6.6 percent above that of a year earlier. Puzzled officials pledged $206,000 for a "more aggressive" publicity campaign. Speculating that they have not successfully reached newcomers and young people, officials are experimenting with new outreach programs, such as using social networking Web sites. This may help, but there is also a human nature factor at work: our perceptions of ourselves dramatically depart from reality. As Garrison Keillor says about Lake Wobegon, "all the children are above average." For water use, many people are simply unaware of how their consumption compares with that of their neighbors. A 2007 Water CASA survey asked, "Would you say that your water use is above, below or about average?" A remarkable 87 percent believed that their water use was either average or below average.

These government-sponsored water conservation programs have taken four forms. Hortatory programs urge citizens voluntarily to reduce their water use in times of crisis, such as drought. These programs usually work because Americans are by nature a generous people, willing to contribute to the well-being of the community. But there needs to be a real crisis, and the duration had best not be too long, for we're also an impatient lot. It's helpful to have public service announcements and a broad-gauge publicity campaign to educate citizens about the reasons to conserve. In Tucson in the late 1970s, the water utility faced a peak water use crisis. Its storage facility was almost tapped out from so many residents watering their landscape plants on late summer afternoons. Tucson Water devised a "Beat the Peak" campaign, complete with a little yellow duck, Pete the Beak, who serves as the mascot for conservation. The community responded, the crisis abated, and thus started a culture of voluntary water conservation that still pervades Tucson.

A second type of conservation involves mandatory restrictions on water use. These restrictions, usually adopted during a drought emergency, typically limit water use to certain hours of the day or days of the week, or prohibit certain activities, such as washing cars, watering lawns, or filling pools. Long-term regulatory programs often mandate low-flow fixtures in new construction or prohibit ornamental fountains or lawns over a certain size or in particular locations. Most citizens readily accept such restrictions, but not everyone does. In 2007 in Brownsburg, Indiana, Rosie Igo complied with lawn-watering restrictions only until her lawn started to turn brown; then she set her timer to come on at 3:30 in the morning so no one could tell. I wonder how Rosie explains to her neighbors her green lawn—the only one on the block. Divine intervention, perhaps? A third program offers rebates, redeemable at a local hardware store, for low-flow toilets, smart controllers, low-flow showerheads, and pool covers. And a final program involves giveaways of low-water-use fixtures, especially toilets. This may be the most effective program of all, but it is very capital-intensive.

Several cautions are in order about water conservation programs, whether voluntary or mandatory. State and local governments should avoid creating a system with so much regulatory red tape that it's ultimately counterproductive. In Southern California, Chuck Carr is president of a landscaping firm that manages irrigation systems for medium- to large-sized homeowners' associations and commercial properties. California mandates that "landscape audits" be performed by "certified landscape design auditors," who must meet "irrigation specialist certification standards." Landscape managers, such as Carr, must pass state-established certification tests that show proficiency in calculating soil moisture, evapotranspiration ratios, and the like. But the workers who actually set the irrigation control timers are often undocumented workers from Mexico who lack the language skills and training to comply with the rules. Says Carr, "You have to have a Ph.D. to understand all the requirements. These irrigation controllers are basically computer systems and the average person out there doing the work in the field is not computer literate."

Or consider Arizona, which prides itself on the 1980 Groundwater Management Act requiring municipalities to lower their per-person, per-day consumption rates. Tucson Water managed to reduce its residential consumption from 115 gallons to 99 gallons between 1997 and 2007, but because the utility's customers swelled from 605,000 to 734,000, actual water use increased. And it's getting harder for cities such as Tucson to make much more progress. The city has already picked the low-hanging fruit through no-brainer programs that targeted obvious waste, such as apartment complexes whose sprinkler systems sprayed as much water on streets as on lawns and ornamental plants. The city now faces what its conservation program manager, Fernando Molina, calls the "hardening of demand." It must now reach for the higher fruit. The cornerstone of Arizona's groundwater act is the requirement that developers demonstrate "an assured water supply" before getting their subdivision plats approved. Developing regulations to implement this program has vexed the Arizona Department of Water Resources. Countless public meetings and innumerable drafts—with debates over every comma and semicolon—finally yielded rules totaling thirty-six pages of single-spaced fine print.

The California and Arizona examples offer a cautionary lesson for advocates of government-mandated water conservation: conservation standards fraught with complexity, thereby requiring elaborate monitoring programs, may not be cost-effective or achieve meaningful results. Mandatory programs require the government to allocate resources to enforce the rules. This demands either "sprinkler police," a questionable use of trained law-enforcement personnel, or a second class of administrative officers who write tickets for violations.

Progressive communities such as Long Beach, San Antonio, Albuquerque, and Tucson have enacted pioneering water conservation programs. Disparate motivations drive conservation efforts and impede them. Not all Americans, even in water-strapped communities, are on the same page when it comes to reducing their water use. Much remains to be done, not only in educating citizens but also in facing hard questions about our collective future. A pious platitude about conservation by government officials in the face of rubber-stamped plats for new subdivisions engenders cynicism rather than citizen commitment to conservation.

CHAPTER 11

Water Harvesting

NEAL SHAPIRO COORDINATES the city of Santa Monica's Urban Runoff Management Plan, an ambitious, multipronged program to harvest rainfall and urban runoff and reduce the pollution that flows into Santa Monica Bay. Cities have traditionally viewed storm water as a problem to be dealt with as quickly and as cheaply as possible. Urban development, with its attendant roads, sidewalks, driveways, parking lots, and rooftops, has replaced permeable land with impermeable hardscape. Water quickly flows off these surfaces, exacerbating floods and carrying with it the detritus of urban living.

Until recently, wastewater posed the biggest pollution threat to coastal waters. But by 1995, says Stephen Weisberg, executive director of the Southern California Coastal Water Research Project Authority, storm water had become the source of as much as 90 percent of the toxic chemicals that washed into coastal waters. According to Bruce Reznik, head of the environmental group San Diego Coastkeeper, 1 *trillion* gallons of urban storm water foul the ocean waters off Southern California each year. This runoff contains industrial chemicals such as chromium, copper, lead, and nickel; pesticides and fertilizers; swimming pool chemicals; oil, gasoline, and grease; dog, cat, and feral animal waste; highway trash; and cigarette butts.

Lots of cigarette butts. According to the California Department of Transportation, 20 percent of the material removed from highway drainage filters consists of cigarette butts. And it's not just oceans that feel the impact of urban slobber. The U.S. Environmental Protection Agency reports that urban storm water ranks among the largest sources of pollution for estuaries, wetlands, lakes, and rivers. Nationwide, the year 2006 witnessed a record number of beach closings and advisories due to contaminated storm water runoff.

Santa Monica, well known for its sandy beaches, city pier, political activism, celebrities, well-educated citizenry, and environmental concern, attracts 3 million to 5 million visitors a year. The city's director of environmental and public works management, Craig Perkins, wryly observed, "It's logical to assume that they would prefer that the beaches and ocean are safe and clean." To that end, the city adopted a management plan that views storm water as a resource to be harvested rather than an annoyance to be disposed of.

In 2001, the city brought online its Santa Monica Urban Runoff Recycling Facility, or SMURRF (a joint project between the city and Los Angeles), a $12 million plant that processes an average of 350,000 gallons of urban runoff every day. The first facility of its kind in the country, SMURRF prevents this runoff from fouling the Pacific Ocean off the city's famous beaches. Still, a question arises. Why does Santa Monica need to operate such a facility full-time? The city gets only fifteen inches of rainfall per year, most of that during a four-month rainy season from November through March, and it enjoys 325 days of sunshine a year. Why not shut the plant down the rest of the year? Santa Monica needs the recycling facility to process what Shapiro describes as "local dry weather runoff." The runoff comes not from rain or any natural storm event but from Santa Monica's residents and businesses excessively watering lawns and gardens, washing cars and equipment, draining pools, and hosing down sidewalks and driveways. The city has an ordinance that prohibits wasting water, but its police department devotes most of its resources to serious safety threats. As Shapiro's three-person staff

Figure 11.1. Photograph of Santa Monica Urban Runoff Recycling Facility. Photograph courtesy of Neal Shapiro, City of Santa Monica.

gamely tries to patrol, observe, and document violations, repeat offenders get fined $250, even up to $500 for runoff violations. But, Shapiro laments, "the problem of wasting water is chronic, daily, 24/7. We are not here on weekends and at night to catch violations. We do the best we can."

Although it seems perverse to process runoff when it hasn't rained, the recycling facility has successfully halted pollution from urban runoff and provided a new source of reclaimed water. Located within sight of the walkway connecting the Santa Monica bluffs with the pier, the recycling facility consists of modernist concrete walls and pillars, red metal cylinders, artist-created sculptures, and red- or blue-tiled spillways and chutes. Although the treatment process is conventional, the city assembled artists, architects, and engineers to add intense colors, unique architecture, dramatic lighting, artwork, and educational information to make the recycling facility a place that would attract visitors. A walkway allows passersby to observe the complete treatment process, and signboards offer educational messages that highlight what citizens can do to reduce runoff pollution. Santa Monica's recycling facility is so sensible that coastal communities throughout Southern California have followed the city's lead and installed their own systems—right? Wrong. Santa Monica's is the only urban runoff recycling plant in the United States, perhaps in the world.

Santa Monica already imposes obligations on new construction and mandates the retrofitting of existing structures in order to minimize runoff from sites and to maximize recharge of the aquifer. Infiltration fields, permeable pavers, catchment basins, retention cells, and rain barrels below gutters reduce runoff at private residences, apartment complexes, office buildings, public facilities, and city parks. These simple and obvious steps make environmental and economic sense. If cities don't eliminate urban runoff, or don't prevent its formation (by harvesting it before it becomes runoff), then they'll pay to treat it.

In 2007, the San Diego Regional Water Quality Control Board announced plans to require local governments to curb bacteria in urban runoff. Under current rules, coastal cities face no penalties if bacteria pollute the beaches when it rains. The San Diego Coastkeeper welcomed the plan as "aggressive and ambitious," while developers complained that it would ruin the local economy and tack on "more than $30,000" to the price of a new home. In Santa Monica, voters approved a second storm water–clean beaches fee in November 2006 to expand storm water management, but not everyone buys into Santa Monica's environmental values. In 2006, voters in Encinitas, California, in northern San Diego County, overwhelmingly rejected a $5 monthly fee to support the city's clean water program. If residents of this affluent city, with a 2005 median household income that exceeded $86,000, refuse to support such a modest fee to help keep their beach clean, the prognosis for reform elsewhere is alarming.

Other communities around the United States are finding creative ways to harvest storm water runoff. In the late 1990s, when Seattle developers began to solicit tenants for King Street Center, a typical downtown office building, the county department of natural resources expressed interest in leasing space but asked the developer to include environmentally friendly, sustainable elements. The center collects rainwater off the roof in three 4,500-gallon tanks. It uses this water, which would normally flow into Seattle's sewer system and require treatment, to flush 105 toilets, saving an estimated 1.4 million gallons each year.

On the other side of the country, in Kannapolis, North Carolina, Tony Brandon, an engineer with a consulting firm, developed a system to reuse water that comes from the urinals, toilets, and sinks at the David H. Murdock Core Laboratory Building on the North Carolina Research Campus. In 2006, Brandon's firm installed biological sand filters to treat the water, added an ultraviolet light disinfection process, and reused the water in the building's toilets. The system was not cheap to install, costing about $90,000, but the payback time is only four years. More important, by reducing the amount of water the laboratory sends to the sewer system, this arrangement allowed the laboratory to get needed building permits in a region with an already stressed local wastewater system.

In New Jersey, Walter J. Mugavin, the owner of Aqua-Mist Irrigation in South Hackensack, is working with the Del Webb division of Pulte Homes, one of the nation's largest home-building companies, on an age-restricted residential community in Manchester, New Jersey. At River Pointe, Del Webb proposed building 504 single-family homes marketed to baby boomers in an effort to encourage them to retire in New Jersey rather than move to Florida. Del Webb faced two problems. Manchester Township, in central New Jersey's Ocean County, prohibits the use of potable water for irrigation. And the township's planning and zoning department initially rebuffed Del Webb's request for permits because it determined that the already burdened wastewater treatment plant could not accommodate the additional storm water and sewer water that would be generated by River Pointe.

Del Webb figured it could obtain potable water by drilling its own groundwater wells, but that would leave the problem of storm water and sewer water unresolved. In examining the plans for River Pointe, Mugavin noticed that two ponds included in the proposed development had a capacity of 9 million gallons, while the irrigation needs for River Pointe were approximately 261,000 gallons per cycle. Mugavin suggested that Del Webb collect storm water runoff in the ponds and use it as landscaping water for the entire project.

The New Jersey Department of Environmental Protection loved the idea. First, capturing the storm water eliminates the runoff problem and the potential burden on the local wastewater facility. Second, storm water collected in the ponds will provide sufficient water to irrigate all of River Pointe's landscaping—front yards, backyards, side yards, common areas, amenities, and the clubhouse area. Mugavin added aerators in the ponds to eliminate mosquitoes, as well as a filtration system on the pump and an automatic backwash on the filter to mitigate the problem of murky water after heavy rains.

River Pointe's irrigation system differs strikingly from those in other master-planned communities. Typically, each homeowner has an irrigation controller that regulates outdoor watering of trees, shrubs, and lawns. At River Pointe, that would mean 504 timers all coming on in the middle of the night, dropping the water pressure and posing a potential firefighting problem. Instead, Mugavin installed a central computerized control system, with all the bells and whistles available, that manages the watering needs of the entire project. It's a win-win solution. Retiree homeowners avoid the hassle of messing with a computerized controller and save money by not paying for landscape water. Del Webb must front the costs for the ponds, pumps, filters, and timer, but it's not a big expense. More important, from Del Webb's perspective, Mugavin's solution allowed the company to build the project. The driving force behind River Pointe's solution was not water conservation. Water was plentiful and cheap. The driver was that, without Mugavin's creativity, Del Webb would not have received the necessary permits, which are worth a great deal to Del Webb. Indeed, the company has decided to adopt this approach at other sites. It began selling River Pointe homes in 2007.

These examples from Santa Monica, Seattle, Kannapolis, and Manchester illustrate "green infrastructure," a new approach to managing storm water that is cost-effective and environmentally friendly. It's an exciting departure from the traditional engineering mentality that embraces massive systems to collect, convey, store, and discharge storm water. From green roofs to rain gardens, and from vegetated median strips to porous pavements, green infrastructure captures, cleanses, and reduces storm water runoff, making it available for reuse.

SINCE THE TWELFTH CENTURY, the Hopi Tribe in northeastern Arizona has harvested water using simple techniques that slow the movement of water to irrigate their fields. Across drainages that flowed only periodically, they built small rock dams that catch soil and water. Soil accumulates behind the dams, providing stable gardens with water supplied by gravity. Terraces, bermed beds, gravel- and rock-mulched fields, and floodplain farming similarly took advantage of natural contours to capture water.

Today, urban evangelists preach the gospel of water harvesting. Perhaps none is more charismatic and persuasive than Brad Lancaster, who lives in Tucson and advocates harvesting rainwater in the desert. Tall, lean, and appropriately bearded and ponytailed, Lancaster exudes a preacher's confidence. In conversation and speeches, he is animated, opinionated, passionate, and inspiring. Listening to him describe how he has lived off the grid for thirteen years in Tucson, relying on 100,000 gallons of water a year that he and his brother harvest on an eighth of an acre in the city, makes one want to dash off to Ace Hardware and buy a cistern. He turns this harvested water "into living air conditioners of food-bearing shade trees, abundant gardens, and a thriving landscape incorporating wildlife habitat, beauty, edible and medicinal plants, and more. Such sheltering landscapes cool buildings by 20 degrees, reduce water and energy bills, and require little more than rainwater to thrive." His message is almost irresistible.

Tucson Water, the local water utility, has bought into Lancaster's message. Its "City of Tucson Water Harvesting Guidance Manual" offers developers, engineers, designers, and contractors detailed, illustrated information about harvesting principles, site designs, and techniques. For homeowners, Lancaster has written a two-volume guide, *Rainwater Harvesting for Drylands and Beyond*, that will get nearly anyone up to speed. As a sign of the green times, Wal-Mart sells Lancaster's guide.

While some folks live to harvest, others harvest to live. On San Juan Island, off the very northwestern corner of the continental United States, excessive groundwater pumping has caused the water table to plummet and salt water to migrate laterally, contaminating some wells. Island residents are hard-pressed for water. Portland, Oregon, jazz musician Teddy Dean and his wife, Alice, retired to San Juan Island in 2000. They built a 1,900-square-foot house and a 1,700-square-foot shop and music studio. A well already on the property supplies water, but it's 750 feet deep and produces only a quart a minute, sometimes erratically. Before the Deans finished construction, they met Tim Pope, who operates Northwest Water Source, a company that specializes in designing and constructing water catchment systems. Working with Pope, the Deans decided to give it a try.

At the bottom of a hill below the house and shop sits a silo that looks like it should be in Nebraska and full of corn. That's because it came from Nebraska. A rubberized liner from Alberta, Canada, tightly seals the silo, which holds as much as 15,000 gallons of water. Gutters and downspouts on the roofs of the house and shop collect rainwater, which flows by gravity downhill into the silo. Teddy Dean calculates that each inch of rainfall produces about 1,600 gallons of water. An electric pump at the bottom of the silo supplies water when needed. Because fir trees surround the buildings and their litter falls into the gutters and decomposes, Pope added a triple filtration system, with two sediment filters and an ultraviolet electronic bacterial filter that brings the rainwater up to drinking water quality. Maintenance is simple. Every nine months or so, Dean changes the filters and the ultraviolet bulb, for annual maintenance costs of roughly $500. "The system just works flawlessly," he says. "Knock on wood; everything has gone really well in the last five years."

It cost the Deans approximately $15,000 for the entire system. It's a lot of money, to be sure, but Teddy Dean notes that without it, they "would be having a pretty precarious existence." They must still husband their supply. San Juan Island, like the Seattle area in general, has a winter rainy season followed by a very dry season between July and October. So the Deans monitor their water use to get through the dry periods. Still, the catchment system adequately provides for their indoor needs, and Alice has a 4,000-square-foot garden where she grows an array of fruits and vegetables year-round. Rainfall adequately waters the garden except during the dry summer months, when she uses a drip irrigation system with water from the silo. Because they use the well only occasionally, they're saving money by not pumping well water from 750 feet, and of course they pay no monthly water bill to a local water utility. The end result is that these homeowners who live on a bed of clay surrounded by ocean water have a fully functioning, sustainable, year-round water supply. This solution is catching on with the other 14,000 residents of the San Juan Islands. Pope has installed 190 rainfall catchment systems.

A water-harvesting system can be much cheaper, especially if the water is not needed for drinking. And it doesn't take an engineer to install such a system. Something as simple as a gutter downspout connected to a fifty-five-gallon drum with a spigot for a hose will provide water for the garden. The city of Portland, Oregon, encourages on-site storm water retention to relieve pressure on the wastewater system, and it pays residents incentives to disconnect residential downspouts from storm sewers. In 2008, Albuquerque began to require newly built homes to have rainwater collection systems.

Inside the home, a variety of new gadgets or modest efforts can save water. Faucets with sensors automatically shut off the water, much like the faucets in many public washrooms. Ever notice that it may take a while before your sink or shower delivers hot water? Some earnest folks have addressed this waste. Doug Pushard, a software company manager in Austin, Texas, places a pail in his shower to capture the water before it turns hot. He then uses that water in his garden. Another way to save the same water is to install a recirculating pump that returns the cold water to the water heater. Ever notice that you waste water prepping vegetables at the kitchen sink? Consider installing a pedal valve on the floor that allows you to turn the water on and off with your foot.

These examples only scratch the surface of options to conserve water. But many of them are expensive; for example, a pedal valve costs between $300 and $500. For some people, such as Pushard, it's not about the money. He also installed a $5,000 rainwater-harvesting system. Although he's saving more than 30 percent on his water bill, roughly $30 a month, it will still take more than fifteen years to recoup his investment. In New York City in 2005, the average water bill for a single-family home was $571 a year. With water priced so low, it takes a very strong environmental ethic, or a government mandate, for most people to install pricey water-saving devices.

Some harvesters have gone even further than Doug Pushard. Laura Allen is one of the Greywater Guerillas, a group based in Oakland, California, that promotes the recycling of gray water, which is water from sinks, showers, and washing machines but not from the kitchen sink, dishwasher, or toilet. Gray water can safely flush toilets and water gardens. Allen and her friends have installed many gray water systems in the San Francisco Bay Area. Once underground and decidedly countercultural, the movement to use gray water has entered the mainstream. In 2007, the state of Arizona, never known as a hotbed of progressivism, enacted a $1,200 tax credit for builders and homeowners who install gray water systems. The state of California became the first state, in 1994, to enact gray water guidelines, but the plumbing code is hideously complicated and, in the opinion of many gray water advocates, an impediment to the growth of gray water systems. Says Steve Bilson, founder of a company that installs gray water systems, "The code is so overbuilt that I'm beginning to think it's better to just have everyone do it bootleg." Laura Allen notes, "We're forced into being guerilla-style because of economics. Permitted systems cost between $2,000 and $10,000. Non-permitted systems can be as cheap as $100."

Brad Lancaster, Teddy and Alice Dean, and the Greywater Guerillas epitomize citizens who are seizing control of their water use in a noble effort to live sustainable lives. They're joined by private sector actors, such as Walter Mugavin, who embrace conservation not for its own sake but as an imperative if they are to stay in business. State and local governments should encourage water harvesting by simplifying the plumbing codes and by creating incentives for home builders and homeowners with tax credits and rebate programs.

CHAPTER 12

Moore's Law

FARHANG SHADMAN, an Iran-born engineer, studied at the University of California, Berkeley, and trained at the General Motors Corporation's research and development center before settling in at the University of Arizona, where he is a Regents Professor, the highest honor the university can bestow on a faculty member. A driven, intense man who strictly applies principles of logic and rationality to his work, he finds room for personal expression in his spare time by playing the hammer dulcimer, an instrument from old Persia. Around 1990, Shadman became interested in the semiconductor industry.

At the heart of every computer, NASA satellite, cell phone, clock radio, and automobile fuel injection system lies a microprocessor—the operating system that runs most modern conveniences. These integrated circuits, or chips, are made in a semiconductor fabrication facility, or fab. "We add value to sand," quips Craig Barrett, Intel Corporation's chief executive officer from 1998 to 2005, referring to the transformation of silicon, the second most abundant element in the earth's crust, into microprocessors. Starting with silicon wafers eight to twelve inches in diameter, the process removes contaminants from the wafer, etches pathways on it, and modifies its electrical properties. It's a bit like constructing a tall building story by story as twenty separate steps deposit different materials atop the silicon base, a painstaking process that takes more than two months for the production of a single microprocessor.

The building blocks of the microprocessor are transistors, the "traffic signals" for electric impulses. In 1959, a single transistor was about the size of a pencil eraser and cost six dollars; by 2007, Intel's engineers had reduced transistors' size so dramatically that 30 million could fit on the head of a pin. Transistors cost 500,000 times less than they did in 1959. In 1965, Intel cofounder Gordon Moore famously predicted that the number of transistors on a silicon chip would double about every two years. Despite the challenge of meeting this geometric progression over forty years, Intel has kept up with Moore's law. The number of transistors on a single chip has mushroomed to unimaginable numbers, from 4,000 in 1970 to more than 1 billion in 2007.

As the width between lines of transistors has decreased, the need for greater attention to pure air and water has increased. At

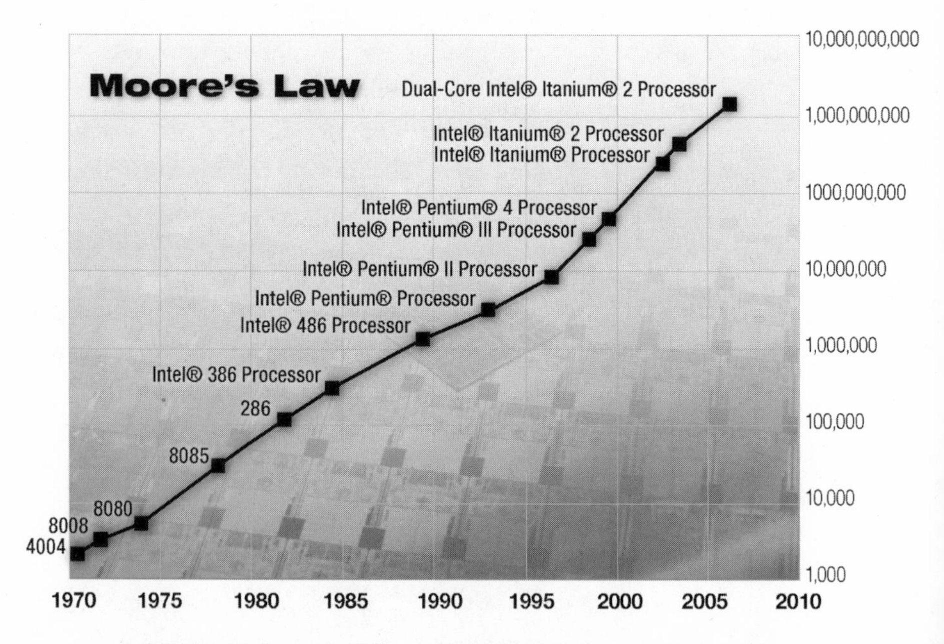

Figure 12.1. Intel's graph of the number of transistors on a silicone chip.

Intel's fabs, sophisticated air filtration and scrubbing systems remove contaminants such as dust, hair, insect mites, and flakes of skin. Workers in bunny suits monitor air-pressurized clean rooms that Intel claims are "10,000 times cleaner than hospital operating rooms." In this process, water washes and cools the chips. Yet, until recently, Intel's scientists and engineers paid little heed to water, either its quantity or its quality. Compared with other components, such as refined silicon, water was a second-class citizen.

Shadman thought that principles of chemical engineering might improve the semiconductor manufacturing process. He persuaded Craig Barrett, then a professor at Stanford University, that super-clean water could shorten the manufacturing time line and increase efficiency. The industry had been moving in a different direction as Intel's scientists tried to perfect a dry-processing system that would sharply reduce the amount of water needed. This strategy made sense for two reasons. First, the semiconductor industry uses large volumes of very toxic materials, such as arsenic, boron, arsine, and hydrogen peroxide as well as nitric, sulfuric, and hydrofluoric acids. Disposing of this nasty stuff has contaminated groundwater and produced Superfund cleanup sites, as the Motorola Corporation found out the hard way. The company spent hundreds of millions of dollars in litigation costs and damages for pollution from its Chandler, Arizona, fab.

Second, many Intel and other semiconductor fabs are located in the arid West, where water supply has become a major issue. In New Mexico, Intel's Rio Rancho plant, north of Albuquerque, drew the wrath of the city and local residents as Intel's groundwater wells lowered the flow in the Rio Grande and caused a drop in the groundwater table. Wouldn't it be better, Intel scientists thought, to eliminate the quality *and* quantity problems by curbing water use in fabs? But it hasn't happened.

In the mid-1990s, Shadman became director of a National Science Foundation center at the University of Arizona. The Engineering Research Center for Environmentally Benign Semiconductor

Manufacturing sounds like an oxymoron, but it isn't. Many semiconductor manufacturers collaborate with Shadman and others in academia to address several problems facing the industry. Tackling the problem of toxic chemicals first, they rather quickly devised ways to strip the toxins from the discharge water and then to reuse the cleaned-up water in fabs. Solving this problem also helped with the water supply problem: recycling used water reduced the need for new water.

In the semiconductor industry, the intelligent use of water is saving water, time, and money. Since 1998, comprehensive reduction, reuse, and recycling programs have saved Intel more than 20 billion gallons of water. In 2007, the U.S. Environmental Protection Agency gave its Corporate Water Efficiency Leader Award to Intel for its Ocotillo Campus in Chandler, Arizona, a suburb of Phoenix. This 700-acre facility houses three fabs and employs 9,500 people. Without water conservation, the high-volume manufacturing processes would require 5 million gallons per day. But Intel has taken steps that dramatically reduce its water use.

First, the company partnered with the city of Chandler to build a state-of-the-art water treatment plant. Using reverse osmosis, this plant has cleaned up and returned to the aquifer more than 3 billion gallons for eventual reuse. Second, Intel purchases reclaimed water from the city's municipal treatment plant, which the company uses for its landscaping, cooling towers, air purification, and mechanical systems. Finally, Intel has its own internal recycling plant, developed in collaboration with Shadman. Using a brine evaporator that resembles a spaceship, Intel reuses fabrication-process water for its ultra-pure water system. These three strategies save almost 4 million gallons per day, a 75 percent reduction in water use.

Intel's use of ultra-pure water in the fabrication process produces a better chip. Fabrication-process water must be ultra-pure because, as Shadman puts it, "cleaning the wafer is not a simple dishwashing process. Water has unique, hard-to-replace properties that chemically interact with the wafers to remove molecular contaminants." A sizable chunk of the energy used at a fab centers on water treatment, at the front end to make ultra-pure water and at the tail end for the internal recycling system and the brine evaporator.

Intel goes to extraordinary lengths to produce ultra-pure water. Matthew L. Brandy, a senior environmental engineer at Intel, explained that "one of the first steps in making ultra-pure water is to run the water through a reverse osmosis system." *A first step?* In most water uses, reverse osmosis produces the epitome of pure water. But Intel goes further. To produce the highest grade of water, known in industrial applications as 18 megohm water, Intel passes the water through ion exchange resins, activated carbon beds, and additional filters. Intel officials were reluctant to provide specific details about their process. Eighteen megohms is 18 million ohms. An ohm is the basic unit of electrical resistance, or resistance to the flow of electrons in a circuit, which means that 18 megohm water has more than a million times the electrical resistance of a typical household electric circuit. Because ultra-pure water contains no inorganic ions to carry electric current, it does not conduct electricity well, which is exactly what Intel wants for rinsing and cleaning the wafers.

Heat has bedeviled the semiconductor industry for years: moving electrons through a greater and greater number of transistors requires more power, which produces more heat. Keeping pace with Moore's law has required the chip industry to confront a technical obstacle. As a chip's transistors get thinner, they tend to leak electricity, which makes the chips run hotter and consume more power. In 2007, Intel announced that it had solved this problem and would begin marketing microprocessors using an alloy of hafnium, rather than silicon, as an insulator. IBM and others quickly issued press releases saying that they too had overcome the obstacle. This new generation of chips offers unparalleled computer power *and* energy efficiency. As an IBM official noted, "It's the difference between can openers and Ferraris."

Before I visited Intel's Ocotillo campus in December 2007, my image of Intel was of workers covered from head to toe in white space suits. As I drove in through the main gate, that image quickly evaporated. I was staring at immense factories with industrial facades, smokestacks, pipes, cooling towers, air scrubbers, and air-conditioning systems. From all appearances, I could have been approaching a Ford Motor Company assembly plant or a Monsanto chemical facility. The men who greeted me, Matthew Brandy and Len Drago, who has responsibility for the site's environmental master plan, were dressed in sport shirts and slacks, not astronaut apparel. I was visiting Fab 32, which, Intel boasts, is "the world's most advanced high-volume digital logic chip manufacturing facility." This $3 billion plant came online in November 2007, but Drago and Brandy never let me into it. I had to gaze at it from a distance. Intel is very determined to keep its processes secret. Fab 32 employs 1,500 people who produce chips, now known commercially as Core 2 Duo, with transistors that are forty-five nanometers wide: that's forty-five billionths of a meter. According to John Pemberton, the plant manager, each microprocessor has more than 400 million transistors. By 2009, Intel expects to market more than thirty microprocessors based on its forty-five-nanometer technology.

To Shadman, the decline of conventional manufacturing in the United States is inevitable, but control over the future—nanomanufacturing—is up for grabs. Partnering with the major semiconductor manufacturers, Shadman's center has led the way in understanding the linkage between water use and energy consumption in the semiconductor industry. As the semiconductor industry moves forward, more challenges await. If the past presages the future, water will play a major role in shaping the industry. Much is at stake for the economy of the United States. Intel is manufacturing chips with forty-five-nanometer transistors in new or expanded plants in New Mexico, Oregon, Arizona, and Israel. A company whose 2007 revenues exceeded $38 billion has many choices for where to locate future plants. In 2009, Intel will open its first fab in China. For new plants to be built in the United States in the future, this country must ensure that high-tech and high-value companies, such as Intel, have access to an adequate supply of water and energy.

To grasp the relative value of the water used by Intel, consider that it takes roughly 135,000 gallons of water to produce one ton of alfalfa, but it takes fewer than 10 gallons to produce a Core 2 Duo microprocessor. A Core 2 Duo chip sells for approximately $400, while a ton of alfalfa fetches up to $110. In other words, each acre-foot used to grow alfalfa generates at most $264. That same acre-foot used to manufacture Core 2 Duo chips generates $13 million.

Part Three

A New Approach ~

CHAPTER 13

The Enigma of the Water Closet

"Civilized people should be able to dispose of sewage in a better way than by putting it in the drinking water."
—Theodore Roosevelt, 1910

IN THE 2006 FILM *Borat: Cultural Learnings of America for Make Benefit Glorious Nation of Kazakhstan*, Sacha Baron Cohen plays Borat, a journalist from Kazakhstan who is making a documentary in the United States. As he and his producer travel across the country in an ice cream van, they meet real-life Americans, not actors. They encounter Americans with deeply racist, sexist, homophobic, and anti-Semitic beliefs. Part of the humor in *Borat* is Cohen's ability to unearth cultural assumptions about how we expect people to behave. Again and again, the hapless people he encounters respond politely to Borat even though he violates deep cultural norms.

In one scene, Borat visits a southern lady, an expert on etiquette. Sitting at an elegant dining room table, Borat receives instructions from his hostess about table manners. At one point, Borat excuses himself to go to the bathroom. When he returns, he presents his hostess with a plastic bag of feces. Her disgust and revulsion prompt her to dispense with her norms of politeness and, unceremoniously, throw Borat out of her home.

In real life, a friend of mine experienced almost the same thing. Members of the Tarahumara Tribe, an indigenous, cave-dwelling community in southern Mexico, were visiting the University of Arizona in connection with the release of a book of photographs about their way of life. My friend hosted one of them. When her houseguest used her bathroom, he was perplexed as to how to use the facilities. In the end, he emptied his bowels in her bathtub rather than her toilet, leaving my friend with a disposal problem of her own. In his culture, water is sacred and no sane person would ever contaminate potable water with human waste.

It is time for us to reexamine this deeply embedded part of American cultural life—how we dispose of human waste—because toilets are responsible for the largest waste of water in the United States. A 2003 study of California water use by Peter Gleick's Pacific Institute found that toilet flushing consumes 32 percent of domestic indoor use. The U.S. Environmental Protection Agency (EPA) reports that Americans use 24 gallons per person per day to flush toilets. That's more than 7 billion gallons. A more recent study, probably more accurate because it considers the move to install low-flow toilets, pegs the total at 5.8 billion gallons per day, or more than 2 trillion gallons per year. That's a lot of water.

Municipalities spend billions of dollars annually to treat the sewage that is produced. And for what? So they can treat it again. It reminds me of the *Peanuts* cartoon strip with Lucy promising Charlie Brown that this time she won't snatch the football away as he's about to kick it. Like a fool, he believes her every time. The managers of municipal water facilities prepare water to drinking water quality, send it to our homes, where we urinate and defecate in it, and send it back to begin the process all over again. Treating all water to potable standards makes no sense given that we use only a small fraction of it, roughly 10 percent, for drinking and cooking. It's an enormous waste of money and energy.

If we want to be serious about water conservation, we need to consider the water we use to dispose of human waste. I have no illusion that my proposal will be an easy sell. When I tried the idea out on my water law class, one woman protested, "I'll clean my laundry with a rock before I'll give up my toilet." In *Godless: The Church of Liberalism*, Ann Coulter anoints the flush toilet as "mankind's single greatest invention."

As innumerable stand-up comedians, as well as Sigmund Freud, have proven over the years, we're fixated on our bowels. We're not alone. In 2006, Bangkok, Thailand, hosted the second World Toilet Expo & Forum, with the theme "Happy Toilet, Healthy Life." The sponsor, the Singapore-based World Toilet Organization, decided that there was a critical need for an independent world body to maintain the highest standards for toilet design and cleanliness. So it started the World Toilet College, which offers courses in restroom design and ecological sanitation for the aspiring restroom training specialist. (I have an image of a conversation in a singles bar in Manhattan: "And where did you go to college?")

Engineering has made the flushing of human waste so convenient, with the mere flick of a wrist, the press of a button, the wave of a hand, or the triggering of a sensor on an automatic unit, that we in the United States take for granted something that 2.5 billion people worldwide don't have: basic sanitation. Our household toilets, connected to 700,000 miles of subterranean pipes, constitute a remarkable technological feat.

Until the mid-nineteenth century, when cities began to build municipal water supply systems, city dwellers deposited their waste in cesspools or privy vaults, either lined or unlined, that were periodically emptied, with their contents often dumped on agricultural land for fertilizer. Unpleasant odors accompanied such concentration of human waste, as did outbreaks of disease. In England, Thomas Crapper, whose name has become synonymous with slang for "toilet," helped popularize the water closet, although he did not actually invent it. He became a plumber to English royalty and shifted Windsor Castle from cesspools to water closets in the mid-nineteenth century.

In the United States, as indoor plumbing became a sign of social status, the installation of toilets spread rapidly up and down the East Coast. The combination of rapid urban population growth and the need to dispose of piped-in water, now contaminated with waste from toilets, created a public health crisis in many cities. The sheer volume overwhelmed cesspools, privies, and even storm sewers. In response, cities installed sewer pipes to transport the wastewater away.

In the nineteenth century, it was widely believed that the decay of organic material caused disease. It made sense to solve the public health problem by dumping it into nearby rivers because, it was thought, "running water purifies itself." With the widespread pollution justified by this theory of disease, rivers functionally became open sewers. And disease, once confined in volume and space to cesspools and privy vaults, spread widely in the medium of water. Flush toilets spread epidemics from upstream to downstream cities.

The superintendent of health for Providence, Rhode Island, Charles V. Chapin, rationalized this method of disposal as follows: "If sewage is discharged into a large enough body of water in such a manner that it quickly becomes diffused through it," he stated, "it is so gradually oxidized as not to give rise to any serious offense." Overseeing sewage systems were public works engineers, ironically called sanitary engineers, who purportedly measured the carrying capacity of rivers to accept raw sewage.

Although notable scientists challenged the idea that rivers could absorb raw sewage and purify themselves, cities found the practice convenient and inexpensive. As a result, epidemics of waterborne diseases, such as dysentery, cholera, smallpox, and typhoid, occurred episodically downstream from sewage outlets. In the early twentieth century, Massachusetts officials went further and endorsed a mischievous notion of "beneficial contamination," claiming that discharging industrial wastes into rivers, such as the Merrimack, hastened the river's natural process of self-purification. Only after the downstream cities of Lowell and Lawrence suffered sharp spikes in typhoid mortality did local officials investigate the link between waterborne sewage and disease. A scientist from the Massachusetts Institute of Technology, William Thompson Sedgwick, documented that the typhoid bacillus had traveled down the Merrimack and into Lowell's water supply.

After eastern cities realized that rivers cheaply and conveniently carry away human waste, midwestern cities followed suit. But when Chicago started dumping its waste into the Chicago River, a problem emerged. The Chicago River, dismissed by one mid-nineteenth-century observer as "a sluggish, slimy stream, too lazy to clean itself," slowly flowed into Lake Michigan near the city's freshwater intake. The city's sewage contaminated its water supply and caused outbreaks of cholera and typhoid that persisted until the early twentieth century. By then, the city had found an engineering solution to manage its waste disposal.

The Chicago Sanitary and Ship Canal, completed in 1900, reversed the flow of the Chicago River, sending it backward through the Des Plaines and Illinois rivers and over a ridge that marks the Continental Divide, and on down into the Mississippi River. The canal disposed of Chicago's sewage and established a navigable waterway that connected Lake Michigan and the Mississippi River. It was a wonder of nineteenth-century civil engineering, and city and state officials boasted of their accomplishment. Illinois governor Richard Yates proclaimed that the Chicago River "purifies itself through natural causes."

In St. Louis, downstream on the Mississippi River, officials had a different appraisal of the canal. By their estimate, Chicago was sending down the Mississippi 1,500 tons of filth and sewage each day generated by Chicago's 1.5 million residents, stockyards and slaughterhouses, rendering plants, and distilleries. The canal dramatically lowered Chicago's typhoid death rate and increased that of St. Louis. Missouri promptly sued Illinois in the U.S. Supreme Court, but lost. The Court required Missouri to prove that the increase in typhoid did not come from actions of Missouri. And some of Missouri's own discharges turned out to be located upstream from the St. Louis intake.

It is impossible to overstate the profound change in water use in the United States that accompanied the rise of indoor plumbing. By one estimate, per capita water use prior to piped-in water averaged between 3 and 5 gallons per day. In Chicago, water use jumped from 33 gallons per day in 1856 to 140 gallons per day by 1872. Most of this was for residential use, which continued to climb dramatically so that by the early 1990s, cities such as Chicago, Pittsburgh, and Philadelphia were distributing more than 200 gallons per day per resident.

The nineteenth-century decision of cities across the country to build combined sewers, carrying storm runoff and human waste in the same pipe, made sense when people believed that rivers clean themselves. But now that we know better, we must also recognize that this practice has profound financial consequences for cities because the greater volume of wastewater from combined sewers drives up the cost of treatment. To this day, 752 American cities have combined sewers. And torrential rains frequently overwhelm sewage treatment plants, resulting each year in the discharge of hundreds of billions of gallons of untreated sewage into our rivers and streams.

In the late nineteenth century, Louis Pasteur and Robert Koch promoted the germ theory of disease, which explained how water-borne bacteria transmit disease. In response, cities slowly began to adopt primitive treatment processes, such as filtration and then chlorination, when they diverted water from rivers. Once the threat of typhoid and other diseases receded, thanks to chlorination, cities found no persuasive reason to treat the sewage before releasing it into rivers. Cities continued to dump raw sewage into rivers until the mid-twentieth century. Upstream cities thrust the responsibility on downstream cities to filter and chlorinate because it was cheaper for the upstream cities not to treat the sewage, and the legal system didn't require them to treat it.

That changed with the 1972 Clean Water Act, when Congress underwrote construction of municipal water treatment plants. The Clean Water Act has been spectacularly successful, dramatically reducing the discharge of raw sewage into our lakes, rivers, and oceans. Since 1972, Congress has appropriated more than $100 billion in construction costs for treatment plants. State and local governments have chipped in another $25 billion. Other federal monies have come from the U.S. Department of Agriculture and the U.S. Department of Housing and Urban Development. In 1987, Congress shifted gears when it tried to put the program on a self-sustaining basis with the Clean Water State Revolving Fund, which makes available low-interest loans. Since 1987, states have used more than $50 billion from this fund on wastewater treatment facilities. Lately, the fund has been limping along as federal appropriations have been slashed, to $850 million in fiscal year 2006. While this is not chump change, it's not nearly enough money to address our crumbling water delivery and treatment infrastructure.

As of July 2007, www.youtube.com offered seventy-three clips featuring broken water mains. Among them, the most popular was a 2006 clip from Girard Avenue in Philadelphia, where a twenty-inch water main ruptured, flooding streets, cars, and houses. It's an impressive display of the force of water. Such scenes frequently occur around the country as the aging infrastructure of our water delivery and sewer system crumbles, prompting cities to issue "boil water" advisories. According to an estimate by the House Transportation and Infrastructure Committee, Philadelphia suffers almost 800 breaks in water and sewer lines in an average year. The American Society of Civil Engineers releases an annual report card on the condition of cities' infrastructure. In 2006, Pennsylvania received a grade of D-minus for its wastewater system because "the useful life of Pennsylvania's wastewater infrastructure is about to expire." The life cycles of pipes range from 15 to 100 years, but in some cities, including Philadelphia, some pipes are almost 200 years old. Philadelphia is not alone in neglecting its water infrastructure.

In 1992, the city of Tucson began to deliver Central Arizona Project (CAP) water. But the water that came out of taps in some homes was brown, prompting an outraged citizenry to demand a halt to using CAP water. The city manager lost his job for this inexcusable betrayal of the public trust. We expect, and should expect, our public officials to deliver safe, clean water to our homes. In hindsight, there was nothing wrong with the water, which is Colorado River water delivered through the CAP canal. The problem was corroded pipes; decades' worth of chemical residues had accumulated on the insides of the main water pipes. The city manager was caught up in the consequences of neglect that long preceded his term in office. I sat dumbfounded at an early morning meeting as I heard his successor describe how Tucson Water replaces 0.5 percent of the pipes each year. Even I could do the math: the city expected pipes to last 200 years! They've since changed that policy.

The water infrastructure in the United States consists of approximately 54,000 drinking water systems, with more than 700,000 miles of pipes, and more than 17,000 wastewater treatment plants, with approximately 800,000 miles of pipes. The useful life of many of these plants and pipes is coming to an end. In 2003, the EPA estimated that over the next two decades the United States needs to spend $276 billion to improve drinking water infrastructure and $185 billion to upgrade and expand wastewater systems. Here's the alarming part: the EPA concedes that this total of $461 billion is a conservative estimate. In 2004, the American Society of Civil Engineers put the gap between needs and funding at more than $0.5 trillion. A 2002 Congressional Budget Office estimate placed the range between $500 billion and $800 billion; the Water Infrastructure Network, a coalition of water industry, engineering, and environmental groups, pegs the future costs at $1 trillion.

I expect that most of us have had the experience of spending a good deal of money on something that gave us no sense of happiness. For me, it was a new garage door. My old one was just fine. It went up and down when I hit the button. Closed, it kept out the elements, people, and critters; open, I came and went. But then suddenly it didn't work and couldn't be fixed. So I plunked down $2,000 for a new door. Did my new door make me happy? Not at all. Every time

I came home, it reminded me of the waste of money. This human reluctance to spend money on things we take for granted explains why cities spend as little money as possible on water and sewer pipes. No politician wants to stand for reelection on the slogan "I overhauled the sewer system!"

Despite the documented acute need for a massive infusion of funds to repair and replace our water system infrastructure, it will not be easy to persuade public officials to spend millions or billions on water and sewer pipes rather than on schools and parks. To be sure, the construction industry's able lobbyists find creative financing options, usually bundles of federal cash, to make it more palatable to politicians to vote for infrastructure repair. But water infrastructure is competing for support with lots of other worthy civil engineering projects, such as roads, bridges, and public transportation. With something as "out of sight, out of mind" as pipes hidden beneath city streets, it's no wonder our system is in a woeful state of disrepair.

Here's the punch line: it makes no sense simply to rebuild the existing wastewater infrastructure. First, our sewer systems represent an enormous misallocation of resources. We have foolishly combined storm and sewer water pipes, causing a massive and unnecessary investment in treating storm water when the water we need to treat is the sewer water. And we treat the combined storm and sewer water to potable standards when we use only 10 percent of the water delivered to our homes and businesses for drinking and cooking. American cities and towns annually spend more than $50 billion to treat water to drinking water standards. A sizable chunk of this money pays for energy to collect, treat, and distribute the water. Second, as pointed out in chapter 9, our water treatment plants do not remove many endocrine-disrupting compounds. As the pharmaceutical and agricultural industries introduce new, more exotic drugs into our environment, we will become engaged in a great chemical experiment on ourselves because we have no scientifically based standards for health risks of exposure to these chemicals singly or, what may be worse, in combination. Emerging chemical contaminants, such as solvents, methyl tertiary butyl ether (MTBE), N-nitrosodimethylamine (NDMA), perchlorate, and pharmaceutical and personal care products, pose unknown risks to public health when combined with one another in our drinking water. As a matter of sound public policy, it is folly to rebuild a system that wastes water, energy, and money; harms the environment; and fails to adequately protect public health. At the heart of this conundrum is the flush toilet.

Ann Coulter dismisses those who seek alternatives to "mankind's single greatest invention" with a really nasty epithet. They are "liberals," who "are worried we're going to run out of something that literally falls from the sky. Here's an idea: Just wait. It will rain." I guess she won't buy this book. Coulter dismisses these "apostles" of composting toilets as people who advocate "living in your own excrement." Harsh words. Freud would have a field day. I won't go there.

But the question remains: are there viable alternatives to the flush toilet? Composting toilets lead the list because they are odorless, user-friendly, and cost-effective. Additionally, they save water and energy, and the composted material consists of organic wastes suitable for fertilizing trees and nonedible plants. You may think that this last point is of no consequence if, for example, you live in an apartment on Manhattan's Upper East Side, but your apartment building could be converted to use a composting system that provides fertilizer for commercial farms. The Bronx Zoo has installed an eco-friendly restroom for its millions of visitors, featuring composting toilets and gray water gardens. With strategically placed educational messages, it has become a huge success and the subject of a charming mini-documentary, "Poop at the Zoo," available at www.youtube.com.

Composting toilets work on an elegantly simple concept: aerobic organisms break down organic matter (excrement) into oxidized humus, just as occurs in forests and in garden compost piles. Aeration is critical; adding a bulking material, such as wood chips, maintains a porous texture. The process destroys human pathogens, reduces the risk of human infection, captures the nutrients in human waste, and doesn't contaminate water supplies.

Introduced in Sweden in the 1930s, composting toilets are used mostly in summer homes, remote locations, recreation areas, and physical settings that preclude hooking up to a sewer system or using a septic system. Let's face it: most people would rather flush and forget than go waterless. Even the environmental writer Florence Williams conceded that point in a 2007 piece in the *New York Times Magazine* about the composting toilet she installed at her summer cabin. "The idea of human waste sitting in one spot—right next to you—for months at a time is difficult to stomach, but I had little choice." Her remote cabin has limited running water and inadequate soils for a septic system. But she's happy with the composting toilet. "Vented by a pipe out of the roof," she notes, "it doesn't smell at all."

A self-contained composting toilet looks just like any other toilet, except it's taller. A collection tray at the bottom allows for removal of the humus. One manufacturer, Envirolet, maintains that this tray needs to be emptied as little as once a year. In homes with basements, the composting unit is located in the basement, below the bathroom. Venting is a critical issue for composting toilets, so most manufacturers use electric fans to circulate air to maintain the aerobic conditions and expel odors. A small wind turbine at the top of the vent pipe further reduces odor. Composting toilets are not cheap, ranging in price from roughly $1,500 to $3,000, depending mostly on the expected number of users. But there is a relatively short payback period, thanks to reduced water and sewer bills.

The use of composting toilets is expanding, even among homeowners with access to sewer systems. With this relatively untried technology, manufacturers have not yet ironed out all the wrinkles, but, as sales increase, a sharply increasing knowledge curve should allow them to improve on what already works pretty well. Venting seems to be the most troublesome problem because most plumbers have little experience in installing composting toilets. And this system requires homeowners to assume some responsibility for maintenance.

Incinerating toilets provide an alternative to composting ones. As the name suggests, these systems use electricity or gas to burn human waste to a sterile ash. If installed properly, incinerating toilets are simple, safe, clean, and relatively easy to maintain. A bowl liner, which, when I saw one, brought to mind a paper diaper, is placed in the bowl before use, and a foot pedal drops the liner and waste into the incinerator chamber. The ash is disposed of in household trash. A friend who has one complains, however, that his uses a lot of electricity. The manufacturer of a popular brand, Incinolet, claims the toilet demands about one and one-half kilowatt-hours per use, which translates to roughly fifteen cents at average electric utility rates. Although incinerating toilets do not use water directly, the electricity consumption involves a trade-off. A third option, portable chemical toilets, are found in airplanes, on buses, and at construction sites and concert venues, where they do the job. But the liquid chemicals involved, usually a form of formaldehyde, are irritating to the nose and throat. This option has not found a footing in residential use.

As another water savings option, some homeowners are installing waterless urinals. Kohler makes a supermodern stainless steel version that looks a bit like a funnel. More likely, male readers have seen a ceramic version at an airport, university, or sports arena. An estimated 100,000 are in use, and sales increased by 50 percent a year between 2005 and 2008. Using a replaceable cartridge or a floating biodegradable liquid seal, waterless urinals allow urine to pass through to the sewer while trapping odors. Even compared with ultra-low-flow urinals, they save more than a gallon per use, which can annually amount to more than 40,000 gallons per urinal.

Waterless urinals have their critics, including the maintenance company for Tucson Water. The company complained that these urinals were prone to staining and took an unwarranted amount of time to maintain. As a result, Tucson Water, a pioneer in water conservation, removed waterless urinals from its headquarters. No press release accompanied this decision. To be fair, the first generation of these urinals encountered problems. Without water to transport urine, it is critical that installers get the right slope to the discharge pipes. Wiping down the fixture, cleaning the filter, and changing the cartridge regularly are essential. At the University of Arizona, which has installed 141 waterless urinals, Chris Kopach of facilities management notes that odor problems generally occur if the maintenance staff fails to follow the manufacturer's recommendations or uses certain cleaning agents that degrade the liquid odor barrier. These problems, says Kopach, "have been few and far between."

In San Diego, the plumbers' union warns that dangerous sewer gases could escape as a result of poor maintenance or during changing of the cartridge. Others see job protection in this resistance to change. In 2004, the Los Angeles city council considered a measure to allow installation of waterless urinals, but the measure died after opposition from the plumbers' union. The *Los Angeles Daily News* believes the union's opposition was based on job security. "Waterless urinals," ran an editorial, "don't have flush valves, which break down a lot. Plumbers make a lot of money repairing them." Yet change is coming; more than two dozen states approve of waterless urinals.

The largest manufacturer, Falcon Waterfree Technologies, argues for improved hygiene as well as aesthetic benefits from waterless urinals. Because they are touch free, there is little chance of bacterial transfer from a prior user. And the company claims that studies have shown that, without water as a breeding ground for bacteria, restrooms equipped with its products have significantly lower bacterial counts. These points have been made in a much funnier way by Ellen DeGeneres in a skit about public restrooms wherein she details all the automated features, from the flushing of the toilet to the dispensing of the soap, but notes that she must grab a dirty door handle in order to leave.

As the debate continues over these waterless alternatives, it's clear that these options hold promise but need some tweaking to become mainstream replacements for flush toilets. To gain widespread public acceptance, we need rigorous, independent studies of the competing options and technologies. In this process, we need to reexamine the role of wastewater treatment plants in handling human waste. Congress served the country well with the 1972 Clean Water Act and with the appropriation of billions of dollars for construction of treatment facilities. Today, without a similar federal commitment of resources, the states find themselves with obsolete plants that increasingly threaten public health. There is now urgent need for a change in direction.

We need a national commitment to find alternatives to combined storm water and sewer water systems, and to treatment plants. This infrastructure uses trillions of gallons of water to transport human waste, requires more than $50 billion a year to operate and maintain, and wastes vast amounts of energy, treating to potable standards an enormous quantity of water that is not used for drinking and cooking. Only the federal government can lead such an initiative. The president and Congress should establish a national commission to address these issues. We need to devote substantial resources to finding sustainable solutions to the problem of human waste disposal.

Fiscal incentives, both credits and subsidies, should encourage the development and use of waterless technologies. Start-up companies should get tax breaks. Homeowners should receive monetary credit for installing waterless toilet fixtures. Presently, the benefits from a homeowner taking such a step inure to the community as a whole because that homeowner uses less water and generates less wastewater to be processed. But the substantial up-front costs must be absorbed by the homeowner. Congress should dig deep and find matching funds to encourage state and local governments to find sustainable solutions to this problem. Cities and counties will still need water treatment facilities to process storm water, but since the volume will be less and the need for tertiary treatment diminished, we can make treatment plants smaller and save on the energy and water now consumed in bringing all water up to potable standards. In the end, Teddy Roosevelt was right: a civilized society should be able to figure out a better way of disposing of sewage than by putting it in the drinking water.

CHAPTER 14

The Diamond-Water Paradox

> "It is scarcity and plenty that make the vulgar take things to be precious or worthless; they call a diamond very beautiful because it is like pure water, and then would not exchange one for ten barrels of water."
>
> —Galileo Galilei, 1632

THE TONY AWARD–WINNING musical *Urinetown*, which opened in New York in 2001 and is still playing in other cities, is an absurdist melodrama about a Gotham-like city suffering a drought so severe that, as one song goes, "it's a privilege to pee." A malevolent corporation, Urine Good Company, controls all bathrooms and uses its monopolistic position to jack up rates for the privilege. The show borrows a theme from *Romeo and Juliet* as the daughter of the company's president falls in love with a fellow who leads a charge against the proposed rate hike. Together, they wrest control of the toilets away from the company and people can pee whenever they want for free. Happy ending, right? Not quite. At the end, the narrator suggests that "the water became silty, brackish, and then dried up altogether. Cruel as the company president, Caldwell B. Cladwell, was, his measures effectively regulated water consumption, sparing the town the same fate." Is the moral that we should pay to use water?

Consider a Las Vegas resident, Sharry Quillin, who circumvented 2007 restrictions on lawn watering with automatic systems by manually watering her lawn with a garden hose. "If you drove by," she proudly noted, "you would not know that we're living in a drought situation." Water conservation programs can significantly reduce consumption, but they'll have no effect on citizens with the conscience of Quillin. Citizens need some other motivation to save water. Such citizens, and the rest of us as well, need a new approach to water scarcity. After decades of experience with both voluntary and mandatory conservation programs, it's time to try another approach. What we haven't attempted in the United States is to encourage water conservation through price signals that create financial incentives to conserve. Quite simply, we must raise the price of water.

Water rates constitute the third rail of water politics. In 1976, the Tucson city council learned this lesson the hard way. When the council imposed a hefty rate hike, it provoked a successful recall election that left a majority of the council members looking for other employment. The political lesson was learned. Since then, all politicians portray themselves as opponents of even modest rate increases. The council subsequently established a citizens' watchdog group, the Citizens' Water Advisory Committee, that carefully monitors rate increases proposed by Tucson Water, the city's water utility. But, as a former member of the committee, I can say from experience that our real function was to shield the council from the wrath of the electorate. Even if the committee signs off on modest price hikes to support infrastructure maintenance and repairs, it is never certain that the council will go along.

Tucson is not the only city with curious attitudes toward water pricing. Fresno, California, has an archaic flat-rate water system that charges all users the same rate regardless of the amount of water used. Local belief in the right to a limitless supply of cheap water is so ingrained that, in 1992, voters amended the Fresno city charter to ban the use of water meters. In 2003, a controversy erupted over the installation of meters in people's homes. In 2004, the California legislature settled the matter when it passed a bill requiring the installation of meters—by the year 2025. Talk about political courage.

Water meters enable a city to insist that residents be responsible in their water use or pay financial consequences. Not surprisingly, residents in cities without meters use considerably more water than do residents in cities with meters. In Fresno, a city without meters, per capita use hovers around 300 gallons per day; in neighboring Clovis, which has meters, use is approximately 200 gallons per day.

Turning to the agricultural sector, western water law allows farmers to divert water from rivers and to pump water from aquifers without paying for it. And farmers intend to keep it that way by resisting the call to install meters on wells. In California's Central Valley, nearly two-thirds of irrigation water comes from wells, but no one has any idea how much water the farmers pump. And the farmers are afraid that if meters are installed, the state may restrict how much they pump. The same concern motivates farmers in Nebraska, where 100,000 irrigation wells pump more water than in any other state, and no one knows how many wells are metered. In both states, groundwater pumping has lowered the water table and caused subsidence, so you'd think that the farming community would be interested in addressing these problems, yet in 2007, proposals to meter wells in both states were defeated. An exasperated Mike Clements, general manager of the Lower Republican Natural Resources District in Alma, Nebraska, complained, "How can you plan for where you need to go when you do not know where you have been?"

Most Americans pay less for water than they do for cable television or cell phone service. Water is ridiculously cheap in the United States. Nationwide, Americans pay an average of $2.50 per 1,000 gallons; that's $0.0025 per gallon, or four gallons for a penny. Prices vary tremendously; typically they are higher in the Northeast and lower in the South and West. It costs the typical American family approximately $20 per month for water. Compared with the situation in other developed countries, water in the United States is a real bargain. Only Canadians, with their modest-sized population and abundant water resources, pay less than we do.

How can water, the earth's most valuable commodity, essential to life itself, be inexpensive? In 1776, in *The Wealth of Nations*, Adam Smith dubbed this phenomenon "the diamond-water paradox." Diamonds then were remarkably valuable as jewels, even though they had no utilitarian use. Water, by contrast, had almost no economic value but immense utilitarian use. Economists explain this paradox by noting that diamonds are scarce, whereas water is the world's most abundant substance. Yet our usable supplies of water are diminishing or even being exhausted.

Farmers get an even better deal because much of their water use is subsidized. In 2004, when California's State Water Resources Control Board began to impose modest fees on diversions and on pumping, one outraged farmer brought suit and described the effect of the board's decision as "kind of like going to jail before you are tried." As this example illustrates, there was immense resistance to modest charges for water even as California staggered through a sixth year of drought. Elsewhere, in 2003 the Nebraska Public Power District made a controversial decision: it raised the rates it charges farmers to $3 per acre-foot. In other words, farmers would pay one cent for 1,080 gallons—and this was controversial!

The U.S. Bureau of Reclamation provides heavily subsidized water to farmers in seventeen western states. The fees charged for water recover only a fraction of the cost of providing it. As a result, in California's Imperial Irrigation District, farmers pay $15 per acre-foot, or $0.0006 per gallon, regardless of how much water they use. Many irrigation districts impose a flat fee, say $90 per acre of land, allowing the farmer to use as much water as desired. This is one reason why so many western farmers grow low-value crops, such as alfalfa and cotton. They have no incentive to conserve.

Most people would be surprised to learn that the water bill they pay—whether to their local municipal water department or to a private water company regulated by the state public utility commission—does not involve a charge for the water itself. This sounds counterintuitive and begs the question: what are we paying for? We're paying the costs of the water distribution system and, in some cases, the costs of a sewer system. Funds raised through water bills reimburse the municipal water department for providing this service to city residents. Private water companies recoup their costs plus a reasonable return on the company's investment—usually 8 to 10 percent. In other words, a water bill typically includes only the costs of delivering the water, including (1) the extraction costs of drilling wells or of digging ditches and conveyance systems; (2) the energy costs of pumping the water; (3) the infrastructure costs of the distribution, storage, and water treatment systems; and (4) the administrative costs of the water department or company. With very rare exceptions, water rates do *not* include a commodity charge for the water itself. The water is free.

The nature of water rights in the United States explains why there is no charge for the water. Under the legal rules governing surface water and groundwater, the state grants its citizens a right to use the public resource. It is up to the individual to divert the river or drill the well to gain access to the resource. The state awards permits without charge, except for modest permit-processing fees. Whether the user is a homeowner, a farmer, a miner, or a municipal water provider, there is no charge for the water itself. This giveaway system presents a problem as we begin to address the issue of water scarcity. Economists suggest that the real cost of water is the *replacement* value of the water. That cost is not simply the sum of the components of providing the water but also includes the cost of obtaining alternative supplies of water.

Raising the price of water would encourage all users—homeowners, farmers, businesses, and industrial users—to examine carefully how they use water, for what purposes, and in what quantity. Economists agree that rate increases would encourage water users to eliminate marginal economic uses and use the water for more productive purposes. An increase in rates might stimulate new water-saving technologies and efforts to harvest water. Effluent would become more attractive to potential users as the price of potable water rises.

Water rates must be crafted in a way that is sensitive to the economies of rural regions and that gives adequate notice that users will pay more. Increasing the water rates for farmers raises other issues as well—issues of equity and history. It is unreasonable to expect farmers to pay much higher water rates without warning. Rate increases can begin slowly and increase over time. Some economists believe that water used by farmers is so heavily subsidized that farmers don't respond to modest price increases. But a 2006 study published in *Water Resources Research* found that farmers do respond by fallowing land, adopting more efficient irrigation practices, and changing cropping patterns to transition from low-value crops to ones with higher economic value.

In many communities around the country, the existing water rate structure perversely rewards profligate water use through flat rates or, even more bizarrely, decreasing block rates. In other words, as a water user, say Sharry Quillin, uses more water, each additional thousand gallons costs less than the earlier thousand-gallon blocks. Is this the right incentive? Wouldn't it make more sense to have *increasing* block rates, where the unit price for water increases as the volume consumed increases? If someone such as Quillin wants to water her lawn in the middle of a drought in Las Vegas, she'd better hope to do well on the *Wheel of Fortune* slot machine.

While we're at it, let's target discretionary uses of water, such as swimming pools, lush landscaping, and sprawling lawns, by imposing a seasonally adjusted rate structure. People use more water in summer than in winter. The great detective Sherlock Holmes would explain that people do not take more showers, flush more toilets, or wash more clothes in summer than they do in winter. If water use increases in summer, it's for nonessential uses. If people want to water their flowers and shrubs, grow a lush lawn, fill a swimming pool, or cultivate herbs and vegetables, they can pay the real cost for these personal choices.

Even if we increase water rates and impose fines, the question remains whether people will change their behavior and reduce their water use in order to save money. Perhaps people so cherish green lawns that they'll water them as before. Eden Prairie, Minnesota, on the southwestern fringe of the Twin Cities, has a median family income that exceeds $93,000. The city has year-round watering restrictions that include a prohibition on lawn watering between noon and five o'clock and an odd-day/even-day watering regimen. In 2007, the city issued more than 800 citations, with fees that started at $25 for the first offense and climbed to $300 for the fifth and each additional violation. City officials noticed that most repeat offenders lived in the wealthier neighborhoods. Even a $300 fine, observes city manager Scott Neal, "is well below the threshold of what it's worth to have a green lawn." Or consider a Palm Beach, Florida, homeowner who used a mind-numbing 11.7 million gallons in 2006. As reported by the *Wall Street Journal*, his water bill was $33,629.

The Eden Prairie and Palm Beach situations suggest that repeated fines and enormous water bills may not alter the behavior of some people. But such outliers do not disprove the basic economic proposition that most people respond to price signals. Consider that, as the price of gasoline crept above $4 per gallon during the first five months of 2008, Americans reduced their driving by 30 billion miles as compared with the same period in 2007. Western Resource Advocates, a nonprofit environmental policy organization based in Boulder, Colorado, has studied water rate structures in Utah. The researchers found that even in communities with increasing block rates, the increase from one block to the next must be substantial enough to get the homeowner's attention. If the increase is minor relative to the homeowner's overall water bill and disposable income, the rate increase is not sufficient incentive for a large-volume user to reduce consumption. Price signals must be aggressive enough to alter behavior. If Eden Prairie wants its residents to obey its watering restrictions, it should increase the fines until people comply.

In Florida, four water management districts recently completed the largest study ever undertaken of how water rates affect single-family residential water use. It found that "water use decreases with increases in water price. The decreases are predictable and statistically valid." No surprise, really. Most people don't want to waste money. The study also found that fixed monthly fees did nothing to encourage conservation and that utilities would be better off eliminating the fixed fee and increasing the charges for water used. In an attempt to find out if wealthy people respond to water price signals, it examined the relationship between water use and the assessed property value of homes. People in more expensive homes not only use more water but also, in response to price hikes, reduce their water use at a greater rate than do people in less expensive homes. Because the affluent use more water for discretionary purposes, such as expansive lawns and lush landscaping, they are better able to cut back in response to price hikes.

State public utility commission rules often constitute a major impediment to using financial incentives to encourage conservation. In some states, requiring private water companies to raise water rates will necessitate a change in state law or even the state constitution. Many states allow private water companies to charge customers only for the "cost of service," which is essentially the cost of providing for the distribution of water. Indeed, some public utility commissions will not even permit private utilities to obtain reimbursement for developing long-term, renewable water supplies, such as effluent or high-cost desalinated water. These restrictions on private water companies are designed to protect the consumer from being gouged, but they discourage water utilities from carrying out the public interest by obtaining access to renewable supplies and building the infrastructure necessary to reuse municipal effluent. At the same time, we wouldn't want private water companies to reap windfall profits from more sensibly priced water. We can capture revenues that exceed the cost of service and use that money to fund other critical public needs. Although one can debate the specific characteristics of a particular pricing structure, the principal point remains that we are not paying the true value of the water we are using.

The idea of charging for water curdles the blood of people who think that would be like charging for air. Both air and water are critical for survival, so it seems unfair and perhaps immoral to extract fees for an essential resource. However, there is a critical difference between air and water; both are public resources, but water is exhaustible. Because water is a public resource, perhaps the government should provide it to citizens as a public service. But should the government allow citizens to use limitless quantities for trivial uses? Precisely because water is a public resource, the government has a stewardship obligation to manage it wisely. The government has failed to meet this responsibility by creating incentives to use improvident amounts of water.

But won't raising water rates harm poor people, especially vulnerable groups such as the elderly and infirm? Because water is essential to survival, should we recognize a human right to water? Absolutely. The richest country in the history of the world can easily make this commitment to its people. The government should guarantee its citizens a safe supply of water for essential human needs. We need to shelter persons of fixed or moderate income from having to pay for water for basic domestic needs.

Maude Barlow, through her books *Blue Gold* and *Blue Covenant*, may be the world's staunchest advocate of the idea that water is an inalienable political and social right and that each person should be guaranteed a "water lifeline," which she calculates as 6.5 gallons per day. Peter Gleick also argues for a human right to water and has concluded that it takes a minimum of 13 gallons of water per day per person for drinking, cooking, bathing, and sanitation. The amount of water needed to supply basic human needs constitutes only a tiny fraction of the water used each day in the United States. Thirteen gallons per day multiplied by 300 million people in the United States totals roughly 3.9 billion gallons of water per day. That's less than 1 percent of the 408 billion gallons used each day. Water for drinking, cooking, and bathing is so critical that it should be free—end of debate. On the other hand, residential use that exceeds a certain threshold often reflects water use associated with discretionary uses, such as swimming pools or lush outdoor landscaping. Water rates can target such discretionary uses and discourage them by imposing graduated (nonlinear) rates for consumption above the threshold amount.

In 2005, I gave the most challenging speech of my career, to girls detained at the Pima County Juvenile Detention Center in Tucson. I was a participant in a speaker series, Inside/Out, that was designed to change how these troubled teens think about themselves by encouraging them to think and write about the vital issues of the day. I was very nervous before my talk. What, after all, could I possibly say about *water* that would interest fourteen- to seventeen-year-old girls? I began by asking them if they liked to take long showers. They did. "How long?" I inquired. "Twenty minutes!" "Thirty minutes!" "Sixty minutes!" Suppose, I probed, that you had to pay to take a shower? Hisses descended on me. And suppose that the amount rose with the length of the shower, say, one dollar per minute? Would you take shorter showers? They all said yes. One added, "I'd use the money for something else."

CHAPTER 15

The Steel Deal

IN JULY 2005, an era of Utah history ended with a literal bang when 3,000 pounds of plastic explosives demolished Geneva Steel's three blast furnaces, nine hot blast stoves, and two stacks. During World War II, the federal government financed construction of the Geneva Steel mill in Vineyard, approximately forty-five miles south of Salt Lake City and just outside Provo, the home of Brigham Young University. The location made sense because of the proximity of raw materials—coal deposits, iron ore, limestone, dolomite, and water. It was also near major railroad lines and an educated workforce. Most important, Vineyard offered an inland location, far from a possible Japanese attack on the West Coast. After the war, the government sold the plant to U.S. Steel. During its heyday, the mill could produce 3,600 tons of steel each day.

By the end of the twentieth century, the United States steel industry was in a tailspin. A prime purchaser of steel, the American automobile industry lost ground to Japanese competitors beginning in the 1970s. Pittsburgh, the hub of the steel industry, came on hard times, especially as Chinese steel producers grabbed an ever-increasing share of the market. Obsolete plants, higher labor costs, and increasing foreign competition caused the demise of American steel production. Geneva Steel faced an added challenge: the plant's distance from any major steel market ultimately made it uncompetitive. These economic forces led Geneva Steel to shut down in 1987. It reopened under new management later that year and managed to stay afloat for another decade, but in 1999 Geneva Steel filed for bankruptcy. Attempts to reorganize ultimately failed, and the company was forced to liquidate its assets to pay off creditors. Fortunately, the company had some major holdings.

First, it owned 1,750 acres of prime, developable land located along the Greater Wasatch Front, a metropolitan area that stretches from Ogden in the north past Salt Lake City to Provo in the south. In 2005, Geneva Steel sold its plant site to Anderson Development for $46.8 million. It then peddled its mill machinery and equipment to, not surprisingly, a Chinese firm, Qingdao Iron and Steel Group Company, for $40 million. Geneva Steel also owned an iron ore mine, which it sold to a mining company, Palladon Ventures, for $10 million. Finally, by closing down, the company paradoxically created another asset: its right to pollute the air. Under the Clean Air Act, companies have an incentive to reduce their emissions because they can sell the right to pollute to other companies. Geneva Steel put on the market its 7,000 tons of emission reduction credits, which fetched approximately $4 million. These substantial assets brought in a total of $100.8 million.

Then, in 2005, Geneva Steel sold its water rights—for $102.5 million. That's right. The water rights turned out to be more valuable than the combined value of the company's real estate, steel mill, iron ore mine, *and* pollution credits. Geneva Steel sold a portion of its water rights to Summit Vineyard for $14 million, for use at its Lake Side Power Plant, a highly efficient natural-gas-fired combined-cycle plant located on sixty-two acres of the former Geneva Steel plant. The Central Utah Water Conservancy District purchased the bulk of the water rights for $88.5 million. Behind the district's acquisition lie rational, savvy businessmen willing to pay more than $100 million for water. They are doing so because they can't do their deals without water.

The Central Utah Water Conservancy District wholesales water for municipal and industrial uses in the Salt Lake City valley. In addition to controlling Utah's share of Colorado River water delivered through the Central Utah Project, the district has assembled a portfolio of other rights to groundwater and surface water based on state law.

In the case of Geneva Steel's water rights, the district saw an opportunity to secure rights to 42,400 acre-feet of both groundwater and surface water to add to its portfolio to satisfy future demand. These water rights are now part of the Central Utah Water Conservancy District Water Development Project (CWP), a $400 million project that will provide not only water but also the infrastructure to move the water where it is needed, which is in the cities northwest of Utah Lake, such as Saratoga Springs, Eagle Mountain, and Lehi. Developers in these communities lobbied the cities to secure access to some of the Geneva Steel water. But the developers will pay for the water, and it will be some of the priciest water in Utah history. None of the $400 million that the district is spending on the CWP will be subsidized by taxpayers. The district will recoup its expenditures by charging the cities an annual fee of roughly $200 an acre-foot to reserve the water, a onetime up-front fee of $6,000 per acre-foot for the water right, and operation and maintenance expenses for the CWP's infrastructure. In turn, the cities will pass these costs on to the developers. For developers, it's just a cost of doing business—fees that they'll bundle into the price of houses. In this way, new users will pay to retire existing water rights.

The Central Utah Water Conservancy District was not initially keen on bidding for the Geneva Steel water rights. It is, after all, a political subdivision, not an entrepreneurial organization. In the past, it has aggressively pushed water conservation and water recycling; in 2006, it opened a $1.5 million education garden at its headquarters to show people how to reduce their water use. "But no matter how hard you push conservation and recycling," notes Rich Tullis, the district's assistant general manager, "it is not enough. We have to continue to develop new water supplies."

Anyone who wants to use water in Utah must file an application for a permit with the state engineer's office. After Jerry Olds became state engineer in 2002, he aggressively moved to enforce the state's prior appropriation system. His predecessors didn't look very carefully at applications for new appropriations. As a result, Utah's water is overappropriated, meaning there are permits to divert or pump an unsustainable amount of water. To Olds, "water is a public resource owned by the state and we have a stewardship responsibility over its use." On Olds' watch, the candy store has closed. He's placed some basins off limits to new permits. If existing users already divert or pump the entire renewable supply, Olds is not going to allow new diversions or wells. Even existing users may expect a careful examination of their water rights by the engineer's office to ensure that they've actually been using all the water they're entitled to. And if they haven't, which is often the case because older claims to water often exaggerated how much water would be used, then Olds cuts back on the right until it reflects the actual use. His actions have annoyed some water rights holders whose rights have been cut back. But in the long term, Olds is helping Utah avoid the plight of other states: too many straws in the same glass. His actions will end up protecting the interests of existing users. "Water," says Olds, "is a valuable, valuable asset." And indeed it is, so long as access to it is limited.

Olds has created a new water paradigm in one of the fastest-growing states in the country. New homes, businesses, and industries have generated a surging demand for water, but the supply is tapped out. The only alternative available to developers is to acquire water from existing users. That's why Geneva Steel received more money for its water than for the rest of its assets.

In Utah, the public owns the water, but the state allows people to buy, sell, and trade water rights. At the moment, those rights are thinly traded, as economists would put it. It's an immature market lacking a sufficient number of transactions to set guidelines for value. At the moment, it's whatever the market will bear.

And the market can be cruel. Daniel Jensen, a partner in a prominent Salt Lake City law firm that represents developers, explained that in the past developers paid little heed to water. Some still acquire real estate only to realize later that the city will not grant building permits until the developer demonstrates that he has locked up sufficient water to supply the development. At that point, Jensen might get a frantic phone call saying, "Gee, I didn't know I had to do that. Where am I going to find some water?" These desperate developers will pay virtually any price to move the deal forward.

While Jensen is a terrific lawyer, acquisition of water rights may take a considerable period of time. It's a two-step process. First, someone has to be willing to sell a water right. And that right is helpful only if it's located near the development. It's not going to do a Provo-area developer any good to find a willing seller in southern Utah. Assuming there is a willing seller, the developer would prefer not to purchase the water right but to acquire an option, exercisable if the deal ultimately goes through. Without some escape clause, the developer may find himself with a water right but without the proper zoning. On the other hand, water rights sellers have little interest in tying up their rights while the developer's attorney navigates the shoals of the rezoning process.

Step two involves securing the approval of the state engineer's office. These transactions almost always involve a change in the purpose of use from irrigation to municipal and industrial, and often a change in the place of use. It's at this stage that Jerry Olds scrutinizes the past use of the right. If the right has not been consistently and fully used, Olds may deem the right forfeited. He also scrutinizes how the transfer may affect other water users, who can enter protests against the proposed transfer. Only after Olds is satisfied does he approve the changes to the original water right.

Meanwhile, time drags on. "If it's a cookie-cutter, zero-controversy situation," says Jensen, "it takes a minimum of six months. More likely, it takes a year and a half or two years." When Jensen advises clients of this time framework, "they freak out. They start asking me, 'Well, what's Plan B?' But there is no Plan B or C." And even if Olds approves a transfer, that does not prevent a disgruntled protester from appealing his ruling through the courts. The large number of uncertainties in the process increases what economists call transaction costs. These are the costs associated with finding willing trading partners, negotiating a price and conditions on the sale, presenting a persuasive case to the state engineer, and defending the engineer's ruling before a court. And all these steps are a mere prelude to taking the water rights to the local planning and zoning commission and seeking its approval for the development.

Into this void have stepped individuals such as Bill White, a Utah attorney who has developed a niche as a water broker. He's knowledgeable as to potential buyers and sellers, the market for water rights, the intricacies of Utah water law, and the administrative practices of the state engineer's office and local planning and zoning officials. By helping to link up buyers and sellers and shepherding them through the administrative process, White lessens the developers' risk that they may end up with water rights to grow alfalfa when they wanted to grow subdivisions.

For an example of how one person creatively responded to this emerging market, consider Christopher Robinson, who is in business with his sister and brother. Under the umbrella of The Ensign Group, based in North Salt Lake, Robinson manages and operates a number of closely held companies engaged in agriculture, real estate investment and development, and manufacturing. The Robinsons have holdings in several states that roughly total 225,000 acres, but their principal focus is the Salt Lake City area, where their father developed the foothills surrounding Ensign Peak, just north of the state capitol. One of these holdings is Timpie Farm, some 1,600 acres set in the Tooele Valley, west of Salt Lake City, with 5,000 acre-feet of irrigation water rights. Robinson decided to sell a fraction of these rights, to "clip a coupon," as he put it. But to maximize their value, he enlisted the help of Dan Jensen.

Timpie Farm abuts Grantsville City, which is in the path of development. In 2006, city officials and Robinson began receiving calls from real estate agents, water brokers, and developers inquiring about the availability of water rights. With help from Jensen, Robinson proposed to Grantsville City officials the creation of a water bank: 600 acre-feet of Timpie Farm's irrigation rights would be dedicated for municipal uses in Grantsville. City officials accepted his proposal, and now when a developer wants water for a project, the city refers him to Robinson. If they reach agreement on the price, Robinson deeds the rights over to the developer, who in turn gives the rights to the city, which supplies water to the development through its municipal water department. This arrangement streamlines the process, as Robinson provides developers one-stop shopping for water rights.

Thanks to Jerry Olds, the state of Utah is squarely confronting the problem of water scarcity. Unlike other states that blindly continue to award permits to divert water from rivers and pump water from aquifers, regardless of the effect on other water users or the environment, Utah is coming to terms with reality. And its solution offers a lesson for the rest of the country. Utah has not drawn a line in the sand and prohibited further development. Instead, it has decided that developers who want to place a new demand on Utah's scarce supply of water must purchase and retire some other water user's right. Without water rights in hand, the state engineer will not grant a permit to use the state's water.

The story of Jerry Olds illustrates a pragmatic water manager in a water-starved region turning to markets as the only practical option to free up water for new demands. Water markets take many forms, including one in high-efficiency toilets in the most picturesque of places: Santa Fe.

Nestled at 7,000 feet in the foothills of the Sangre de Cristo Mountains, Santa Fe, New Mexico, was founded in 1607 by Spanish explorers. The 1848 Treaty of Guadalupe Hidalgo ceded the New Mexico territory to the United States. Lacking a major airport and with a modest-sized population of 75,000, Santa Fe surprisingly has more art galleries than any other city in the country, save New York and Los Angeles. Tourists flock to Santa Fe for its mild climate, its art scene, the sweeping vistas that Georgia O'Keeffe painted, the world-renowned Santa Fe Opera, and the robust cultural traditions of Native American pueblos and Spanish and Mexican settlers. A rich culinary scene with creative chefs, inspired by the flavors of the Southwest, has created nuevo New Mexican cooking. While the Coyote Café, Santacafé, Geronimo, and Café Pasqual continue to produce superb southwestern cuisine, a new generation of chefs offers fine dining with few nods to the bold flavors of the Southwest. Anchored by a central plaza, Santa Fe exudes a charming ambience, created in part by a fifty-year-old ordinance insisting that new or remodeled buildings, especially in the historic district, have Spanish Territorial or Pueblo architecture, characterized by flat roofs, adobe construction, and vigas.

For water, Santa Fe relies on runoff from the Sangre de Cristo Mountains stored in reservoirs on the Santa Fe River, on groundwater from wells located along the Santa Fe River and west of town along the Rio Grande, and on an allocation of San Juan–Chama Project water, diverted from the Rio Grande. These sources have been adequate until recently. A drought in 2002 demonstrated that the city had run out of water. The city responded aggressively with a comprehensive strategy to tighten annexations, regulate new construction, and conserve water. A stage-based drought ordinance imposed harsh limits on water use while other innovative programs educated residents to detect leaks, convert landscaping, and change behavior. The result was spectacularly successful: the city estimates that in 2007 residents used approximately 110 gallons per person per day—one of the lowest figures in the Southwest.

The city wasn't finished. No longer would growth occur on the basis of "paper" water rights or unregulated groundwater pumping. Santa Fe would dovetail land use decisions with water supply. This policy may seem commonsensical, but western cities have gotten along quite nicely for decades, thank you, without paying the slightest attention to where water will come from. Santa Fe looked in the mirror and concluded there must be another way. It decided to allow new construction only if the development would not create a demand for additional water: all new construction must offset the water required for the development. The easiest way to achieve this goal is to retrofit existing homes with low-flow toilets. To obtain a permit to build a single-family home, the city requires a developer to retrofit eight existing toilets. This sensible system imposes the costs on those who would place an additional demand on a tapped-out water supply. It does not halt development, as a "no growth" policy would, but instead places the burden on developers to underwrite water conservation measures.

Developers nimbly responded, and Santa Fe plumbers saw a chance to make some money. A veritable cottage industry developed as plumbers sought out homeowners in older homes and offered to replace their high-water-use toilets for free. Plumbers documented the swapped-out toilets with the city and bundled together credits for water conservation, which they then sold to developers. In short order, most existing toilets were replaced with efficient ones. The conservation program worked beautifully, but demands for new water continued.

In 2005, the Santa Fe city council enacted a water rights transfer ordinance requiring developers of projects over a modest size to bring water rights with them to the negotiating table. A developer must find a willing seller of water rights, acquire those rights, and then tender them to the city for review and approval before obtaining a building permit. Instantaneous shock waves reverberated through the development community. Phones in the offices of water lawyers rang off the hook as desperate developers scrambled to find water rights. But the city held firm. Once again, entrepreneurs emerged, this time serving as water brokers who find prospective sellers, evaluate their water rights, and link them up with developers. As frantic developers scrambled to find water, prices for water spiked from $7,500 an acre-foot in 2005 to $35,000 an acre-foot in 2007. Developers were bidding against one another.

From the developers' perspective, the ordinance operated harshly to give the city unfettered discretion to reject water rights, for which the developer may have paid millions of dollars, and to place projects in limbo for years as developers acquired, tendered, and transferred water rights. In 2006, the council softened the ordinance, but make no mistake about it. This is a new way of doing business in the West: Santa Fe expects developers to secure water rights for their projects before their projects are approved.

The developers adapted but did lobby for change. Why wait, they asked, for us to propose a development before we tender water rights to the city? Let us acquire the rights in advance so that, when the development gets rolling, we won't be bogged down in the water transfer process with the state engineer's office. The city saw the wisdom in that and created a water bank that allows developers to deposit water rights with the city without reference to a specific project. As a project firms up, the developer can withdraw the water for the project or can sell the rights to another developer whose project is further along. This sounds reasonable enough, but it undercuts a central tenet of western water law: prior appropriation.

The prior appropriation doctrine arose in the agrarian West of the nineteenth century. States rewarded farmers with water rights if they invested hard labor to dig a ditch and made an actual diversion from the river. The rights did not vest until the farmer used the water. Because they had to use it, this system discouraged speculation and water hoarding. A farmer got rights only to the water he could use at the time, not to an amount of water that he might use in the future.

Santa Fe's water bank allows developers to transfer water without even having a construction project. The rights are held in the bank until needed by a developer. The fear that people might hoard water rather than use it made sense when appropriators acquired rights for free. But because entrepreneurs must spend hundreds of thousands or even millions of dollars acquiring and tendering water rights to the city, the risk that they'll merely sit on those rights is almost nonexistent. Perhaps some speculators will acquire water rights hoping that the price of water will rise. We've seen that strategy in the housing market, accompanied by some serious losses for speculators. But gambling on future price rises is like wagering in Las Vegas; it's a bet, not an investment.

Santa Fe's wet growth regulations link land use decisions with available water supply. No longer can new growth feed off water supplies paid for by existing residents. No longer can construction occur without water to support it. In Santa Fe, developers bear the entire risk of acquiring water for growth.

Why didn't the developers try to scuttle this reform? They did, and they failed. But in the end, it's no big deal for developers. A prominent developers' lawyer told me, "The only answer that is unacceptable to my clients is 'no.' The answer 'yes, but' is merely a cost of doing business." If a deal is worth doing, developers pass the costs along to the residential, commercial, or industrial purchaser of the property. Although this system may encourage growth, it's growth that pays its own way. In many parts of the West, developers are riding the coattails of a water supply paid for by existing residents. Santa Fe has broken the relentless cycle of overuse and integrates land use planning decisions with available water.

CHAPTER 16

Privatization of Water

THE 1988 Oscar Award–winning film *The Milagro Beanfield War*, directed by Robert Redford and based on the novel of the same name by John Nichols, portrays Joe Mondragon in a heroic struggle to save his humble bean field from corrupt politicians and greedy businessmen. Filmed in Truchas, New Mexico, located between Santa Fe and Taos on the scenic Taos High Road, the movie chronicles a fictional fight over water. But in reality, that fight is taking place today in northern New Mexico.

By the year 1700, approximately 60 acequias (Spanish for "irrigation canals") operated in northern New Mexico. That number had jumped to almost 500 by the end of the 1800s, and today more than 800 acequias deliver water to modest-sized farms, most less than twenty acres and some smaller than one acre. Unlike modern irrigation canals, which are straight as an arrow and lined with concrete to reduce seepage losses, acequias meander around trees and large boulders to take advantage of gravity flow. These unlined canals, originally dug with wooden spades and shaped with knives, allow subsistence farmers to grow native seed crops for local consumption.

Acequias are not only the physical ditches; they are also community organizations managed by elected commissioners and a majordomo, who opens and closes headgates as appropriate.

Landowner-members receive water according to their needs rather than in quantities determined by prior appropriation. "El agua es la sangre de la tierra" (Water is the blood of the land). In these close-knit communities, everyone helps maintain the ditch or acequia. Each spring on an appointed day, acequia members gather to clean the ditch, from the acequia madre, or mother channel, down to each field. At Indian pueblos, special songs keep the rhythm of shovels hitting stone. Special blessings with sacred cornmeal celebrate the opening of the compuertas, or floodgates. At Hispanic Roman Catholic acequias, priests bless the water and processions honor the patron saint of agriculture, San Isidro Labrador.

Water marketing threatens the rich cultural and spiritual tradition of acequias, which foster community cohesion through communal water management, local self-governance, and cooperative labor. These economically challenged villages and neighborhoods face an uncertain future if wealthier individuals and businesses transfer acequia water rights for urban and commercial use elsewhere. If farmers in acequias sell their water, fewer people will be available each spring to maintain the canals. As water is diverted upstream, less water will arrive from the acequia madre to purge the ditch of sediment and allow gravity flow to deliver water. At some point, these centuries-old institutions will collapse.

But this will not happen without a fight that would make Joe Mondragon proud. "El agua es la vida!" (Water is life!) proclaimed the Congreso de las Acequias in its 2006 campaign to defend water as a community resource. Under the umbrella of the New Mexico Acequia Association, acequias from around the state are resisting the commodification of water. The New Mexico legislature has acted to protect the viability of acequias by enacting a statute that allows acequias to require that any transfer of water be approved by the acequias' commissioners. That approval will be hard to come by. Kenneth Salazar, a commissioner of an acequia near Española, vowed, "I'm not going to let any water leave."

If putting a price on water is a good mechanism for getting people to take its use seriously, then should we abandon the acequia system? From a strictly dollars-and-cents perspective, an entrepreneur could generate more cash flow if acequia water were transferred to Santa Fe. But what of the cultural costs? Don't they dwarf any benefit from such a transfer?

Isn't water a public resource, essential for life, not a commodity to be bought and sold like pork belly futures? We let people sell other natural resources, such as oil, which is largely privatized. But oil has only an economic value. Water has cultural, spiritual, religious, environmental, and economic value. To conceive of water as private property—owned by someone who can unilaterally decide whether to sell it, to whom, for how much, and for what purpose—raises profound philosophical and moral issues as well as troubling political questions about the role of corporations, especially multinational corporations, and about the ability of local communities to be independent, autonomous, self-sufficient, self-determinative. To critics of privatization, a society is bankrupt of values if it treats water as simply a marketable commodity, no different from video games or kitchen faucets.

In a world of 6 billion people, where more than 1 billion lack access to safe, potable, and affordable water, the issue of privatization of water resources poses an immense challenge to the international community. The context is etched sharply by recent strife in Cochabamba, Bolivia. In 1998, the World Bank insisted that the Bolivian government turn over its public water utility to the private sector, or else the bank would refuse to guarantee a $25 million loan for improvement of the water system infrastructure. The bank required that infrastructure costs be passed on to consumers. The company that received the concession, a subsidiary of the Bechtel Corporation, increased water rates by 35 percent at the instruction of the bank. A series of escalating protests resulted in seven deaths and spurred Bolivia's president, Hugo Banzer, to place the country under martial law. Other demonstrations against similar conditions have occurred in Argentina, Ecuador, Panama, and South Africa.

To many observers, these third-world incidents share one thing in common: multinational corporations are exploiting the dire economic situation of poor people. Corrupt political regimes, often bribed by these companies, pay no heed to citizens' complaints because the citizens have no political power. These episodes have led some opponents of privatization, such as Maude Barlow and Tony Clarke, authors of *Blue Gold*, to draw a line in the sand by stating unequivocally that "the move to commodify depleting global water supplies is wrong—ethically, environmentally, and socially." They think privatization of water resources allows allocation decisions to be made by corporations that desire to maximize profits and ignores the environmental and social consequences of water allocation policies. These companies, focused only on the bottom line, are unlikely to invest in new technology or water conservation. To Barlow and Clarke, privatization interferes with citizens' ability to allocate and manage their own water, concentrates power in the hands of monopolist corporations, and makes it difficult for local governments to reclaim control over the water system.

So is privatization a good thing or a bad thing? As with so many other things in life, it depends. The devil is in the details. To some, what happened in Cochabamba epitomizes what is wrong with privatization. But since the uprising, the cooperatively run water system that replaced Bechtel is in shambles, possessing neither the capital to overhaul the infrastructure nor the experience to run a public utility. To analyze the situation in Cochabamba or elsewhere, one must know the state of affairs *before* the private company arrived. What was the condition of the infrastructure? Was it decayed and neglected? Was everyone in the community receiving water? And what exactly did the company do? Did it build, repair, or replace the infrastructure; deliver water to people; charge people for water delivered; respond to the demands of local politicians to divert resources to their pet projects? Except to those who believe that privatization is ideologically unacceptable, only a before-and-after comparison allows an accurate appraisal of how well privatization works.

One oft-heard criticism accuses the multinationals of reaping profits from the sale of water. If a company invests tens of millions of dollars in rebuilding a decayed infrastructure, restoring and expanding water delivery to poor urban and suburban communities, and putting in place a competent water administration system, it quite justifiably expects the return of its capital and a reasonable profit. Unless the profits are excessive, the fact that the company may earn a profit is not sufficient reason to condemn the corporation as exploitative or privatization as a bad idea. The passion generated by water privatization is nicely captured by an exchange between an Argentinean opponent of privatization, who argued that water "is a gift from God," and the president of Veolia Environnement (which supplies water to 100 million people throughout Europe, Asia, Africa, and the Americas), who responded, "Yes . . . but he forgot to lay the pipes."

Despite the intense debate, privatization is an elastic concept that embraces many different scenarios involving the transfer of the assets or operations of a public water system into private hands. Most water systems in the United States are publicly owned and operated, but things were not always this way. In the early nineteenth century, most citizens received water from a private water company. At the end of the nineteenth century, municipalities began to assert control over these services because they recognized that private companies were not providing adequate service to all citizens. Private companies often failed to invest sufficient capital in the system, and sometimes they supplied water to the wealthier sections of a city but not to the poorer sections. Any concern over the quantity and quality of the water delivered took second place to maximizing the company's profit on its investment. By the year 2000, private companies served only 15 percent of the American public.

Recently, the pendulum has swung back again. The drivers behind water privatization reflect several impulses. First, financially strapped municipalities are eager to have a private corporation put forward the huge amount of capital necessary to update the crumbling infrastructure of municipal water and sewer systems. Water utilities are more than twice as capital-intensive as other utilities. Second, many economists argue that private businesses are more cost-effective in providing services than the public sector. And third, in some quarters political ideology favors downsizing government and outsourcing things to the private sector.

There are several forms of water privatization. A limited, often uncontroversial form involves a local government contracting with a private company to operate the municipal water system or the wastewater treatment system. Municipalities regularly request bids from the private sector to design, construct, operate, and maintain public facilities. Even though public employee unions prefer that the jobs remain in the public sphere, few other people would object to a private company administering the billing and revenue collection services, or the payroll obligations, of a municipal water department—so long as they do it well. Since 1997, the number of publicly owned water systems operated by private companies under long-term contracts has jumped from 400 to approximately 1,100. Even so, 94 percent of water systems in the United States are publicly controlled.

A more contentious type of privatization involves selling or transferring the assets—the pumping plants, treatment facilities, headquarters building, and distribution systems—of the municipal water system to a private company. Often this exchange is the quid pro quo for the company's agreement to infuse the system with a major dose of new capital. If this occurs, the municipality avoids the need to rely on its municipal bonding and financing system to generate new monies for the project, which would saddle residents with higher taxes. Whether this form of privatization is desirable depends on the understanding between the municipality and the private contractor. Does the municipality have adequate oversight on issues of water quantity and quality? Will the company undertake sufficient water conservation efforts, such as attempts to increase the use of municipal effluent? The answers depend on the specifics of the contract between the two parties. The most bitter international controversies have involved situations in which local governments have entered into contracts with the private sector to operate the system for a substantial period of time, often fifty years. The time period necessary to recoup the heavy initial investment by the private corporation justifies the length of the contract, but the duration may effectively cede control over a public resource to a for-profit corporation. Moreover, the company has no incentive to invest money to maintain the existing infrastructure. At the end of the contract, the city may find itself with a system of pipes held together by duct tape.

But the form of privatization that really gets the ideological juices flowing, that most resembles a red flag in front of a bull, involves actual ownership of water. If a state allows a private corporation to own the municipal water supply, may the company sell the water at whatever price it wishes to whomever it wishes? If a company owns the water, it may distribute the water unequally, favoring the wealthy, who can pay more, and the politically powerful, who can help in other ways. To prevent these abuses, the contract between the government and the private corporation must specify which sections of the city will receive water and at what price.

Privatizing any aspect of a municipal water system risks shutting out the public from participation and insulating a company's practices from transparency and accountability. Environmental consequences, water quality, and dam safety are three other issues raised by privatization. A private corporation has little incentive to protect the environment from the adverse effects of providing water. Surface water diversions and unsustainable groundwater pumping may have horrible environmental consequences yet may be of little concern to the private corporation that delivers the water. The company does not internalize these environmental costs but shunts them off onto society generally, which often means they are not addressed until years or decades later. As for water quality, private companies often resist undertaking expensive monitoring programs for low levels of pollutants. Corporate executives fear, often reasonably, that it will be difficult to recoup these costs through rate increases, which are subject to both consumer acceptance and public utility commission approval.

In the United States, when economic conditions produce natural monopolies of scarce resources, we have frequently created regulated industries, such as for electricity, telephone service, cable television, or natural gas. Whether regulation of such monopolies is successful depends on the strength of our political institutions. The United States has a strong tradition of democratic oversight of private utilities through public utility commissions (PUCs). Every state has a PUC with appointed or elected commissioners whose mandate is to protect the public. Most PUCs carefully regulate utilities to prevent them from gouging consumers. I do not mean to suggest that PUCs always provide the most efficient, democratically accountable regulation of private utilities, but they do a pretty good job.

That is why the many valid criticisms of water privatization in third-world countries do not transfer very well to the United States. Although privatization appears to make the most sense in those countries where weak governments have failed to provide basic human needs for their people, it is unlikely to be successful precisely because the ineffective or corrupt governments lack the capacity or will to adequately regulate the private sector. Yet global financial institutions, such as the World Bank, often pressure third-world governments to privatize their water resources.

In the United States, corruption and incompetence plagued the privatization of the water and sewer system of Atlanta, Georgia. In a triumph of hope over experience, the city thought it would save money, and United Water, a subsidiary of Suez Environnement, based in France, thought it would make money, from privatization of these city systems. In 2003, after four disastrous years of burst water mains, interrupted service, broken meters, and "boil water" advisories, Atlanta's mayor and United Water's chief executive officer announced the "amicable dissolution" of the contract. Political corruption indictments soon followed, further tarnishing the largest water privatization in American history. Atlanta's sour experience has dampened the appetite of other cities contemplating privatization.

For privatization to be successful, governments must regulate water as a social good, ensuring access to all. PUCs must carefully monitor the financial returns to the private company and link any rate increases to agreed-on improvements in service, conservation programs, or environmental stewardship. Cities that go down this path should do so cautiously, for privatization is no panacea for financial woes and decayed infrastructure. Privatization can benefit a city, as it has Indianapolis, where a private company administers the water delivery system but does not own the water supplies. In any event, government should retain ownership of the water resources.

In the United States, there is no reason to surrender the ownership of a municipal water supply to a private corporation. But, in our democracy, the answer to the question, whose water is it? is not so simple. As we shall see in the next chapter, the experience of three irrigation districts in California and Arizona offers surprising insights into political control over water.

CHAPTER 17

Take the Money and Run

AMONG THE IDIOTIC QUESTIONS tourists routinely ask park rangers at Grand Canyon National Park, my favorite is "Where did they put the dirt?" The assumption that human beings dug the Grand Canyon is ludicrous, but the question of what happened to the dirt is not. Forces of nature created the Grand Canyon, as we know from the science of plate tectonics. Large pieces of the earth's crust moved together and forced the uplifting of the mountains that form the Colorado Plateau. As the mountains slowly rose up, the Colorado River cut through them, much as in cutting a cake by raising the cake into the knife. Geologists now believe that the Grand Canyon did not gradually evolve by erosion. Instead, a series of abrupt events carved the canyon.

Each winter for millions of years past, the Colorado Rockies have received a heavy fall of snow. In the spring, tributaries to the Colorado River capture the snowmelt, causing floods. During summer in the Grand Canyon region, the monsoon creates flash floods, triggering debris flows in side canyons that empty into the main stem of the Colorado River. These floods and debris flows widened and deepened the Grand Canyon. Before construction of dams on the river, the force of the water flow transported the sediment downstream. Most of it ended up in the Sea of Cortés, lying between the

Baja Peninsula and the mainland of Mexico, making the sea the earth's shallowest ocean; but the river also deposited some sediment alongside the main stem in the Lower Colorado River region. Below Yuma, Arizona, the river meandered and occasionally turned west into what is now California's Imperial Valley before discharging its sediments. The alluvial soil from the Grand Canyon has created fertile land for growing crops. It needs only one thing: water.

The Lower Colorado River region—from Wellton, Arizona, west through Yuma, and on to El Centro and Brawley, California, and Mexicali, Mexico—is blisteringly hot. Summer temperatures frequently reach 115 degrees. The region receives only a few inches of rain a year and has little natural vegetation. Settlement of any kind is relatively recent in the Imperial Valley. Spanish explorers described their encounters with the area as "el viaje de los muertos" (journey of the dead).

In 1901, the California Development Company began diverting Colorado River water into a canal that began in the United States, ran for most of its length through Mexico, and then recrossed the international boundary into the Imperial Valley. By 1904, the Imperial Valley had 7,000 settlers irrigating farmland. However, the California Development Company's main canal often became clogged with sediment. To bypass this obstruction, the company cut a new canal in the river's banks, but it failed to install an adequate control gate. In the spring of 1905, heavy rains on the Gila River in Arizona flooded the Lower Colorado River and overwhelmed the modest headgate for the diversion. The entire Colorado River flowed into this makeshift canal, flooding the Imperial Valley, enlarging the New and Alamo rivers, and creating the Salton Sea.

By June 1905, 1,000 cubic feet of water per second—or 27 million gallons per hour—was pouring into the canal. The California Development Company begged E. H. Harriman, president of the Southern Pacific Railroad Company, to corral the river. A two-year fight, costing more than $3 million, eventually forced the river back into its usual channel. In the meantime, the river had washed four times more earth into the Salton Sea than had been excavated for the Panama Canal. The flood wiped out a quarter of a million acres of farmland and deposited enough water in the Salton Sea to create a body of water thirty-five miles long and fifteen miles wide. In return for its help, Southern Pacific took over the assets of the now-bankrupt California Development Company. In 1911, the Imperial Irrigation District acquired the assets from the railroad. By 1922, the IID was delivering water to nearly 500,000 acres in the Imperial Valley, making Imperial Valley farmers the first people to make serious use of the Colorado River's water.

As drivers travel west on Interstate 10 from Yuma, Arizona, into California, a bridge takes them over the Colorado River. Glancing down, one might think it's not much of a river, a pale contrast to the untamed Colorado, which the explorer John Wesley Powell successfully navigated in the 1870s. Upstream, major diversions have siphoned off most of the river's flow. A few miles farther west, drivers come upon a much larger river, wide and deep, moving at a steady clip. Despite appearances, this watercourse is not a river at all; it's the All-American Canal. An engineering marvel when it was built by the federal government in the 1930s, it begins at Imperial Dam on the Colorado River, approximately twenty miles northwest of Yuma, and then flows eighty-two miles, propelled by gravity, into the Imperial Valley, where it provides water to the IID and, farther northwest, past the Salton Sea to the Coachella Valley. The federal government built the canal at the behest of farmers in Imperial Valley, who lobbied for a canal to be built entirely on United States soil; hence the name All-American Canal. Completed in 1942, the canal delivers approximately 3.1 million acre-feet of water, more than 20 percent of the Colorado River's annual flow.

A visitor to the Imperial Valley is struck most by what is missing. A vast horizon surrounds this perfectly flat landscape where farmers grow alfalfa, Bermuda grass, and Sudan grass as feed grains for animals, along with some winter vegetables. Surveying the valley, one slowly realizes that there is no sign of farm life. There are no farmhouses with white picket fences, shade trees, and long driveways. That's because most Imperial Valley farmers are absentees who lease their land to locals, who live clustered in the towns of Brawley, El Centro, and Calexico. Families may farm in Imperial Valley, but there are few family farms. This is large-scale agribusiness, with almost a half million acres under cultivation by five hundred farming entities organized as partnerships, trusts, real corporations, and dummy ones. Dryland farming is not possible in the Imperial Valley; each and every acre is irrigated with Colorado River water. Imperial County is one of the top agricultural counties in the United States, producing roughly $1 billion worth of crops annually and supporting more than 1,000 jobs in the Imperial Valley. Against this backdrop of wealth, Imperial County has one of the highest poverty rates in the nation. Approximately 90 percent of the farmworkers are Mexican Americans or Mexican nationals, though virtually none of the owners are.

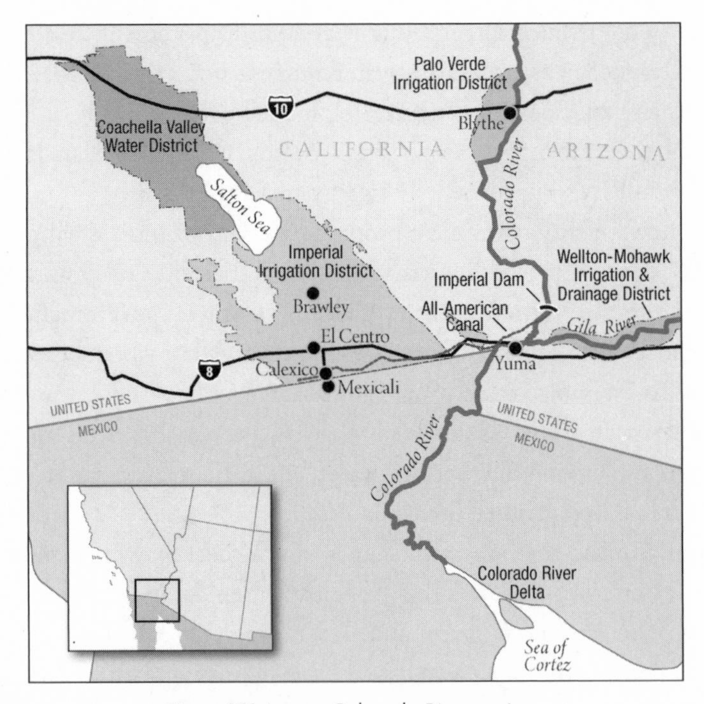

Figure 17.1. Lower Colorado River region.

IID farmers confront a challenging salinity problem. Salinity is the natural enemy of farmers because too much salt at the root zone of crops will kill them. Colorado River water, salty to begin with, becomes 30 percent more so by the time it reaches IID fields as evaporative losses along the All-American Canal concentrate the salts. The irrigation process picks up additional salt from the soil. To address the problem, IID farmers use flood irrigation to leach salts from the soil, but this presents farmers with another problem: how to dispose of this supersaturated saline water. The answer lies in elaborate drainage ditches that capture the excess water and dump it into the New and Alamo rivers, which eventually flow into the Salton Sea.

When Imperial Valley settlers began to draw on the life-giving force of the Colorado River, most farmers in the West used flood irrigation. Today, IID farmers use more water per acre than is typical in the West because they must leach salts from the soils, the growing season is long, and the temperatures are high. But Imperial Valley farmers developed their irrigation techniques when water was plentiful and no one else needed it.

Over the past generation, IID farmers have found their way of life under attack: critics claim that the IID uses enormous quantities of water to grow marginally productive crops. In 1984, California's State Water Resources Control Board ordered the district to reform its irrigation practices. Driving the board's decision was one remarkable fact: of the 3 million acre-feet of water diverted by the IID each year, "approximately one million acre-feet per year of Colorado River water enters the Salton Sea as irrigation return flow from IID."

Southern California's urban interests think they can make better use of that water. The Metropolitan Water District of Southern California, the largest public utility in the United States, is a consortium of twenty-six cities and districts that provides drinking water to 18 million people (about the populations of Illinois and Indiana combined) in Los Angeles, Orange, San Diego, Riverside, San Bernardino, and Ventura counties. Metropolitan delivers 1.7 billion gallons of water every day to an immense service area that covers 5,200 square miles. Set in a desert, Los Angeles has managed to accommodate population growth that has turned it into the nation's largest and most populous metropolitan region by importing water. Lots of it. The city's sprawl could not have occurred without imported water.

Some IID farmers have modernized their distribution systems. Ronnie Leimgruber, whose grandfather came to the Imperial Valley from Switzerland in the 1920s, has expanded his family's operation and now grows several thousand acres of alfalfa and Sudan grass as well as some fruits and vegetables. He bought his first farm in 1983 for $3,800 an acre; today he would not be able to sell the land for much more than that. In the IID, irrigation system improvements are expensive relative to the cost of the water and the value of the crops. Nevertheless, Leimgruber has installed a pump-back system that takes tailwater from the low end of his field and pumps it back up to the higher end for reuse. Quite revealingly, it costs him almost as much for the energy to pump that water from the bottom of his field to the top as it does for the Colorado River water delivered to his farm by the IID.

In 1931, California agreed to limit its rights to Colorado River water to 4.4 million acre-feet annually, yet the state regularly diverts 5.2 million or even 5.3 million, thanks to annual declarations by the U.S. secretary of the interior that there is a "surplus" of water available. If California actually had to live within its allotment of 4.4 million acre-feet, Southern California's cities would face a stark reality: they would be cut off first, while IID farmers could continue to grow alfalfa. The cities' fears were realized in the 1990s when Secretary of the Interior Bruce Babbitt, after years of futile efforts to get California to be more "judicious in its use," used his discretionary authority to declare a surplus as leverage to force California to reduce its consumption of Colorado River water.

By 2002, thanks to Babbitt's intervention, the stage was set for the consummation of the largest water transfer in American history. The parties had agreed on the terms of the Quantification Settlement Agreement, or QSA, a collection of some forty contracts and side agreements that, when stacked on top of one another, are several feet tall. The Metropolitan Water District and the San Diego County Water Authority would provide the IID with a major infusion of cash, almost $7 billion over seventy-five years, more than enough money to overhaul its inefficient irrigation system. To free up water in the short term, the IID would pay farmers to fallow portions of their land. After improvements make the distribution system more efficient, a process expected to take fifteen years, the farmers can bring the fallowed lands back into production. This agreement would solve the cities' water problems with enough additional water for hundreds of thousands of homes. By 2020, one-fifth of San Diego's water would come from the Colorado River. Finally, the agreement dedicated water to protect the Salton Sea.

It seemed as though the agreement would benefit all parties. The only piece missing was a majority vote by the IID's five-person board of directors. But in a stunning and unexpected turnabout, the IID board voted 3–2 against the transfer. The parties had spent an estimated $50 million, all for naught. Much finger-pointing revealed the bitterness among the parties. Many people found it perplexing that the IID board would vote against an agreement that involved selling water for $250 an acre-foot, for which it had paid $15 an acre-foot.

So why did the deal collapse? Community leaders urged the board to reject the deal because they believed it would wreck the region's largely agricultural economy. "Nobody wants it. Nobody," said Larry Bratton, a local jeweler and business leader. "All the heavy weight is on the Imperial Valley. There are all these threats that if we don't do this deal, all these bad things could be done to us." According to Andy Horne, a member of the IID's board of directors and a former president of the Imperial chamber of commerce, the business community opposed the transfer because it would bear the brunt of any adverse effects to the local economy from a reduction in agricultural production. Horne eventually voted against it because he couldn't "sign something that was not in this community's interest."

Many IID farmworkers live in Mexico and cross the border each morning at Calexico. There, they wait at appointed corners for a bus to pick them up and take them to a field, where they work for $6.75 an hour. Workers such as Antonio Paez and Feliz Baragas, whose families live in Mexicali, stay at a shelter in Calexico "because we can get to work so much faster." They follow the crops from Imperial to Yuma to Salinas for weeks at a time. Through an interpreter, Baragas stated, "There are a lot of people working in the farms. There will be lost jobs for everybody." Baragas works the fields so that his daughter and two sons will have better opportunities. He thinks that the water should stay in the Imperial Valley, not be transferred to support golf courses, tract homes, and Wal-Mart and Home Depot stores. Opposition also came from Imperial County's board of supervisors; one board member described it as "the great water rape" and another as "a shotgun wedding."

Even IID farmers were divided. Don Cox, a farmer and former IID board member, opposed the deal because he thinks that the farmers were not offered enough money. But his son Larry, an organizer of the Imperial Valley Water Users Association, favored it because he would like to benefit financially from fallowing some of his fields. Don's other son, Mike, who was president of the Imperial County Farm Bureau and who leases some fields, opposed the deal, fearing that the landowners might take the money and run, leaving him without fields to farm. "One of my concerns," says Mike, "is whether all of the money will go to the landlord."

Not all agricultural interests are compatible. If one farmer sells out, that drives up the remaining farmers' costs of maintaining the infrastructure and paying the overhead of the irrigation district. Even farmer-oriented organizations such as the Farm Bureau and the Agribusiness Council objected to the sale. Members of these groups include not only farmers but also owners of farm-related businesses, such as the local John Deere dealer, the cotton gin operator, and feed, fertilizer, and pesticide suppliers. If a farmer sells out, these businesses lose a customer. Municipal and county governments within the Imperial Valley also opposed the sale, fearful of a decline in municipal or county sales, property, and income tax revenues.

In the middle of the dispute sat Bruce Kuhn, a fifty-one-year-old Imperial Valley native and IID board member who runs a construction firm and an agricultural land-leveling business. His family has been in the valley for several generations; his grandfather helped to dig the All-American Canal. A volatile fellow, he blasted U.S. Senator Dianne Feinstein in spring 2002 as a "bureaucratic gasbag" after she urged the IID to go along with the water sale. He later apologized for his remark, but not for the emotion behind it. Newly elected IID board president Stella Mendoza reacted defiantly to pressure from both state and federal officials. "If you push me around, I'll push back. We'll see them in court. Without water, the Imperial Valley is nothing."

On the other side, the IID's rejection of the transfer agreement annoyed officials in Los Angeles, San Diego, Sacramento, and Washington, DC. *The Los Angeles Times* editorialized, "It has come to this: a one-vote majority of an obscure elected body in a remote desert community with 140,000 residents has put the water future of 18 million Southern Californians in jeopardy." The *Times* understood why the deal collapsed. "The plan might have passed if the farmers ran the irrigation district. In Imperial County, unlike most of the state, the water district directors are chosen by all voters, not just the farmers. Only one of the five is a farmer."

To the *Times*, this voting system was an irritant that impeded movement of the water from the farms to the cities. In many irrigation districts, only landowners are eligible to vote for the board. This system has an odd ring to modern ears, but, in the early twentieth century, when many irrigation districts were formed, the only residents were farmers. They banded together to create a government organization to build the infrastructure to irrigate their lands. The farmers shared the cost of these improvements through taxes on their lands. Only landowners paid these taxes, and only landowners determined membership on the board.

To this day, many irrigation districts restrict voting to property owners and sometimes base voting rights on the number of acres a farmer owns, or even the assessed value of the property. But in the IID, any registered voter may vote to elect board members. As a consequence, the board is sensitive to a much larger political constituency that includes elected officials of municipal and county governments, owners of local businesses, tenant farmers, farmworkers, and citizens in general.

A month after the IID board of directors rejected the transfer agreement, Secretary of the Interior Gale Norton and Assistant Secretary for Water and Science Bennett Raley announced that, in 2003, the IID would receive 2.8 million acre-feet rather than the 3.1 million acre-feet it expected. The president of the IID board, Lloyd Allen, complained to Norton: "Your action sends a message to all water right holders throughout the West: Comply with the desires of the Department of Interior and the urban populations, or your water rights will be confiscated." Norton and Raley justified this decision on the ground that the IID was not using water beneficially.

The concept of beneficial use traditionally has been a state law doctrine, administered through the courts in each state. Never before had the U.S. Department of the Interior under any administration, Democratic or Republican, claimed authority to determine whether any given use of Colorado River water was beneficial. Norton and Raley, conservative Republicans who disfavor federal intervention and regulation, asserted unprecedented authority for the department to regulate water users. In an interview, Raley described himself as "bordering on being a libertarian" but noted that as assistant secretary he was not free to implement his own ideology. To Raley, what happened was a natural culmination of the process begun during Bill Clinton's administration under Bruce Babbitt. The process had such momentum, according to Raley, that he and Norton had no choice but to follow through.

In 2003, the IID board finally ratified the agreement and agreed to transfer more than 30 million acre-feet over seventy-five years. Secretary Norton described the largest transfer of water in American history, involving on average 300,000 acre-feet per year (the same amount as the state of Nevada's entire Colorado River allocation) as "an historic event" and as "a roadmap for future water trades in the West." Municipal officials, particularly in San Diego, were overjoyed with the agreement. "I knew the day would come. I was just hoping it would come in my lifetime," said Maureen Stapleton, general manager of the San Diego County Water Authority.

Some people may object that the transfer agreement resulted in a huge windfall to the IID. And perhaps it did. But if one is serious about modernizing an incredibly inefficient irrigation system, there was no other choice. The largest irrigation district in the country, with rights to 3 million acre-feet of Colorado River water per year, distributed water through unlined canals, had no meters to measure water use, charged a pittance for the water, and had lax enforcement of regulations against wasting water. IID farmers had no incentive to modernize their customs. Their wooden gates hemorrhaged water.

This intolerable waste had to end. But how? The state of California could have imposed new conservation standards, but these rules would have affected *all* farmers in the state, a politically powerful constituency that the state legislature will not alienate. To be sure, a few IID farmers will make fortunes by selling water they never paid for. But who's actually hurt by that? Not Los Angeles or San Diego, surely, for their developers benefit from the security of a new supply of water with a priority that protects the cities during times of drought.

What will happen in the Imperial Irrigation District? Adverse effects to the community have turned out not to be as substantial as residents of Imperial County feared. The fallowing involves only a small percentage of the irrigated acres in the IID: 14,000 out of 465,000 acres, or 3 percent of IID land. Farmers are not fallowing their fields of lettuce or broccoli, high-value crops that require labor-intensive harvesting, picking, packing, and shipping. Instead, they are fallowing the fields of alfalfa, a low-value crop that requires the labor of only a few workers to cut and bale. Farmers, as businessmen, are taking out of production their marginal, less productive fields rather than the fields that make them a lot of money.

Some adjustments will be stressful. The contentiousness and bitterness among the different groups of farmers, as well as a spate of litigation in the wake of the agreement, reflect this fact. But the IID can, must, and will make adjustments. Considering the small percentage of land involved, it's surprising that the transfer generated so much angst. Given the financial windfall that IID farmers will receive, why didn't the IID just take the money and run the first time? Because transfers strike fear in the hearts of all agricultural water users, who, perhaps correctly, anticipate that agreeing to one transfer will pave the way for additional, larger transfers. Still, the IID will be around for a long time to come, and when it modernizes its distribution system, the fallowed lands can once more be productive.

What caused the board to reverse course? Pressure from the federal government obviously was an influence. But two other critical factors, both unprecedented, help explain why the board changed its mind. The agreement provides funds to compensate workers and businesses harmed by the transfer, and it dedicates water to protect the environment. The lesson to be drawn from the Imperial Valley's experience is that large-scale transfers of water can be successful, not just for the acquiring cities but also for the rural agricultural communities, provided that the process protects innocent third parties and the environment.

Of course, that's not always what happens. With stakes as high as they are in water-scarce regions, there will always be the risk that a powerful few will just take the money and run, with little care for the public good. Take the example of the IID's neighbor, the Palo Verde Irrigation District (PVID), just off Interstate 10 northeast of the IID.

Each year, more than 8 million cars pass through Blythe, California, on Interstate 10. But drivers stop only long enough to fill their gas tanks or to grab a quick night's lodging at a motel. Despite the city's best efforts to plant new landscaping along the main drag, Hobson Way, Blythe is quite a depressing place, with many buildings boarded up and others in various states of disrepair. Surrounding Blythe is the PVID. In contrast with the IID, few people have ever heard of the PVID, which is fine with the farmers. The district seeks no publicity; until recently, it didn't even have a Web site. But it is well known to the Metropolitan Water District of Southern California. Metropolitan is so enamored of the PVID that, in 2002, it actually purchased more than 16,000 acres of land in the PVID—10 percent of the entire district.

Relations between PVID farmers and the city of Blythe are rocky, which is surprising, given that agriculture dominates the local economy. But in the 1980s, in an attempt to diversify the economy, local officials attracted two state prisons, which were built seventeen miles west of town. The farmers were bitterly opposed because the prisons pay employees pretty well and the farmers did not want to compete for labor. More than 1,800 people work at the prisons, which house some 8,000 inmates—about 40 percent of the total population of the area. Local farmers have long memories, and they harbor grudges, according to Mark Fulton, an officer with the Palo Verde Valley Economic Development Corporation. Twenty years after the fight over prisons, some farmers still refuse to purchase goods from people who supported their construction. Commercial dealings between the farmers and town businesses have declined as the farmers have become more self-sufficient, developing field-based packing plants instead of sending their produce to Blythe.

The PVID is a very large irrigation district, covering 189 square miles of land, about the size of metropolitan Philadelphia. The district has only eighty farmers; five landowners control more than half the acreage. From 1998 to 2002, the total value of the crops averaged $94 million. Cattle feed (alfalfa, Sudan grass, and Bermuda grass) and cotton constitute about three-quarters of the acreage. In 2003, the market value for PVID farmland hovered around $3,000 per acre.

In 2004, the Metropolitan Water District of Southern California and the PVID consummated a water transfer agreement. Participating farmers fallow between 7 and 29 percent of their land, on a rotating basis, for as long as thirty-five years. Metropolitan makes a onetime payment of $3,170 for each acre of land enrolled in the program and an annual payment of $602 per acre for each acre that is fallowed in a given year. It's a terrific deal for the landowners. As an up-front bonus for signing up for the program, farmers receive essentially the fair market value of their land, yet they retain title to the land. Then they receive $602 an acre for each acre they fallow, which is more than farmers could make if they farmed the land! PVID farmers Bob Hull and Danny Robinson had hoped that the day would come when the water used to irrigate their crops would be worth more than the land itself. That day has arrived.

Jill Johnson, a fifth-generation PVID farmer, irrigates 1,500 acres of alfalfa and wheat. Her sign-up bonus was approximately $1.5 million, and the annual payment level is attractive, given that her crops net a maximum of $400 an acre. The program involves no change in water rights; Metropolitan is leasing her water. At the end of the thirty-five-year term, she may negotiate a new lease, sell the water to someone else, or use it to farm once again.

From Metropolitan's perspective, the deal makes good sense. The late Dennis Underwood, a Metropolitan vice president in 2004, emphasized that the high-priority water rights offer security; the arrangement's flexible terms allow Metropolitan to decide each year how much water it needs, ranging from 25,000 to 111,000 acre-feet; and the costs, between $150 and $200 an acre-foot, were attractive relative to other options in Metropolitan's water portfolio.

As for potential third-party losses, Metropolitan will provide $6 million to offset them. The impetus for the $6 million fund, surprisingly, came not from the PVID but from Metropolitan, which took responsibility for the effects of the transfer on the community. Said Underwood, "We wanted to raise a bar of high standard and that's why we put up some upfront monies for community improvements." A committee representing a cross section of the community will determine where the money is to be spent. Metropolitan wants to be seen as acting fairly; the utility understands that it will return, seeking more water from the PVID in the future.

But not everyone is happy with the deal. A report on the consequences of the transfer for the community, commissioned by the chamber of commerce, concluded that, after the initial sign-up bonuses are spent or sent to out-of-state absentee owners, the effects would become negative—very negative: $58 million over the thirty-five-year length of the program. The most outspoken critic of the transfer is Mark Fulton, who believes that while the farmers did quite well for themselves, the larger community was shortchanged. In Fulton's opinion, the $6 million ponied up by Metropolitan does not begin to compensate for the third-party effects. Four days after he made a presentation to that effect before the PVID's board, he found a snake in his mailbox. It was only a black racer, not a venomous snake, but no one would mistake it as a goodwill gesture. Instead of rewarding Fulton for pushing the issue of third-party impacts, the chamber of commerce fired him.

Nevertheless, the PVID-Metropolitan deal went through easily and quickly. The stark contrast with the IID transfer bears careful attention. In the PVID, only property owners may vote to elect members of the seven-person board, and votes are weighted depending on the assessed value of the property. Each $100 of assessed valuation entitles a landowner to one vote. The difference in voting rights between the IID and the PVID is not lost on the *Imperial Valley Press*, which editorialized that the "PVID pretty much does what its landowners tell it to do, since they are the only people in the Palo Verde Valley with a right to vote in PVID elections. As a result, [the PVID-Metropolitan agreement was] never subjected to rigorous public input because the larger community was effectively shut out of the process. If the agreement sounds like a pretty good deal for PVID landowners, that's because it is. If it sounds like a bad deal for everybody else, well, how could it not be?" The *Imperial Valley Press* got it right.

PVID farmers are completely in control of water rights. And what the farmers wanted was money; they could not have cared less about mitigating the third-party consequences. Ed Smith, general manager of the PVID, sharply distinguishes his board from that of the IID, which makes decisions based on "what the mayor of El Centro says or the supervisor of the county thinks," because a popular vote elects them. To Smith, IID decisions have nothing to do with the agricultural industry.

The chance of nonfarmers electing a member to the PVID board is negligible, explained Smith: "If every single non-farming landowner in the Valley got together, they might control a few thousand votes. But it is going to take 40,000 votes to win a seat." As a result, the board has a clear sense of who elected it. Smith thinks this is how it should be because "the people that use the water should be the ones to have the say on it." If they want to sell it and people in the community object, "it's none of their business," said Smith.

When I asked about Fulton's role in this process, Smith bristled. "Mark's an asshole. He stepped on a lot of toes on this water deal and the chamber people knew that he had all the farmers upset. And most of them have a more moderate view of this thing." He denied that there was any pressure from farmers to fire Fulton. "Because we don't give a shit. Excuse the language, but we don't care. We can do what we are going to do and we are going to do it. If Mark doesn't like it, that's too bad."

This voting arrangement has dramatic consequences for the Blythe community. In both the IID and the PVID, water rights are held by the board in trust for the membership. But the PVID board, responding to the interests of farmers, has allocated the water to permit individual farmers to sell off discrete amounts. Metropolitan is paying mitigation money in spite of—not because of—the PVID board. From Metropolitan's perspective, the lesson is obvious: there is no incentive to undertake a cumbersome process to deal with the IID board when it can easily and quickly conclude transfers with the PVID board. Although the long-term effect on the Blythe area remains uncertain, one thing is clear: Metropolitan will need access to additional supplies in the future. Ron Gastelum, chief executive officer of Metropolitan at the time of the Quantification Settlement Agreement and the PVID deal, told me that he doubts the QSA will provide a model for the future because it took too long, cost too much, and is too cumbersome and complicated. Instead, Metropolitan will continue to seek smaller, shorter-term transfers that are easier to pull off. Metropolitan has no interest in burning any bridges, whether to the PVID farmers or to the residents of Blythe. In 2008 and 2009, Metropolitan spent nearly $30 million buying 100,000 acre-feet of water from PVID farmers.

Across the Colorado River in Arizona, people in the Wellton-Mohawk Irrigation and Drainage District are fearful of water transfers. Charlie Slocum, a lawyer who manages the district, thinks it makes more sense to let growth occur where the water is rather than move the water to support urban sprawl. And he's concerned about the future of the area left behind. Farmers may profit from a water transfer, but other residents will not. "The major problem," says Slocum, "is that the haves are attempting to have it all and they are going to leave some of the rural haves as have-nots later on in the future." Wellton-Mohawk officials fear that the metropolitan area of Phoenix, in Maricopa County, has its eye on their water supply. Indeed, they sarcastically refer to the Arizona Department of Water Resources as the "Maricopa Department of Water Resources" and to the state of Arizona as "the state of Maricopa." When the Arizona Department of Water Resources recently asserted an ability to determine the "beneficial use" of water, it struck fear in the hearts of Wellton-Mohawk officials, who think that the state of Arizona may be trying to do to it what Bennett Raley did to the IID.

The folks in Wellton-Mohawk are nervously watching developments across the river in California. And what they see scares them. It's prompted them to devise a simple strategy to protect their water. According to Slocum, "for a rural district to remain viable, it must gain political clout. And one way you gain political clout is to look very closely at the value added by your use of water." Slocum scratched out a back-of-the-envelope calculation of the total value of alfalfa grown in Wellton-Mohawk divided by the number of acre-feet of water used. Then he compared the amount of money that a power plant could generate at five cents per kilowatt-hour of electricity. He figures that the water would have ninety-two times more value producing kilowatts than growing alfalfa.

Slocum explained that using water for a power plant would make it exceedingly expensive for anyone to buy the water. It's not exactly political clout, Slocum explained; it's economic clout. As a consequence, Wellton-Mohawk is actively trying to attract a power plant to the district.

Few people would consider an electric power plant to be a good thing for the neighborhood; most would deem it a threat to the quality of life. But, for the farmers in Wellton-Mohawk, it would insulate their water rights from a hostile takeover. Attracting a power plant would certainly create employment opportunities for residents in the Wellton-Mohawk area; a large coal-fired power plant typically employs 500 workers. Construction costs would be likely to run into the billions of dollars, giving a real boost to local businesses.

The power plant has been stalled since 2004 as the company attempts to navigate the shoals of the environmental permitting process. But the mere fact that Wellton-Mohawk is pursuing it says a lot about the stakes involved, and the need for equity, in water transfers. It matters who gets to approve or reject a proposed water transfer. In the PVID, the board responded solely to the interests of farmers and quickly approved a transfer that benefited farmers while harming the larger community. In the IID, the board struggled long and hard before grudgingly approving the agreement to transfer water to Los Angeles—and then only with huge strings attached to protect the Salton Sea and compensate injured workers and businesses. In Wellton-Mohawk, irrigation district officials figure that they can defend against a raid on their water by shifting some water from growing alfalfa to supporting a power plant. And if the environment suffers along the way, that's still better than losing the water to a heartless city.

The chronicles of these southwestern irrigation districts—Imperial, Palo Verde, and Wellton-Mohawk—offer lessons for the rest of the United States as other regions struggle with water shortages. Control over water must remain with the state or with a broadly representative elected body; otherwise, parochial interests may encourage the crude commodification of water without regard to how transfers may harm workers, other businesses, or the environment. The QSA was not only the largest transfer of water in American history; it was also the messiest process imaginable. No one would suggest it as a template for other transfers, yet a lesson for municipal interests is apparent: share the wealth. Urban interests that come to rural areas seeking water had better work with such areas to craft mutually beneficial agreements. Otherwise, as we've seen from the Wellton-Mohawk board's fear of a raid on its water by Phoenix, rural areas will take steps to keep urban areas from getting their water, even if it harms their own quality of life.

CHAPTER 18

The Future of Farming

BETWEEN 1987 AND 2005, there were 3,232 sales or leases of water rights in the twelve western states, involving a staggering 31 million acre-feet of water. That's more than twice the annual flow of the Colorado River. Plotted on a graph, the recent trajectory looks like a rocket ship taking off. The greatest number of transfers occurs between and among farmers, but the largest amount of water is transferred from farmers to cities. The agricultural-to-agricultural trades usually involve modest quantities of water exchanged within irrigation districts on a short-term, as-needed basis. For example, a corn farmer might take advantage of a price spike and lease water from a neighbor who grows alfalfa. The trades that involve the largest quantities of water, however, involve farmers selling or leasing water for municipal and industrial use or for environmental protection or restoration.

Let's be clear about one thing: the water for new demands, whether refining ethanol or processing semiconductor chips, will mostly come from agriculture, because farmers use 70 to 80 percent of each state's water. Another driving factor is money. In many states, a high percentage of agricultural water is used to grow crops that return a relatively low value. Consider *Medicago sativa*, commonly known as alfalfa, a perennial flowering plant native to Europe and Asia. In the United States, we garnish salads and sandwiches with alfalfa sprouts and feed rabbits with processed alfalfa pellets. But most alfalfa, by far, is cut in the field, allowed to dry, and then baled as hay for beef cattle, dairy cows, horses, sheep, and other farm animals. If an early frost prevents the last cutting of the season, farmers allow their animals to forage in the fields. When harvested before it blooms, alfalfa contains high protein and low fiber, making it an excellent feed. Its high levels of potassium and calcium make it well suited to the needs of dairy farmers.

As a nitrogen fixer that is drought and salt tolerant, alfalfa grows well in poor soils and with low-quality water. But during summer, when temperatures are high, alfalfa grows slowly and blooms quickly, producing a harvest that is low in quality or quantity or both. A diet of lower-grade alfalfa can reduce milk production in a dairy cow by more than 50 percent and can reduce the weight gain of beef cattle from 1.85 pounds per day, while feasting on prime alfalfa, to only 0.06 pound, when fed poor-quality alfalfa.

Known as the queen of forages, alfalfa is the fourth-largest crop in the United States, after corn, wheat, and soybeans, respectively. Wisconsin, thanks to the needs of its dairy industry, leads the nation in alfalfa production, followed by California. Production of alfalfa, which is grown in every state in the country, exceeds 23 million acres. With an average yield of three tons per acre and a price of slightly less than $100 per ton, the 2004 alfalfa crop generated just under $7 billion. Alfalfa also sustains the production of honey, an annual $150 million market.

Seven billion dollars is a substantial market, but the production of alfalfa consumes more water than most crops and generates less revenue per acre. The California Alfalfa and Forage Association, a trade group representing growers, concedes that some California growers use more than eight acre-feet of water per acre of alfalfa per year. In 2000, the market value of the California agricultural industry exceeded $27 billion, but only $730 million came from alfalfa, which ranked tenth, just ahead of broccoli. Parents beg children, often in vain, to eat their broccoli, yet this commodity barely lagged behind alfalfa in market value, even though farmers planted only 120,000 acres of broccoli, compared with more than 1 million acres of alfalfa.

In the Imperial Irrigation District, where alfalfa is the largest crop, with more than 159,000 acres planted in 2004, farmers get ten cuttings of alfalfa each year. The cooler season from November through May produces high-quality alfalfa, but from June through October, high temperatures drive down yields and increase water consumption. During summer, when temperatures in the Imperial Valley regularly reach 115 degrees, farmers use the most water to raise the lowest-quality alfalfa.

Critics such as David Pimentel, Lester Brown, and Fred Pearce focus on the use of huge quantities of water to grow alfalfa merely to feed cattle. It takes 2,500 gallons of water to produce one pound of beef (another reason to reconsider the amount of red meat we eat). Water used to grow alfalfa, critics suggest, produces food only indirectly. That may be true for beef cattle, but it begs the question for alfalfa used in the dairy industry. If dairy farmers feed Bessie alfalfa, she gives them milk twice a day. That's a pretty direct production of food, including yogurt, butter, cheese, and ice cream. Alfalfa as an input to dairy production isn't going to go away; it makes economic sense.

But it's not nearly as efficient as it could be. If alfalfa production is to be made more efficient, farmers will need incentives to turn off their irrigation pumps and sprinkler systems during the dog days of summer. Some growers have begun to experiment with "summer dry-downs," letting the alfalfa go dormant during summer and then resuming irrigation in fall, when the temperatures moderate. Other growers are switching part of their farms from alfalfa to other crops as they try to maximize returns by growing higher-value produce. The disparities in value create tremendous opportunities for trade in water rights. In California, one acre-foot used to grow alfalfa generates approximately $60 in revenue, but the same amount of water used in the semiconductor industry generates almost $1 million.

The same pattern holds true for cotton, another water-guzzling crop grown in Arizona and other areas with desert climates thanks largely to federal commodity support payments. Cotton, which grows like a weed in humid regions (as anyone from Mississippi knows), must be irrigated in the West. Agricultural economists at the University of Arizona recently examined the relationship between irrigation patterns and cotton yields in Arizona. They found that water applied toward the end of the growing season, in late August and September, does not produce a better crop or generate greater returns for the farmer. Late-season irrigation has "a marginal negative value of water." In other words, it ironically costs the cotton farmer more to irrigate the crop than the crop is worth. Surely we must encourage these farmers to save that water.

The economic value of this water for municipal and industrial uses dwarfs the value of the same water to farmers. California growers consume 80 percent of the state's water yet contribute only 2 percent to the gross state product. This observation is not a criticism of farmers, who have several good reasons for using a lot of water. First, they have rights to the water, either under state law or through contracts with the U.S. Bureau of Reclamation, which provides water at heavily subsidized rates. Second, federal farm policy encourages farmers to grow water-intensive crops, especially rice, wheat, corn, soybeans, sugar, and cotton, by doling out handsome subsidies, more than $20 billion a year. Third, farmers operate on razor-thin profit margins and can't afford to pay for expensive infrastructure modernization, a fact that critics often overlook.

Here's the surprising reality. While cynics accuse farmers of milking the government, farmers are suffering from their own remarkable productivity. Few Americans realize that the United States has the cheapest food supply in the world—even taking into account the run-up in food prices in 2008. According to the U.S. Department of Agriculture, in 2005 Americans spent just under 10 percent of their disposable income on food, less than the citizens of any other country. Even more remarkably, that percentage has sharply declined over time. Twenty years ago, consumers spent 11.7 percent; 30 years ago, they spent roughly 15 percent; in the late 1940s, the number was 20 percent; and in 1933, Americans spent 25 percent of their disposable income on food. Increased production has lowered the cost of food to the public. Better equipment, mechanization, hybrid seeds, fertilizers, pesticides, and irrigation explain the so-called green revolution. Between the 1950s and 1980s, crop yields per acre doubled for wheat, rice, and corn. That was good news for farmers, right? Not exactly.

Increased productivity did not bring increased incomes for farmers. In 1950, farmers received 41 cents of every dollar spent on food by consumers. By 1980, that number had dropped to 31 cents. Today, it's less than 20 cents. The farmers' share has dropped by more than 50 percent because processors, wholesalers, and retailers have taken a larger portion of each food dollar. Farmers could adjust to receiving a smaller percentage if prices were to rise enough to offset the declining share. But the prices that consumers pay for food, held constant for inflation, are about the same today as they were during the era of the New Deal.

Consider what has happened to wheat prices. According to the U.S. Department of Labor's Consumer Price Index, the price of wheat in 1955 was $12.73 per bushel; by 2000, the price had dropped to $2.62. To make matters worse, it's much more expensive to run a farming operation today than it was fifty years ago. The inputs needed to achieve such remarkable productivity are costly, ranging from tractors and combines to high-priced genetically programmed seeds and to pesticides and fertilizers based on fossil fuels. Despite declining prices and increasing productivity, farmers have not benefited. No wonder some farmers are angry and others are selling out.

The decline in the number of American farms, to roughly 2 million from 2.2 million in 1993, is a very disturbing trend because it devastates rural economies, upsets community cohesion, stimulates the creation of ever-larger corporate farms, and may threaten the nation's food supply. Won't water transfers just make this worse?

The answer is no. Satisfying new demands for water requires the transfer of only modest quantities of water from relatively unproductive agricultural uses, including the production of nonfood crops such as cotton. Since 1985, farmers have sold or leased millions of acre-feet of water to other users, yet aggregate farm revenue, held constant for inflation, has not declined. This startling conclusion confirms the commonsense perception that farmers adjust to using less water by being more efficient. When farmers sell or lease water rights, they fallow the least productive fields on the farm, shift the crop mix, or change the irrigation system.

Sometimes farmers lease or sell water rights in conjunction with the sale of their property. The decision to sell is usually made either because the farmland is located in the path of suburban or exurban development or because the next generation does not want to take over the operation. Sprawling cities and endless suburbs are displacing more than 2 million acres of farmland each year. "Every minute of every day," according to the American Farmland Trust, "we lose two acres of agricultural land to development." California lost 90,000 acres of farmland between 1998 and 2000; in Michigan, 300,000 acres were converted to other uses between 1997 and 2002. Nationwide, farmland dropped from 986 million acres in 1990 to 933 million in 2006, a loss of 53 million acres.

Protecting existing farmland is critical for the nation's economy and food supply, our national security, the fiscal stability of local governments, and even the environment because farmland provides open space, food and cover for wildlife, flood control, and wetlands protection. The American Farmland Trust is spearheading reform of tax policies, land use regulations, and economic incentives to preserve existing farmland. Conservation easements, agricultural zoning, and preferential tax treatment for farmland offer viable strategies for saving farmland from development.

Farmers are responding nimbly to these challenges by finding higher-value niche markets, engaging in vanguard agriculture (identifying new ways of growing and marketing produce), and moving to value-added products. In 1999 in California's Central Valley, Daniel Errotabere switched from annual crops such as garlic, cotton, and sugar beets to four varieties of almonds. It was a risky decision because in a drought he may lose a season's produce with the annual crops, but his almond trees would die if they went without water for a year. Yet almonds are a lucrative crop, and growers such as Errotabere have increased the worldwide demand for almonds, confounding academic theorists. Richard Howitt, an agricultural economist at the University of California, Davis, concedes, "The almond growers have done a fantastic job of keeping the demand growing despite all the predictions of experts like me." Changing conditions and chronic water shortages forced Errotabere to adapt. A Central Valley water lawyer, Tom Birmingham, notes that his clients "were going to have to grow higher-value crops to pay for the water to stay in business."

In the Coachella Valley Water District, a Southern California agricultural district that includes upscale communities such as Palm Desert, farmers have shifted from flood to drip irrigation, and they grow dates, fruits, and vegetables that fetch approximately $8,000 an acre. With their farms generating this kind of money, they wouldn't dream of growing alfalfa. To the east, across the Colorado River in Arizona, there are signs of change. In the Yuma Irrigation District, Tim Dunn, a corn grower, recently switched from flood to sprinkler irrigation. He also uses Global Positioning System (GPS) links to laser-level his fields, which eliminates runoff. He's saving water not to save money—the price of his water is ridiculously low, about $15 an acre-foot—but because using too much water cuts into his yields. "And the land is so valuable and the cost of production is so high," he says, "we have to have high yields." A fellow farmer, Doug Mellon, made the same choice. "We found that the celery does great with sprinklers. We've found that using less water, we can grow crops better." Another neighbor has gone even further, installing drip irrigation with plastic tubing, which delivers a precisely measured amount of water to each plant. It's costly, to be sure, but the profits from his organic vegetables made the investment worthwhile.

Thirty miles east of Yuma, in the Wellton-Mohawk Irrigation and Drainage District, Robby Nickerson changed his crop from iceberg lettuce to baby lettuces. Until recently, most Americans' lettuce of choice was a head of iceberg, broken into chunks and served with ranch dressing and a maraschino cherry on top. I still think that nothing quite matches the crispness of iceberg lettuce. But a consumer craze has taken over the United States for baby lettuce and exotic species from arugula to radicchio. People are even eating dandelion greens, before they have been sprayed with Roundup, I hope.

The growing and processing of lettuce have changed profoundly in recent years. At the retail end, many consumers now buy precut lettuce in vacuum-sealed plastic bags rather than as individual heads of iceberg, red leaf, or romaine lettuce. Several companies are leading this industry, including Fresh Express, a publicly traded company, and Earthbound Farm, a privately held company specializing in baby organic lettuce that has seen its sales grow by an average of 55 percent annually since 1995, to more than $200 million in 2002.

Nickerson has responded to new consumer tastes by switching to these exotic lettuces. The growing, harvesting, and packaging of these lettuces testify to farmers' ingenuity. Nickerson plants baby lettuce in raised rows. When the lettuce is ready for harvest, a tractor pushes something that looks like an enormous electric razor. The wheels of the tractor go down each trench and the razor shaves the lettuce off about an inch above ground. The shaved lettuce leaves are conveyed up a belt immediately behind the razor's edge. At the top of the belt, the lettuce falls into boxes that are moved by another conveyor belt and placed as needed by a worker. Another worker then moves the filled boxes from the tractor to a trailer truck in an adjacent row. When the trailer is full, the driver immediately heads off to the processing plant. Four workers can harvest an entire field in a few hours, whereas it takes a score of workers better than a day to harvest heads of iceberg.

This lettuce operation has no need for weed killers because the lettuce is planted so closely together that weeds cannot grow between the individual leaves. A sophisticated method of planting the seeds ensures an even distribution that covers all areas of the row. Once the row is harvested, it is immediately fertilized and watered; the trimmed plants soon begin to sprout new leaves. Nickerson's operation is a clear response to the market and, perhaps just as important, to the recognition that as the value of water increases, growers need to take advantage of what they can do well in order to secure a greater return for themselves and a greater value for their crops' use of water.

From harvesting in the field, the boxes of lettuce are moved to refrigerated trucks and rushed to the packaging plant. At the plant, it takes only twenty minutes to chill the lettuce down to thirty-six degrees; if the processor maintains that temperature until the lettuce reaches grocery stores, it will stay fresh for about fifteen days. Inside processing plants, the lettuce is washed, in some operations as many as three times. The exact washing process is a carefully guarded secret, as each company wants a competitive edge on the freshness of its product. From the time the baby lettuce is shaved from the row, it takes only six days to wash, bag, and transport the lettuce around the country.

Consumer demand for tomatoes that taste like tomatoes has created a significant marketing opportunity for Eurofresh Farms, located seventy-five miles east of Tucson in Willcox, Arizona. The operation presents a bizarre sight: Eurofresh grows tomatoes in a greenhouse, a very big greenhouse: 318 acres, or about half a square mile. The tomatoes are grown hydroponically, in an inert medium called rock wool, and fertilized by sixteen nutrients carried by the irrigation water. Even though Eurofresh Farms uses no pesticides, the tomatoes are not organic because the system uses synthetic nutrients instead of manure. But Dwight Ferguson, Eurofresh Farms' chief executive officer, says that business is booming. The company ships more than 3 million pounds each week, year-round, all over the country.

Hydroponics has taken off thanks to two things: computers and plastics. Sensor-equipped computers determine how much water and nutrients these plants need and when they need them. As the plants grow, they climb plastic latticework nets, which are lowered as needed for workers to harvest the crop at eye level rather than stooped over, as in traditional field harvesting. Eurofresh Farms has not always enjoyed untroubled labor relations; a 1999 strike led to the workers being represented by the United Food and Commercial Workers International Union, the same one that's pressuring Wal-Mart. On the other hand, hydroponics production offers year-round employment, thus eliminating the need for workers to migrate from state to state over the growing season. Starting a hydroponics operation involves a substantial capital commitment. Gene Giacomelli, director of the University of Arizona's Controlled Environment Agriculture Center, notes that it takes only a few thousand dollars per acre to prepare conventional fields, but "in a greenhouse you may invest up to $500,000 an acre." But perhaps most important is that the water Eurofresh Farms uses was previously allocated for cotton. It made economic sense to shift production, says Giacomelli, "because tomatoes have greater market value than cotton."

Vanguard agriculture, finding new ways to grow and market produce, has established a foothold in Montana. A pioneer of organic agriculture, David Oien grew up on his family's farm in Conrad, Montana. Beginning in 1987, he and four other farmers took a stab at organic farming when they created Timeless Seeds. Despite his father's warnings that the only person who'd buy his products would be his mother, Oien persisted. After all, the alternative was to continue the cutthroat competition of commodity agriculture. After using black medic, a hardy legume, to fix nitrogen in the soil, the Timeless Seeds farmers planted wheat, barley, and flax. Crop rotation restored the soil as it reduced weeds and insects. Their hard work is paying off. In 2005, Timeless Seeds sold 66,000 packages of organically grown lentils to Whole Foods stores in the United States,

Canada, and England. A niche product, black beluga lentils, has become popular with fine chefs worldwide.

Elsewhere in Montana, Bob Quinn, a native Montanan with a doctorate in plant biochemistry, experiments on his farm with specialty grains. He grows organically for a simple reason: organic farmers get twice the money for their crops and, Quinn says, the expenses are lower. In the early 1990s, Quinn came upon a strain of Egyptian wheat, called Khorasan, that thrives on Montana's high, dry plains. Quinn marketed his wheat as Kamut, the Arabic word for wheat, and promised his buyers an organic, high-protein, nutty-tasting wheat. Cereal producers loved it, and now consumers can find Kamut cereal in most health food stores and many supermarkets. The 110 Kamut growers in Montana ship more than half of their product to Europe, mostly to Italy. "When it comes to food," Quinn observes, "the Italian imagination never quits."

Dean Folkvord, chief executive officer of Wheat Montana Farms, still farms his family's wheat fields, but he's also expanded into other aspects of wheat production. Just off Interstate 90 in Three Forks, he opened Wheat Montana Farms Bakery and Deli, where wheat goes from seed to sandwich under one roof. The facility, dubbed a "wheatplex," contains a processing plant that sends the product to the kitchen, which in turn sends baked goods to the retail store out front. The bakery sells more than 100 types of breads, rolls, pancake mixes, and wheat berries. Wheat Montana, with 110 employees, from farmers to bakers, opened its sixth retail outlet in 2005. The Wheat Montana approach—marketing so-called value-added products—is the fastest-growing segment of the agriculture industry.

These innovative approaches represent just a small fraction of total agricultural output in the United States today, and a great many of the bulk commodities grown in this country cannot be shifted to such value-added production. But the pressures on agricultural land and water will ultimately force commodity production to adapt. And the good news is that farmers can do more with less water. Not only can they remain productive; American farmers can capture some slice of the profits previously enjoyed by the processors. One way they're succeeding is by switching to value-added products. It's not an option for all farmers, as it involves stepping out of familiar territory to tackle a new aspect of food production. And, in the case of fresh-cut lettuces, it's not an endeavor for the undercapitalized: a regional processing plant might cost $20 million. These new developments in farming prove once again the adaptability of farmers to changing circumstances. There is reason to be optimistic about their future as these developments coincide with the emerging "eat local" campaign inspired by Alice Waters, Michael Pollan, and Dan Barber.

CHAPTER 19

Environmental Transfers

NATURE PROVIDES ecosystem services that most of us never think about. Bees and other pollinators fertilize orchards, enabling apple and almond trees to bear fruit. The trees in your front yard trap dust, dirt, and harmful gases. And water in a stream transports nutrients, providing nourishment to aquatic species. Diversions of water from a stream may disrupt or impair these surprisingly fragile processes. Efforts to rehabilitate degraded riparian habitat turn on the willingness of diverters to reduce their water use. But, as we shall see, even seemingly sensible water conservation efforts may yield ironic unintended consequences.

Janet Neuman, an energetic mother of three and an environmental law professor at Lewis & Clark Law School in Portland, Oregon, is constantly juggling schedules and commitments. Beyond teaching courses in natural resources law, Neuman is involved in community-based conservation efforts. In 1994, she assumed the helm of a fledgling organization, the Oregon Water Trust. When the state of Oregon realized that hundreds of miles of its rivers suffer diminished or depleted flows each summer as a result of irrigation withdrawals, it began to allow water rights to be dedicated to instream flows. If a farmer reduces his diversion and leaves more water in the river for fish, wildlife, or recreational users, an instream flow permit can protect that water from diversion by another farmer or water user. It's a nice idea, but why would a farmer leave the water in the river rather than use it to irrigate her land? After all, it's her water right.

In fact, there are several reasons to do so. Many farmers have such deep ties to their land that their sense of stewardship prompts them to act nobly to preserve the environment. Farmers who act this way deserve our praise, but it's asking a lot to expect them to shoulder the financial burden of giving up part of their water rights to enhance a river's flow. A less idealistic motive is to avoid running afoul of the Endangered Species Act, for a court may find that a farmer's diversion has harmed the critical habitat of a protected species in violation of federal law and that the farmer must decrease her diversion. But while farmers may collectively cause severe harm to a species, as the dewatering of the Klamath River did to salmon, it is highly unlikely that a court would find any single farmer in violation of the act and order that farmer to reduce her diversion. The third and most compelling reason is if she receives a financial incentive to do so.

The conservation community has discovered that land trusts offer intriguing options for protecting sensitive land. Consider a third-generation farmer who faces escalating taxes on the farm as rising real estate values make it a prime target for developers, but who wishes to continue farming and protect the land as open space. The Trust for Public Land and The Nature Conservancy have pioneered the use of land trusts, often in the form of conservation easements, as a flexible tool that allows the farmer to keep farming and protects the land from development. Land trusts have recently mushroomed. A 2005 survey by the Land Trust Alliance, an organization based in Washington, DC, that supports land trusts, found a dramatic increase during the previous five years in the number of land trusts, up by 32 percent to 1,667, and the amount of land protected, up by 54 percent to 37 million acres, an area more than sixteen times the size of Yellowstone National Park.

Neuman and her Oregon Water Trust decided to use the legal

model of the land trust to protect rivers. The first water trust in the country, the Oregon Water Trust is a private nonprofit organization that acquires water rights in order to increase river flows. The idea has spread to other states, including Washington, California, Montana, Nevada, and Colorado. At its core, a water trust uses incentives to reallocate water from irrigation to instream use. In Colorado and Oregon, the state water departments can hold instream flow water rights, and both states have recently appropriated funds to buy water rights from willing sellers. The Oregon Water Trust relies on government support augmented by private funds, membership fees, donations, endowments, and foundation grants to acquire water rights from willing sellers.

In the ten years Neuman served as president, the Oregon Water Trust used permanent purchases of water rights, short- and long-term leases, exchange and forbearance agreements, and conservation projects to protect a total volume of more than 124 cubic feet per second in basins across Oregon. In absolute terms, this is not a large volume of water, but, because the trust focuses on small tributary streams, where less than 1 cubic foot per second can make a difference, the trust's activities have had a major effect.

Doing the deals proved to be more difficult than Neuman expected. At the end of her term as president, Neuman reflected that "the Trust's bank balance of acquisition money remains quite healthy, as it has turned out to be harder than expected to spend the money." Although it took time to develop a harmonious relationship between the trust and the farmers, both sides see tremendous benefits in working together. Voluntary agreements, rather than litigation or contentious regulation, have succeeded in putting water back in the rivers. Faced with the choice of being sued for wasting water or receiving a check for that water, farmers prefer the latter option.

The longest undammed river in the Pacific Northwest, the John Day River rises in the Strawberry Mountains in northeastern Oregon, picks up flow from four branches, and runs 280 miles before joining the Columbia River. Its North Fork supports the largest population of salmon and steelhead in the entire Columbia River Basin. In 1988, Congress placed 148 miles of the main John Day and 101 miles of the North Fork and South Fork in the National Wild and Scenic Rivers System. White-water enthusiasts especially enjoy rafting and kayaking the North Fork through a remote canyon, while the main John Day provides an easier and very popular float. Anglers love the upper river for the chance it provides to catch wild salmon and steelhead and the lower river for its terrific smallmouth bass fishing. Farmers rely on the John Day and its tributaries for water to grow alfalfa to feed cattle. The Middle Fork of the John Day is a small but critical tributary because it supports as much as one-third of the spawning salmon and steelhead in the entire river basin. A five-year study by the Oregon Water Trust found that the Middle Fork could support many more fish if it had a little more water, especially late in the summer, when irrigation diversions are the heaviest.

Located on the Middle Fork is the sprawling Austin Ranch,

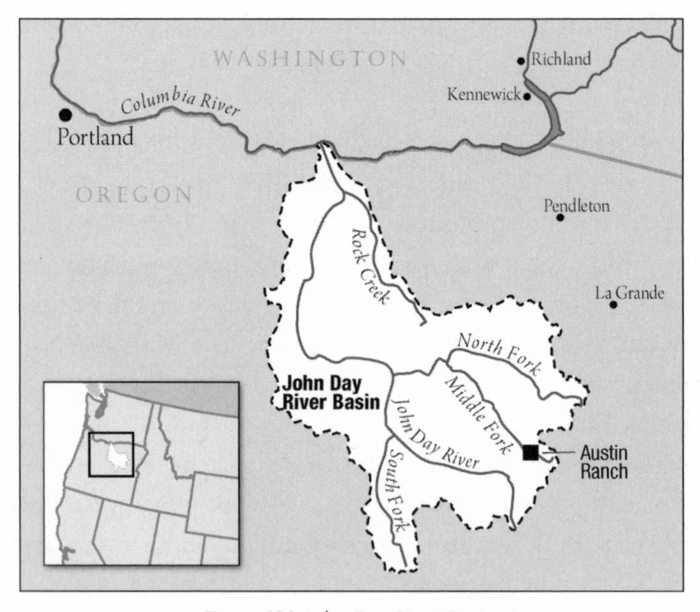

Figure 19.1. John Day River Basin.

owned by third-generation ranchers Pat and Hedy Voigt. The Voigts run cattle and use Middle Fork water to grow alfalfa. In 2006, they agreed to sell a portion of their water rights to the Oregon Water Trust. The agreement is a landmark in creative approaches to water shortages. It's a win-win solution that will increase fish habitat and improve water quality by adding cool water to the Middle Fork during the critical late-summer spawning season.

The Voigts agreed to turn off their irrigation system from July 20 each year until the end of the growing season. This arrangement will enable the Voigts to harvest one cutting of alfalfa each season and then turn the fields into pasture, allowing cattle to graze after July 20. Halting irrigation means the Voigts will forgo a second cutting, which involves a substantial loss of income.

For this permanent, voluntary relinquishment of part of their water rights, the Voigts received $700,000 from the Oregon Water Trust. The Voigts are using that money to modernize their irrigation system, which will provide better yields per acre for the first cutting. "This deal's working for us," says Pat Voigt as he turns the crank on a diversion head gate. "And I believe OWT thinks it's working for them. How does it get any better than that?"

The Oregon Water Trust is thrilled because the agreement will leave up to 6.5 million gallons of water *per day* in the Middle Fork and two small creeks that flow into it. No one else can divert this water, so it will provide 10 cubic feet of water per second to seventy miles of river during the critical low-flow spawning season. This water will have enormous benefits for wild chinook salmon and steelhead trout.

"I give OWT a lot of credit," says Pat Voigt. "They came to us with an attitude of wanting to help and they displayed a great deal of respect for agriculture." Mutual respect and benefits are essential if agreements such as the one between the Voigts and the trust are to become commonplace.

Major funding for this agreement came from an unlikely source: the Bonneville Power Administration (BPA), the U.S. Department of Energy agency that runs most of the dams on the Columbia River system. These dams have extirpated, endangered, or threatened many salmon species—and now the BPA is trying to save the salmon? You bet. The Endangered Species Act has placed enormous obligations on the BPA to protect those species still extant. The BPA loves its dams and really, really doesn't want to remove them. So it's spending $145 million per year in efforts to protect endangered species. Some efforts are comical, like the ones discussed earlier that involve barging salmon fry around dams or, even better, trucking them. But the agency has become more creative, some would say more biologically sensitive, in recent years. In 2002, it established the Columbia Basin Water Transactions Program, a mouthful, to be sure, but nonetheless a critical commitment to finding market-based solutions to environmental problems. The BPA is spending as much as $4 million per year to underwrite projects such as the agreement between the Oregon Water Trust and the Voigts.

The Austin Ranch represents the tip of the iceberg for voluntary water transfers that benefit the environment as well as farmers and ranchers. Between 1998 and 2005, environmental organizations and government agencies spent approximately $300 million on more than 1,000 voluntary water rights transfers in ten western states.

LAURA ZIEMER, a mother of two young daughters, directs Trout Unlimited's Montana Water Project in Bozeman. Even though she works for Trout Unlimited (TU), Laura doesn't fish. Given her limited time for recreation, fishing is too tame a physical challenge. So she climbs mountains. Big mountains. Laura is an accomplished mountaineer mentioned in Jon Krakauer's *Into Thin Air* as part of a rescue team on Mount Everest. On weekends, she and her female climbing buddies scale class 11c routes. When I gasped and said that was like climbing a vertical cliff, one of them nonchalantly replied that these routes actually involve an overhang. In her office, instead of using a chair at her desk, she sits on a fitness ball, bouncing away with boundless energy.

Despite her youth, this University of Michigan Law School graduate has become one of the most creative and effective water lawyers in the American West. This is no small accomplishment because many of her successes have come from developing trusting relationships with crusty old Montana ranchers, a breed not easily cowed by spunky women. But Ziemer's style is part of her substance. It's her mission to figure out the common ground in order to persuade ranchers that saving water for fish is in their best interest. This conclusion does not come readily to Montana ranchers, but they've rarely met a person with Ziemer's infectious enthusiasm. Once a rancher agrees to protect a creek, it's up to Ziemer and a TU colleague, attorney Stan Bradshaw, to finesse legal obstacles and recalcitrant state agencies to make it happen.

Ziemer uses financial incentives to persuade ranchers to participate in stream restoration efforts. Let's consider two examples. Poorman Creek rises in the Helena National Forest and flows fourteen miles to where it meets the Blackfoot River—renowned for its fly-fishing and made famous by the book and movie *A River Runs Through It*. Poorman Creek has suffered badly from placer mining, undersized culverts, dewatering for irrigation, and excessive grazing. Nevertheless, westslope cutthroat trout and endangered bull trout sometimes spawn in its upper reaches. TU biologists believe that more fish would spawn in Poorman Creek but for one simple fact: in the last mile, just before it enters the Blackfoot, the creek suffers from low flows or even dries up in late summer. A cattle and hay operation diverts water through an earthen ditch and flood-irrigates the fields. As water flows through two ditches that meander for a mile, much of the water seeps into the ground along the way. With flood irrigation, the farmer simply opens the gate from the ditch to the field and lets the water run out. It's a notoriously inefficient use of water.

Bradshaw approached the rancher, Eddie Grantier, who was amenable to talking about flexible, commonsense solutions. His 420-acre ranch, begun by his parents, is a small operation by Montana standards, running about 100 cattle. He'd like to be more efficient in his use of water, but the alternatives demand substantial capital. Grantier had a dilapidated sprinkler system that he towed around his fields, but it still worked and a new one would be expensive. Bradshaw proposed that TU, the U.S. Fish and Wildlife Service, and Montana Fish, Wildlife, and Parks support the efficiency improvements. Good idea. Who's going to pay? A grant from the BPA's Columbia Basin Water Transactions Program helped. Then Ziemer and Bradshaw approached the Natural Resources Conservation Service (NRCS), an agency of the U.S. Department of Agriculture that helps farmers and ranchers install efficient center-pivot irrigation systems. A backlog exists for NRCS help, but TU persuaded the service to give priority to funding center pivots when the water saved goes toward stream restoration. When I asked Ziemer how she pulled this off, she replied, "It involved copious amounts of beer."

In 2003, the parties reached agreement. TU raised $110,000 for a new center pivot, an irrigation pipe and pump, and a screen to keep fish from being trapped by the intake pipe. Grantier contributed $20,000 in labor. "This agreement," Grantier notes, "has provided me with a chance to make some improvements to my operation, and hopefully will save me some money in the long run." The new system has also saved Grantier a lot of hard labor. Thanks to the pipe, he no longer needs to dig out the ditch each season, and he can operate his sprinkler system with the push of a button. It's also reduced the amount of water he needs. Grantier previously diverted as much as 18 cubic feet per second; he now diverts 3.3 cubic feet per second. The difference is approximately 7,000 gallons per minute, about the same as the flow in Rock Creek in Washington, DC.

In 2005, Bradshaw did the legal work for Grantier to convert a portion of his water right—the conserved water—from an irrigation right to an instream flow right. That conserved water now flows in the last mile of Poorman Creek, and TU has documented bull trout moving upstream to spawn. The environmental benefits should be long-term because Grantier installed fences to keep his cattle out of the riparian area and planted trees and shrubs to restore the habitat.

Grantier gets another side benefit: with the influx of fly anglers to Montana, ranches with trout streams command a premium, and his ranch has suddenly become more valuable.

As the Poorman Creek story illustrates, we need to provide incentives for farmers to move from flood to sprinkler irrigation, or from sprinkler to drip or micro-irrigation, and to laser-level their fields. In 2000, American farmers irrigated some 63 million acres of land, an area roughly the size of Oregon. Farmers used the most efficient form of irrigation—drip irrigation and laser-leveled fields—on only 4 million acres; flood irrigation and sprinklers, about equally divided, irrigated the rest. These practices present an excellent opportunity to conserve water by adopting more efficient irrigation systems, but the associated costs are immense, well beyond the means of most farmers. As a result, farmers continue to use inefficient forms of irrigation because the cost of water is low, the returns on their investment are marginal, and they have the legal right to continue doing what they're doing. We must find ways to encourage those who need or want the water, be they developers, cities, or environmental organizations, to compensate farmers for the cost of installing more efficient irrigation systems.

Defenders of the status quo, including some environmentalists, maintain that flood irrigation doesn't actually waste water because the excess either percolates into the ground and recharges the aquifer or slowly drains into a nearby river or creek. In some locations, such as Silver Creek in Idaho, a legendary fly-fishing spring creek, less groundwater was discharged into the creek after farmers switched from flood irrigation to sprinklers. In other locations, it is argued, flood irrigation so saturates the fields that late in the growing season the nearby river receives subsurface drainage. Thus, there is more water in the river in late summer, the critical season for fish.

Laura Ziemer and the Montana Water Project are sensitive to these concerns. Before they commit to any potential project, scientists calculate whether reduced return flows would offset the potential environmental benefits of moving from flood to sprinkler irrigation. In the case of Silver Creek, the water for flood irrigation actually comes from the Big Wood River, in another basin. In this unusual case, flood irrigation actually adds water to the system. However, the aberrant occasions in which flood irrigation has environmental benefits should not obscure the larger point: it's wasteful and harmful. Flood irrigation encourages increased evaporative loss, causes irretrievable percolation, promotes salt buildup, leaches fertilizers and pesticides into rivers and aquifers, and consumes energy in diverting or pumping excessive quantities of water.

Two hundred years ago, Meriwether Lewis and William Clark explored the Yellowstone River on their way west. The river rises at the Continental Divide in what is now Yellowstone Park, flows through the majestic Grand Canyon of the Yellowstone, heads north out of the park into Montana, runs through Paradise Valley, and eventually joins up with the Missouri River. The longest undammed river in the continental United States, the Yellowstone enjoys a reputation for having some of the largest brown, rainbow, and native cutthroat trout in the country. The breathtakingly beautiful Paradise Valley, dotted with ranches, framed by snowcapped peaks, and crisscrossed with tributaries to the Yellowstone, is home to multigenerational Montana ranching families and newer well-to-do arrivals, often from the entertainment industry, such as Tom Brokaw, Peter Fonda, Michael Keaton, and Dennis Quaid. To fly anglers, the valley's spring creek tributaries—Armstrong's, Nelson's, and DePuy's—offer some of the best spring creek fishing in the world.

Some Yellowstone River tributaries, including the North Fork of Fridley Creek, go dry each year. Trout Unlimited and the Gallatin Valley Land Trust worked with the Murphy family, the owners for four generations of the Ox Yoke Ranch, a working cattle ranch on the North Fork, to change the ranch's irrigation system. Native Yellowstone cutthroat trout spawn in summer, during the high point of the irrigation season, but a large irrigation ditch built across the Ox Yoke Ranch in the 1930s literally cut off the North Fork's connection to the river. For this project, Ziemer used a private lease of water to facilitate the agreement. TU raised the money to pay rancher Sean Murphy to replace his irrigation canal with small center-pivot systems, install culverts, and drill a groundwater well. In return, the Murphy family dedicated the conserved water to instream flows. As with Poorman Creek, the results were spectacular. For the first time in seventy years, Fridley Creek reaches the Yellowstone. "I've wanted to do this my whole life," said Murphy. "It's a project that keeps us ranching. We get what we need and the fish get what they need." Fridley Creek provides another example of how voluntary collaboration between ranchers and conservationists can improve ranch operations and restore fish habitat.

DRIVERS WHO CONTINUE west on Interstate 10 shortly after passing over the All-American Canal come upon an unexpected sight—the Imperial Sand Dunes. Extending more than forty-five miles and reaching a height of 300 feet, these dunes resemble the Sahara in North Africa. The All-American Canal cuts its way through fourteen miles of these dunes on its way to the Imperial Valley. Here's the shocking part: the engineers did not line the canal with concrete. The All-American Canal is a simple earthen ditch. Nothing prevents water in the canal from percolating through the sand into the ground, which water does in immense quantities. The U.S. Bureau of Reclamation estimates the loss at 67,500 acre-feet per year just for the section that runs through the dunes. In other words, 22 billion gallons of Colorado River water never reach the Imperial Valley. That's enough water for a half million people. Every year. Forever.

If we're serious about water conservation, shouldn't we line the canal? Here's an opportunity to save billions of gallons by taking a step that should have been taken in the 1930s. But at that time, no one lined ditches or canals because water was plentiful. If a large quantity percolated into the ground, the solution was to divert more water, not to line the ditch. Today, water is scarce, especially in the Colorado River Basin. But not in the Imperial Irrigation District, whose rights to more than 3 million acre-feet are ten times those of the state of Nevada.

The 2003 Quantification Settlement Agreement provides a mechanism for San Diego to get more Colorado River water by finally lining the sand dune section of the All-American Canal. Lining the canal seems like a perfect water conservation program. The farmers continue to farm, the city gets more water, the municipal and state interests pay for the conservation improvements, and waste ends. The canal lining makes the conveyance system more efficient. However, a long-overdue conservation measure ironically has become a diplomatic bone of contention between the United States and Mexico. And environmentalists oppose the project because seepage from the canal supports an important agricultural area and a rare wetland in the Mexicali Valley, in the northern Mexico state of Baja California.

In the 1940s, after completion of the All-American Canal, Mexican farmers across the border from the Imperial Sand Dunes found new springs appearing on their property. The water continued to pour out of the ground until, in the 1950s, the Mexican farmers, aided by their federal government, dug drainage canals to capture the seepage. Then they drilled fields of wells to pump this newly available groundwater. Fifty years later, the Mexicali Valley has more than 50,000 acres of irrigated land and has become one of Mexico's most important agricultural regions.

Raul Garcia, who grew up in Mexicali, works as an agricultural engineer for an irrigation district that depends on the water that seeps from the All-American Canal. "Agriculture is the only source of employment in the valley," he says, "and if there's no water, there will be no agriculture. It is a very serious problem." The loss of 67,500 acre-feet saved by lining the canal will have a substantial effect on more than 14,000 Mexicali farmers. The local groundwater is rather salty, and higher-quality seepage from the All-American Canal dilutes it, making it an excellent source for irrigation. No one knows what kind of salinity spike will occur if the seepage flow is cut off. Baja California's government predicts that the lining will affect almost 3,000 acres of prime agricultural land in the northeastern corner of the Mexicali Valley.

In the late 1990s, the environmental community became interested in the Colorado River Delta. Upstream diversions and dams usually dry up the Colorado River before it reaches the Sea of Cortés. However, above-average flows on the Colorado in 1983, 1988, 1993, and 1998 produced more water than cities or farmers could use and more than the Bureau of Reclamation could store behind Hoover and Glen Canyon dams. Millions of acre-feet flowed into the Delta, transforming almost overnight a barren wasteland into fertile riparian habitat. It must be saved, asserts Ed Glenn, a bespectacled plant scientist at the University of Arizona who doggedly pushed to bring attention to the Delta. While conducting low-level aerial vegetation surveys of the Delta in 2002, researchers came upon an unexpected sight: wetlands at the northern edge of the agricultural land in Mexicali and just south of the United States border. Comprising more than 9,000 acres, the Andrade Mesa wetlands support more than 100 species of birds, including some rare and endangered species such as the Yuma clapper rail, large-billed savannah sparrow, gull-billed tern, and California black rail. These unique wetlands consist of marshes, vegetated dunes, and salt grass beds surrounded by stands of cattails.

An unlikely coalition of farmers, environmentalists, and business interests in the city of Mexicali, home to maquiladoras (assembly plants owned by companies such as Sony and Honeywell), formed to challenge the lining of the All-American Canal. The national government of Mexico has also protested the lining, broaching the subject with the U.S. Department of State and the International Boundary and Water Commission, a binational body that deals with an array of border issues. "The responses from the United States," said Alberto Cárdenas Jiménez, Mexico's secretary of environment and natural resources, "were not very encouraging." So the coalition brought suit in 2005 in United States district court against the United States government, the secretary of the interior, and the commissioner of the Bureau of Reclamation.

The suit alleged that the lining would deprive Mexican interests of water rights and that adverse effects on the environment would violate the National Environmental Policy Act and the Endangered Species Act. The major Colorado River water users, including all seven Colorado River Basin states, the Metropolitan Water District of Southern California, the San Diego County Water Authority, and the Imperial Irrigation District, quickly intervened. The water users expected to prevail. Mexican interests had no right to the seepage water, claimed Daniel Hentschke, a lawyer for the San Diego County Water Authority. Lining the canal, said Hentschke, was merely fixing "a leak in the hose that delivers California water to California." The district court agreed and, in July 2006, dismissed the suit.

The Colorado River water users rejoiced, and construction on the $300 million project commenced. Although the plaintiffs appealed, the water users weren't worried until the U.S. court of appeals issued an injunction prohibiting further construction pending a hearing, an ominous development based on the court's finding of a likelihood that the plaintiffs will ultimately prevail. With millions of dollars on the line, 9 percent of San Diego's future water supply at stake, and a construction project halted, although San Diego still had to pay the contractor, the water users decided to explore other options. None of these included settling with the plaintiffs. Instead, they went to Congress.

The Tax Relief and Health Care Act of 2006, 279 pages long, contained tax relief measures and health savings account options for millions of Americans. The act had nothing to do with water until December 2006, when California senator Dianne Feinstein, Nevada senator Harry Reid, and Arizona senator Jon Kyl attached a last-minute rider to the act declaring simply that "notwithstanding any other provision of law, [the government] shall, without delay, carry out the All-American Canal Lining Project." President George W.

Bush signed the law three days after the U.S. court of appeals heard arguments on the plaintiffs' appeal.

The 2006 act effectively changed the rules in the middle of the game. Plaintiffs were outraged. Their lead lawyer, Gaylord Smith, bitterly attacked the act as a repeal of the National Environmental Policy Act and the Endangered Species Act "by midnight legislation that no one in Congress knew they were voting for." Plaintiffs vowed to challenge the act. But they faced an enormous hurdle. Congress writes federal laws, and if Congress doesn't like a federal court's interpretation of a federal statute, it can rewrite the statute. In April 2007, the U.S. court of appeals dismissed the plaintiffs' appeal. Lining of the canal is proceeding.

The All-American Canal saga resonates with other water conservation efforts around the West; increasing the efficiency of water delivery systems may degrade or destroy wetlands and riparian habitat that depend on seepage. Conserving water by lining the canal paradoxically harms other water users and the environment. It's a quandary, admits Jennifer Pitt, an Environmental Defense Fund analyst in Boulder, Colorado. "I'm in a really awkward position having to argue against efficiency projects, but we need to get water dedicated to the environment."

CHAPTER 20

The Buffalo's Lament

BETWEEN 1800 AND 1900, the number of buffalo in North America plummeted from an estimated 30 million to less than 1,000. Everyone shares blame in this morality tale. Native Americans used buffalo not only for their daily needs but also in trade for goods. European settlers decimated the herds to make room for cattle ranching and large-scale farming. The United States Army and the U.S. Department of the Interior favored buffalo eradication as a way to confine roaming bands of Plains Indians to reservations. Eliminating the buffalo became part of America's Manifest Destiny of subduing the frontier and establishing "civilization."

But one group bears special culpability for the demise of the buffalo: commercial hunters. As killing machines, they were admirably efficient in responding to a rising market for tanned buffalo skins used for clothing, rugs, and industrial machine belts. Discard your romantic notion of the solitary hunter perched on a ridge knocking off a few bison and replace it with the hard reality of a cadre of hunters, skinners, gun cleaners, and cartridge loaders, backed by a wagon train serviced by blacksmiths, wranglers, teamsters, and cooks. In the 1870s, almost a thousand commercial hunting outfits harvested buffalo. With no restrictions on the seasons or on numbers, size, or gender of buffalo hunted, they killed more than a million buffalo a year, bringing them to the brink of extinction.

Similar practices of unrestricted grazing, logging, hunting, and fishing decimated the grasslands of the Great Plains, the eastern hemlock forests, the lobster fishery in Maine, the cod and haddock fisheries in the North Atlantic Ocean, and the salmon and halibut fisheries in Alaska, as well as the once-vast populations of whales, swordfish, alewives, shad, green sea turtles, tiger sharks, giant tortoises, sea otters, fur seals, elephant seals, flounder, oysters, porpoises, swordfish, abalone, and bluefin tuna in the open oceans. A recent victim of relentless fishing pressure is the Chilean sea bass (Patagonian toothfish). Whether for oil, skins, wood, fur, or meat, the hunters, ranchers, loggers, and fishers responded to open access with entrepreneurial zeal, seizing the opportunity to make money off natural resources. Commercial buffalo hunters, for example, had no incentive to limit their kills because buffalo not killed by them would be killed by other hunters. It was literally open season. Unrestricted access to a public resource encourages its overuse and potential depletion or extinction, whether that resource is buffalo, grassland, a forest, salmon, or water.

Ultimately, Alaska's halibut fishers saved their livelihoods and the fishery by mutually agreeing to limit their catches. The first efforts along these lines were fragmentary: a five-month season in the 1970s gave way to "derby style" fishing in the 1980s. At a certain starting time, boats raced off to the fishing grounds for forty-eight-hour open periods. The pressure on the fishery was short but brutal; captains hauled in as many fish as possible; overloaded boats sometimes sank; fish size and quality plummeted; and fishers risked their lives by fishing in whatever challenging weather conditions the Bering Sea and the Gulf of Alaska dished up. In a typical year during the 1990s, thirty-four boats capsized and twenty-four fishers lost their lives in Alaskan waters—twenty times the national average for fatalities in industry in general. The boats that did return dumped tons of halibut on the docks all at the same time, causing prices to plummet, as most halibut had to be frozen. This system needed an overhaul.

The International Pacific Halibut Commission, a joint United

States–Canada association established in 1923, oversees the fishery. On the basis of scientific research, with input from fishers, the commission has begun to estimate the sustainable catch for each season and to recommend quotas for implementation by the two governments. In 1995, Alaska began awarding halibut permits to captains of vessels on the basis of their past halibut catches, allocated as a percentage share of the annual harvest. These individual fishing quotas, or IFQs, revolutionized commercial halibut fishing. With percentage shares of the catch, captains no longer need to worry about beating other boats to the fish. Without frantic forty-eight-hour seasons, captains schedule their trips according to weather conditions. And without all the boats returning to the dock at the same time, halibut fishers receive higher prices for delivering fresh fish over an eight-month season. Because anyone who wants to become a commercial halibut fisher must purchase a permit from a current captain, a robust market for IFQs has developed and the value of permits has steadily increased. Dockside prices for halibut have more than tripled since the advent of IFQs, from less than $1.00 per pound in 1992 to more than $3.00 in 2006. In 2007, when prices at the dock soared to an unheard-of $5.50 a pound, halibut fishers wore broad smiles.

By restricting new entrants and establishing catch quotas, Alaska has broken the relentless cycle of overfishing and has invested fishers with a financial stake in the future health of the fishery. Having property rights in the fishery has increased the income of commercial halibut fishers, improved their safety, spread out their working hours, and stabilized their future earning power. In 2007, the Environmental Defense Fund released an in-depth analysis of ten United States fisheries, before and after implementation of catch shares, which concluded that such programs reduced bycatch (unwanted fish and marine mammals caught in nets) by more than 40 percent, increased revenues per boat by 80 percent, doubled safety rates, and ended the race for fish.

Alaska's halibut fishery offers a lesson for how we should regulate water in the United States. We must break the relentless cycle of overuse by restricting new access to the public resource, by protecting existing users with quantified water rights, by making these water rights transferable, and by insisting that new users purchase and retire existing water rights in exchange for permission to place a new demand on the resource.

In eastern states, the doctrine of riparian water rights allows diversions of surface water from lakes and rivers by any owner of land abutting the lake or watercourse. The owner shares the right with fellow property owners. In the nineteenth century, when the law restricted the owner's diversion of water to the amount that could be used on the land next to the lake or river, the riparian doctrine had a self-contained limit on how much water property owners could use. But twentieth-century courts began to allow riparian water rights to be used on other lands, thus vastly expanding the potential for overuse of the resource. Today, most eastern states have moved to a regulated riparian system, with permits issued by a state agency that possesses authority to set limits on the quantity of water used by riparian owners. However, most states exempt from regulation any use under a certain threshold, often 100,000 gallons a day—cumulatively a lot of water—and most states grant considerable discretion to their departments of water resources to implement the system. When asked to put teeth into lax standards, many state agencies, not surprisingly, lack the fortitude to deny permits to citizens, companies, or municipalities that want or need more water.

In the West, states adopted the prior appropriation doctrine, a "first in time, first in right" rule that encourages diversions of water from rivers by rewarding the earliest diverters with senior rights. In theory, the seniority system prevents harm to senior diverters when flows are low by requiring juniors to halt their diversions. In reality, some state agencies continue to grant permits even for rivers that are fully appropriated. Again, state agencies are loath to deny permits. This puts the burden on existing diverters to persuade a judge to order juniors to cease their diversions, a costly, time-consuming process involving complicated factual claims about priority dates, differences between amounts diverted and consumed, forfeiture and abandonment for nonuse, and return flows. Many people think that the priority system creates precise, quantified rights; in practice, however, seniors often have only *claims* to water, not legal rights confirmed by a court decree. Under this unruly system, juniors merrily proceed to use water without regard to priority.

While the riparian and priority systems of surface water rights do set some restrictions on water use, groundwater rules in most states allow unlimited access to a finite resource. The majority of states recognize the "reasonable use" doctrine, an ill-conceived rule that encourages exploitation of the resource. Even those states that restrict groundwater pumping still allow exemptions from regulation for domestic wells. These wells, it is thought, extract so little water and are so critical for homeowners that it makes sense to give homeowners unbridled permission to drill domestic wells. But 15 percent of Americans get their water from such wells, and 800,000 new wells are being drilled *every year*. New Mexico allows domestic wells to pump 1 million gallons per year; in Arizona, it's 3 million gallons. In Washington State, developers have skirted the priority system and provided water to entire subdivisions by using multiple wells, each of which serve only a few homes in the development. We must close such loopholes and end the calamitous tradition of encouraging unsustainable use of our nation's water resources.

This byzantine system needs a major overhaul. As we enter the era of water reallocation, what will bring about this needed change? Perhaps reform could come from government rules and regulations that impose conservation requirements, eliminate subsidies, encourage investment in modernization, and require "full-cost pricing," which would require the beneficiaries of U.S. Bureau of Reclamation projects to pay the actual cost of the water they receive. But these proposals, each desirable in the abstract, would be extraordinarily difficult to execute in the real world. Several involve very expensive system improvements, such as lining canals with concrete or laser-leveling fields, which can cost hundreds of thousands of dollars for a single farm. Where is the money to come from, given that many farmers operate on paper-thin margins? It is not feasible, reasonable, or equitable to require farmers to undertake massive expenditures in order to make their irrigation systems more efficient. It's no fault of farmers that the world has changed around them and that their customary irrigation methods now seem obsolete to outsiders. In general, they still work quite nicely for the farmers.

Neither state legislatures nor Congress is likely to mandate such reforms, for one very practical reason: farmers wield immense political power. Legislators would imperil their political futures were they to require their farmer constituents to shoulder the burden of these huge expenses. Heavy-handed government mandates would generate bitter political controversy and endless litigation over whether the government has the authority to act so cavalierly and whether the United States Constitution prevents the confiscation of water rights.

Perhaps other constituencies could pressure legislators to impose conservation standards on farmers. The rise of the environmental movement, the growing demands of cities, and the increasing role of recreation on public lands in the West have created groups of voters whose interests in water differ from those of farmers. Even though these constituencies consist of a large number of voters, they are geographically diverse and lack focus on any particular issue involving farming practices. For example, a politician contemplating how to vote on a bill to require farmers to laser-level their fields can be confident that 100 percent of the farmers will be strident opponents, but he would be uncertain about the views of his other constituents. This is a good example of a common political phenomenon in which a small number of deeply committed voters wields inordinate political influence over the legislature. Politicians listen carefully to voters who represent the dominant economic interests in the state. Accepting this political reality, we must recognize that state legislators will not impose costly changes on the farming community.

It would be far better to encourage voluntary transfers between willing sellers and buyers. Let them decide what the water is worth to each of them. The best way to reform agricultural water use in the United States is to give farmers a financial incentive to use less: let them sell the water they save to Google for its server farms or Trout Unlimited for fish habitat. The funds generated will enable farmers to modernize their irrigation systems, and make a profit on the side.

Market-based transfers can take many forms, from sales to leases, from forbearance agreements to dry-year options, and from land fallowing to conservation measures that save water. Each offers the prospect of a win-win result. The seller secures a price that she finds attractive and the buyer secures a water supply at a price that he finds attractive. The case for water marketing rests on the assumption that ownership of an item invests the owner with an incentive to take care of it. While this is surely not a universally valid proposition, it is still generally true that ownership changes one's behavior.

Ask yourself, for example, whether you treat your new car the same way you treat a rental car. When was the last time you washed a rental car? The same point might be made about hotel rooms, public parks, and parking lots. I have never seen a cigarette smoker dump his ashtray in his driveway, but some smokers do not hesitate to do so in parking lots. Others dump trash on public highways, something they would never do in their own yards. Of course, not all people treat public property recklessly. Some of us pick up trash in parks and on hiking trails; others tidy up hotel rooms. Whether driven by the golden rule, feelings of guilt or shame, concern for the labor of chambermaids, or the belief that civility ennobles us as a people and a culture, many Americans do take responsibility for public places. Others, alas, habitually degrade public spaces. Whether the habit of littering is cultural or rooted in status and class, people act differently toward things they care about.

The ability to transfer ownership creates an incentive to shepherd the resource wisely, to use property more productively, as we saw in the stories of agricultural innovation and ranchers collaborating with environmentalists. This is the core idea of markets. Owners of property assess its value to themselves and part with it if they will realize a profit. Buyers seek to change the use of property and capture the value added by the new use. In this process, both sellers and buyers may profit, and society benefits from increased efficiency. Water markets have other benefits, including conservation. If a farmer can profit from the sale of water, he will be motivated to invest in conservation practices that save water and thus free it up for other users.

Most defenders of markets rely on economic arguments, but, as my colleague Carol Rose has shown, markets can also promote democracy and liberty. Property in the hands of citizens rather than the state diffuses political power, holds government in check, and expands the range of individual autonomy. It may even foster civility because merchants will be solicitous of their customers. Rose notes that, when McDonald's opened its first restaurant in Moscow, it trained employees to smile at customers, which amazed the patrons.

If water markets are to flourish, there must be a system of quantified water rights that are transferable. Water markets can develop only if current water users have known and fixed rights that they can sell or lease. Without a property right that is quantified and transferable, there will be no voluntary reallocation of water use.

If we could turn the clock back and start over in allocating our water resources, we might not create a prior appropriation system or permissive groundwater rules. But instead of wistfully thinking about "what if," let's acknowledge reality. In the United States, people have legal rights to *use* water. The prior appropriation system has recognized surface water rights for more than 150 years. Entire communities have arisen that depend on reasonable use groundwater rules. In irrigation districts, generations of farmers have relied on contractual rights with the Bureau of Reclamation for water for their crops. And federal law recognizes water rights held by Native American communities. Courts will protect these legal rights, as they should; and states may not unilaterally abrogate these property or contractual interests.

It bothers many people to think of water as a commodity, but water in the United States is both a public good and an economic good. In 1982, the United States Supreme Court decided a case involving a Nebraska law that tried to prevent people from pumping groundwater from beneath their land and exporting it to Colorado. The Court struck down the law as an interference with interstate commerce: Nebraska could not hoard groundwater through rules that discriminate against out-of-state sales because simple protectionism of this kind violates the free market principles in the U.S. Constitution. As shocking as this result may seem, Nebraska has several other options to protect its resources. The state could prohibit the export of water from beneath Nebraska-owned lands. Or it could enact stringent regulations on groundwater use and impose them on all users, including those who would like to export the water. Or Nebraska could protect its environment with regulations evenhandedly imposed on all users. Of course, Nebraska has done none of these things because its farm sector, the dominant economic interest in the state, depends on the state's relaxed groundwater regulations.

Before we get too caught up in the wonders and glories of the market, we need to remember that markets produce winners and losers. Markets have great utility, but I don't worship at the shrine of the free market. At the core of economic theory is the bedrock conviction that markets make everyone better off; when individuals act out of enlightened self-interest, it encourages innovation, promotes specialization, and results in the movement of goods to those persons who can make the best use of them. This belief rests not on immutable principles of science but on faith, as shown by Duncan K. Foley in his 2006 book, *Adam's Fallacy: A Guide to Economic Theology*. Foley, an economist, attacks the idea that it's possible to separate an economic sphere of life, guided by self-interest, from the rest of social life. Adam Smith's fallacy, to Foley, is that selfishness in the service of capitalist market relations helps our fellow human beings. This convenient rationalization allows us to ignore the harsh consequences of a cruel market, ranging from large-scale unemployment to environmental degradation to the destruction of cultures and communities.

For any market to operate, the state must establish and protect property rights and enforce contracts. The choices made by the state about when and how to enforce contractual arrangements involve resource allocation decisions. In other words, there is no such thing as a "free market." Markets involve political decisions made by the state that define property rights and influence the division of labor. The state is responsible for the consequences of these choices. That's why we should insist on state oversight of the process and the outcomes for markets in water rights. I envision a regulated market with state responsibility to protect third parties from potential harm caused by water transfers and to ensure that transfers do not harm the environment.

A regulated market makes eminent sense in the case of water, a resource with cultural, spiritual, religious, environmental, and economic value. Water is a shared resource, widely but unevenly distributed, used and reused, in constant movement through the hydrologic cycle. As a shared resource owned by the state and used by its citizens, water requires stewardship by the state. A state-regulated market makes sense for a public good such as water in order to prevent externalities, an example of what economists call market failure. If a farmer diverts the entire flow of a creek into his canal, the resulting death of the fish downstream is a cost of his action but not a cost paid for by the farmer. Instead, he externalizes this cost, which is paid (or suffered) by his neighbors or society generally. Markets have difficulty internalizing environmental values. The same is true for a farmer who sells water to a distant city and then fires his workers. The lost wages of the farmworkers are a consequence of the farmer's transaction but are not costs absorbed by the farmer, unless we insist on it. And we should. We must also address the concerns of communities harmed by water transfers when people move away, school enrollments decline, and local businesses suffer. The lesson to be learned from the story of the Imperial Irrigation District is that democratic oversight by a popularly elected board of directors protected the community.

Let me respond to three likely objections to water markets: the sellers make windfall profits; markets encourage sprawl; and environmentalists can't compete with the deep pockets of developers. Some people object to the idea that farmers who paid nothing for the right to *use* water under the prior appropriation or reasonable use doctrines can turn around and sell the water for a huge profit. Even worse, tens of thousands of western farmers and ranchers get water through heavily subsidized Bureau of Reclamation irrigation projects. The federal government built the infrastructure and then asked the beneficiaries to repay only a fraction of the costs with zero-interest loans stretched out over fifty years. The government, it is argued, should reap the benefit from its largesse rather than allow farmers and ranchers to realize windfall profits.

Perhaps. But existing water rights holders will not sell their water unless they profit from it. Instead, they'll continue growing alfalfa, cotton, or other low-value products. It's that simple. If we want to encourage low-value agricultural producers to use less water, we must give them an incentive to do so. And that incentive is money. Given a choice between making farmers and ranchers rich from the sale of their water rights and enabling them to continue to use huge quantities of water to grow, say, cotton, I think the choice is easy. The beneficiaries of these trades will usually be cities, whose developers can easily afford to pay for the water rights. Although the farmers paid nothing for the rights in the past, these rights nevertheless have enormous present value. Our focus should be on the current value of the water, not on what the farmers originally paid for it. For that matter, many of the original recipients of the windfall are long gone. Subsequent purchasers paid higher prices for the farms because of the value of the subsidy.

That leads to the second objection: sprawl. Transferring water from farms to cities and suburbs may provide liquid nourishment to feed growth. I'm neither a fan of mindless sprawl nor a shill for developers. As a resident of Tucson, I sometimes feel that the dominant bird call is the peeping sound that heavy construction equipment makes in reverse gear. Losing an acre per hour of pristine Sonoran Desert habitat to support population growth is a tragedy that we must halt. Around the country, developers are replacing the open spaces of pastureland, cotton fields, and rice paddies with red-tile-roof subdivisions, 7-Elevens, Walgreens, and the occasional Wal-Mart or Home Depot. I'd much rather see a green alfalfa field dotted with snowy egrets than row after row of tract housing.

What role does water play in encouraging this growth? In situations in which limited water supplies might constrain development, water marketing could facilitate growth. But lack of water seldom halts development. Despite limited supplies, growth perversely marches on. Our existing water law tolerates both wasteful irrigation *and* mind-numbing sprawl. We have not chosen one over the other: we supply water to serve both. We should require those proposing new development to purchase and retire existing water rights in order to break the relentless cycle of overuse and move toward sustainable water use. If developers must purchase water rights, at least some of them would shelve their plans.

The third objection to water markets is that environmental interests cannot compete in a market-driven system. It is feared that developers will always outbid environmental organizations. This need not be the case if those of us who care about the environment pony up to protect it. If we want farmers to leave more water in the river for fish or habitat, we must give them an incentive. In fact, environmentalists have often prevailed in market settings. In the American West, Forest Guardians, the Western Watersheds Project, and other conservation groups have outbid ranchers for USDA Forest Service grazing leases. The Sonoran Institute and other environmental organizations have spearheaded efforts to reform the sale of state trust lands to protect sensitive habitat. We've seen the Oregon Water Trust, the Montana Water Trust, Trout Unlimited, and other environmental groups use leases and sales of water rights to restore and maintain stream flows.

As a final example, consider The Nature Conservancy, whose main strategy for decades has been to acquire parcels of land and accompanying water rights that its scientific analyses have determined are worthy of protection. No other environmental organization in the world owns as much land as The Nature Conservancy, whose members continue to support its market-based approach to environmental stewardship.

Environmental interests have sometimes succeeded through the political process in securing government funds to protect instream flows, anadromous fish, and riparian habitat. For example, in the 1992 Central Valley Project Improvement Act, Congress dedicated funds to acquire water rights to 800,000 acre-feet of water to protect the environment. Finally, the reallocation of water from agricultural to urban use is unlikely to harm the environment directly because farmers are already diverting or pumping the water to irrigate their fields. A simple change in use poses little risk of further environmental damage. If any particular transfer is environmentally objectionable, environmentalists can raise objections during the government review process.

The environmental community should embrace water transfers as a critical element in a comprehensive strategy for protecting the environment. If we don't require developers to purchase existing water rights, then we're back to business as usual. From an environmental perspective, water marketing lessens the pressure to build new dams, divert additional surface water, and drill more wells. Water marketing avoids these environmentally destructive alternatives.

Resistance to water marketing is visceral in some quarters, an ideological response rooted in opposition to markets, especially for water. I understand the argument, even if I don't agree. Opponents of marketing must offer an alternative for dealing with the immense demand for more water. If we're not going to use markets to reallocate water, there is really only one alternative: the public sector. Such allocations would occur at the direction of elected politicians or at the discretion of bureaucrats. No economist thinks that's an efficient way to make decisions about the allocation of public resources. Indeed, that system already exists in the American West. Water laws routinely favor agricultural interests in states such as Nebraska. Elsewhere, the Bureau of Reclamation for generations has curried favor with important members of Congress and influential agricultural organizations by distributing federal project water to irrigation districts. These cozy relationships prove that water allocated through the political process inevitably goes to the most powerful economic and political interests in the state. Faith in a benign public interest bureaucracy is unwarranted.

Voluntary transfers offer an alternative to a politically driven system of allocating water. Water marketing can provide water for valued new demands and break the relentless cycle of overuse. The lament of the buffalo and the lessons learned by halibut fishers offer guideposts to the future.

CONCLUSION

A Blueprint for Reform

"Anything else you're interested in is not going to happen if you can't breathe the air and drink the water. Don't sit this one out. Do something. You are by accident of fate alive at an absolutely critical moment in the history of our planet."

—Carl Sagan

"I THINK WE CAN SURVIVE THE YEAR," said Barbara Emley, "but I'm afraid it will go on." Emley, age sixty-four, has fished commercially for salmon off the California coast since 1985. She was referring to the total ban on commercial and sport salmon fishing imposed by the Pacific Fishery Management Council in March 2008, after only a small number of salmon had returned to spawn. It's the first time that the council has mandated such sweeping restrictions, which extend from northern Oregon to the Mexican border. The ban will idle approximately 1,000 commercial vessels and countless sportfishing boats, force wholesalers and retailers to reassess their futures, challenge restaurateurs to find alternative supplies, and harm small coastal communities that depend on fishing and tourism. "This is going to be devastating to the economy," notes Paul Johnson, president of Monterey Fish Market, a seafood wholesaler at San Francisco's Pier 33. "It's put everyone on edge. A lot of small-boat fishermen are going to go out of business."

The health of coastal fisheries in the United States—whether salmon off the Pacific coast, shrimp off the Gulf coast, or oysters off the Florida coast or in Chesapeake Bay—depends on an adequate supply of fresh water in rivers. But upstream diversions, dams, and water pollution have caused fish stocks to plummet. The plight of Pacific salmon illustrates the profound connection between water flows and the economic health of coastal communities. Using water for one purpose may preclude using it for another.

We can't grapple with these trade-offs unless we redefine the role of water in our society. We have to understand that choices made about land use, population and immigration, farm policy, and energy policy profoundly affect our water supply. To begin with, we need land use reform that couples zoning decisions made by local governments with the available water supply. In order to overcome the understandable reluctance of ordinary citizens to conserve water when boards of supervisors and town councils indiscriminately rubber-stamp new development proposals, elected officials must insist that developers demonstrate a supply of water adequate to serve the new demand.

The United States' policy on population growth badly needs an overhaul, from federal tax laws that encourage large families to immigration laws that are incoherent. The social policy issues related to population growth and immigration are incredibly complex, yet, without reform, we will add 120 million people by midcentury—a staggering number—and the increased demand for water will have immense consequences for our culture, economy, and environment.

The supply of water must become a critical element in our energy policy. The ethanol story offers a sobering lesson about the unintended consequences of creating incentives for ethanol production, which will require shifting billions of gallons of water from food production to fuel production. Energy policy makers have never taken into account the consequences of their decisions on our water supply. That folly must end.

We must begin to treat water as a valuable, exhaustible public resource. Water is a basic commodity for which there is no substitute, regardless of price. In other areas, we can shift from coal to oil, or oats to wheat, or hydroelectric power to power generated by fossil fuels. But there is no substitute for water. This redefinition of the role of water should involve pricing water in a way that reflects its importance, its uniqueness, and its finite limits.

Throughout this book, I have advocated a new way of thinking about water that involves numerous changes in how we value it. These reforms include the following:

- Encouraging creative conservation
- Using price signals
- Creating market incentives
- Reexamining how we dispose of human waste
- Requiring developers to pay their own way
- Reconsidering the location of wastewater plants
- Separating storm water from sewer water
- Creating infrastructure with dual pipes to supply potable and reclaimed water
- Abandoning business as usual (more dams, diversions, and wells)
- Recognizing the link between water and energy
- Appreciating the critical role played by water in the economy
- Removing barriers to water transfers while providing for government oversight of them
- Creating incentives for homeowners and others to harvest water
- Stimulating alternative waste disposal technologies
- Metering water use
- Securing water for the environment.

These reforms offer several options that move us toward sustainable water use, free up water for critical new demands, protect the environment by reallocating water from low-value agricultural use, ensure a bright but more water-efficient future for farmers, demand that government protect third parties from harm as a result of water transfers, and begin to recognize water as a valuable and exhaustible resource.

Individuals, businesses, and all levels of government in the United States have their roles to play, but the most profound change must come from an invigorated federal role in water management. Congress has generally deferred to the states in matters of water allocation and policy. In large measure, the states have defaulted on their obligations (or opportunities) to craft sensible water policy. The disarray, conflict, litigation, scarcities, and economic dislocations I have described in this book testify to the appalling job performed at the state and local levels. Consider the city of Sacramento, California, which in 2003 rejected a proposal to install water meters. Or the Carefree Water Company in Arizona, which in 2007 repealed a water rate surcharge intended to require high-volume water users (more than 50,000 gallons a month) to bear a larger share of the utility's costs. It returned to a flat-rate system that will discourage conservation.

We need leadership at the federal level to move the country toward recognizing the value of water and to clarify our national policy concerning water. And in 2007 we started to see members of Congress champing at the bit to get involved. For example, the House Oversight and Government Reform Committee held hearings on local complaints about bottled water companies tapping into critical groundwater supplies. The U.S. Senate Committee on Environment and Public Works considered establishing a presidential commission to develop a national plan to evaluate water resources for the future. And five members of the U.S. House of Representatives established a bipartisan Congressional Water Caucus to promote dialogue about national water policy and provide a forum for discussion of water issues.

Congress undoubtedly has constitutional authority to enact sweeping legislation regulating water use in the United States. As early as 1982, the United States Supreme Court held that Congress has "affirmative power to implement its own policies concerning groundwater regulation. Groundwater overdraft is a national problem and Congress has the power to deal with it on that scale." Existing legislation such as the Clean Water Act and the Endangered Species Act reflects the scale and scope of congressional power to regulate local water use. When low flows cause factories to close their doors, employers to lay off workers, state regulators to deny permits to nuclear and coal-fired power plants, endangered species to die, federal regulators to close commercial fishing seasons, and cargo ships to lighten loads, we have an urgent national crisis that justifies federal action.

What exactly should these energized senators and representatives do? First, they should find out what's going on. We are pathetically ignorant about how much water we have and who is using how much and for what purpose. Many states allow citizens to drill wells without even notifying the state, grant exemptions from permitting requirements for commercial wells that pump less than 100,000 gallons a day, and do not require permitted wells to have meters that measure water use. If we don't have accurate data about water use, it's tough to craft sensible regulatory policies. Congress should support the mission of the U.S. Geological Survey by providing adequate funding to gather data about the nation's water supply and uses; funding cuts have forced the USGS to cut back on the scope of its National Water-Use Information Program and surface water gauge measurements. The USGS's current budget to collect groundwater data is $10 million, a paltry sum given that groundwater constitutes one-quarter of the nation's water supply and that roughly half of us drink groundwater every day. Legislation introduced by New Mexico senator Jeff Bingaman in 2008 would fund such data collection efforts.

The intolerable situation of wells pumping unknown quantities must end. At a minimum, Congress should insist that, as a condition to receiving water from federal projects, every recipient must meter its water use or submit annual water use reports. Congress should also instruct the Environmental Protection Agency to give high priority to assessing the dangers posed by emerging contaminants to humans and the environment, and to setting standards for the most dangerous ones, including perchlorate.

Second, Congress should get the federal "water house" in order. No effective mechanism currently ensures that federal agencies work together toward sensible water quality and water supply outcomes. Interagency activity on water is an appallingly low priority.

As a national priority, Congress should embark on a major initiative that focuses on the disposal of human waste and the management of storm water and sewer water. All experts agree that there is pent-up demand for overhauling the nation's wastewater and storm water infrastructure. Given current awareness that climate change may profoundly alter our water supply, the time is right to reexamine how we dispose of human waste in the United States. In addition, Congress should amend the Clean Water Act to encourage municipal reuse of effluent and better management of storm water.

Congress can provide incentives for states to meet federal goals, rather than simply impose regulation and cause conflicts with state statutes. For example, Congress can encourage states to address water problems by making federal funding for water infrastructure and environmental enhancement projects contingent on meeting certain federal expectations related to the reforms I have identified.

Sensible federal farm policy would gradually phase out price supports, commodity purchases, and other subsidies whose roots trace to the New Deal but whose current existence hampers incentives to use water wisely. Without subsidies, water will approach its true market value. Congress ought to use both sticks and carrots to reform agricultural water use. First, the carrots. Congress should encourage farmers to switch to more efficient watering systems by allowing them to sell or lease conserved water. Funding for such improvements should combine grants, interest-free loans, and tax credits.

Now the stick: a tax on water. A phased-in, graduated tax on all surface water and groundwater use will not only encourage conservation but also generate funds to underwrite expensive infrastructure repairs, conservation programs, and environmental restoration. Congress should give water users, especially farmers, ample lead time to adjust their water practices before the tax kicks in.

Despite the imperative for federal action, the states needn't wait for nudging from Congress to get their houses in order. Data collection, including a mandate to install meters, is an obvious first step. States should fulfill their stewardship obligation to recognize water as a valuable commodity and a public resource by ending the madness of allowing new diversions and withdrawals from an already stressed public supply. To free up water for valuable new uses, states must first recognize and quantify existing rights in water. The legal rules in many states purport to recognize water rights but, on closer inspection, prove to be vague or ambiguous, leaving users of water the unattractive option of suing to enforce rights, uncertain whether they'll prevail. It does no one any good to have a veneer of rights that can be easily stripped away by new users.

A lesson from Arizona may help other states. Until 1980, Arizona adhered to the reasonable use doctrine to govern groundwater withdrawals. In practice, this rule set no limits on the drilling of new wells, and as a result wells proliferated and the water table plummeted. Reform came in the form of the Arizona Groundwater Management Act, which did three things essential to sensible water management. First, the act prohibited new wells in already stressed areas. Second, it quantified the rights of existing pumpers. Third, it made those rights transferable.

States can sponsor an array of interesting water conservation programs, from gray water reuse to water harvesting, and can use innovative mechanisms, such as tax credits, to advance them. Beyond that, states should encourage current users to reduce their consumption by giving them incentives to transfer to third parties the right to the conserved water. States should remove impediments to water transfers, such as Colorado's rule that a farmer who makes conservation improvements doesn't get to keep (or sell) the freed-up water. That makes no sense.

As the market in water rights matures, the states should play a critical role in regulating the process. Water remains a public resource even as states allow citizens and businesses to use it. States must monitor water transfers to prevent harm to third parties or the environment. A California statute could serve as the template for other states. California's State Water Resources Control Board oversees most water transfers, and the board can approve a transfer only if it finds that the transfer will not "unreasonably affect" the environment or the economy of the area that is giving up the water.

States should require local governments to link land use decisions with available water supply. In turn, local governments need to insist that developers bring new water to the table. In this way, local government can fulfill its stewardship obligation to protect water resources, the economy, and the environment.

Local government has control over water use through ordinances that regulate plumbing fixtures, swimming pools and fountains, and outdoor landscaping. Although it seems painfully obvious, the local governments in Southern California should recognize that the region is a desert. Asking residents to landscape their yards with low-water-use plants would be a good start. For inspiration, all local governments in the country might turn to San Antonio, Tucson, or Albuquerque, all of which have innovative water management programs designed to educate citizens about conservation and, where necessary, to forbid certain wasteful practices.

A critical role for local governments, particularly those in fast-growing regions, involves the choice of where to build wastewater treatment plants. The conventional approach is one gigantic plant sited at the lowest point in the basin, or even on an island in the ocean, in order to get rid of the effluent as quickly and as easily as possible. But, to make reclaimed water available for reuse, it makes sense to decentralize treatment plants so that local municipalities will have operational flexibility in using this supply.

In the end, democracy gives us the government we deserve. If we as citizens don't demand that government live up to its stewardship obligation, we have no one to blame but ourselves. And don't expect elected officials to get out in front on raising water rates! An active, involved citizenry is essential for ensuring the future of our nation's water supply.

Even those of us who are not politically engaged can take many steps toward solving our water crisis. A good place to start is by taking a fresh look at your monthly water bill. How much water are you using? Your current usage can serve as your baseline for future use. A terrific list of water-saving tips is available online from the Water Conservation Alliance of Southern Arizona (Water CASA), at www.watercasa.org/genwatersavingtips.php.

To save the largest amount of water, examine your outdoor water use. Not only does that category constitute the largest percentage of use for most homeowners, it's also 100 percent consumptive. What you plant and how you water will substantially affect your total water use. Gray water reuse and water-harvesting systems offer two innovative ways to reduce outdoor use.

For indoor water conservation, the options are virtually endless but notably include low-flow faucets, shutoff valves, composting toilets, and waterless urinals. Simple steps, such as turning off the water while brushing your teeth or cleaning dishes, can save a considerable amount of water. One household device that uses a surprising amount of water is the garbage disposal. Water CASA explains that even with a low-flow kitchen aerator, running the disposal for two minutes uses five gallons; if you do that once a day for thirty days, you're using 150 gallons a month to get rid of food scraps that you could easily throw in the garbage. If you're really hardy, try a navy shower: after getting wet, turn the water off, soap up, and then turn the water back on to rinse off. Gloria Schleicher, the eighty-four-year-old woman in Maine, whom you may remember because her well was contaminated with salt water, takes navy showers. She's my mother-in-law and I'm proud of her, even though I prefer to take regular showers.

Finally, energy-conscious citizens have another reason to conserve. Think of it this way: reducing your carbon footprint will also reduce your water footprint. That Prius you're considering buying or the compact fluorescent bulbs you're thinking of installing not only will save gas, oil, and electricity but also will save water.

WATER PRESENTS A surprising riddle. We can neither make nor destroy it, so our supply is fixed; yet it's exhaustible because, as a shared resource used repeatedly, some uses preclude future reuse. Water policy suffers from a profound discontinuity between science and law. Although we know from hydrology that surface water and groundwater are interconnected and that excessive pumping will diminish or deplete surface flows, the legal rules often allow unrestricted diversions and pumping. The result epitomizes the tragedy of the commons: limitless access to a finite resource.

Water is not only a critical environmental issue; it also has immense public health implications and profound consequences for the nation's economy. As the ethanol boom illustrates, a close connection between energy production and water consumption makes it imperative that water figure prominently in the country's energy policy.

Alas, our political system largely ignores the problem of water, except when a squeaky wheel needs a dab of grease. Even then, attention wanders once the squeak is gone. It will be a major political challenge to find the will (and money) to modernize our water and wastewater infrastructure. We humans have a limitless capacity to ignore reality. Many proposed "solutions" to our water crisis involve quick fixes that place an almost religious faith in uncertain technologies and favorable long-term weather conditions.

Before the crisis turns into catastrophe, we must embark on a new course, with firm resolve and steady confidence. Our water use poignantly reflects our culture and values. At the moment, it's not a pretty reflection. More than thirty-five states are squabbling with their sisters over water, and multiple federal agencies have conflicting mandates. We have become accustomed as a nation to wasting a precious commodity and to allowing unrestrained growth and consumption to drive our policy decisions without regard to the availability of water. There is no silver bullet, no black-and-white answer to managing a natural resource as uniquely valuable and as essential to human life as clean water. We have abundant tools available, however, to alter these old patterns of use by acknowledging the true value of water in our lives and our economy. All we need now is the will and the commitment to confront the water crisis.

EPILOGUE

The Salton Sea

HEADING SOUTH ON California Highway 86 from Interstate 10, one comes upon a beautiful, shimmering, dark blue body of water, oddly set amid the sparsest vegetation imaginable. California's largest lake, the Salton Sea is thirty-five miles long and nine to fifteen miles wide, with a maximum depth of fifty-one feet. As one gets closer, it's obvious that this body of water is unlike others. The shoreline has neither trees nor shrubs nor dwellings. It's as if both plants and humans have decided they wanted no part of the place. Yet it's a haven for birds migrating along the Pacific flyway, including hundreds of thousands of ducks and tens of thousands of geese. The sea supports incredibly diverse waterfowl and shorebirds, including American white pelicans, eared grebes, double-crested cormorants, black skimmers, long-billed dowitchers, and cattle egrets—almost 400 species of birds, including two endangered species, the brown pelican and Yuma clapper rail. Because 90 percent of California's wetlands have been lost over time to development, the sea provides critical avian habitat. "The Salton Sea is probably the most important inland aquatic habitat for birds in the Southwest," says Stuart Hurlbert, director of San Diego State University's Center for Inland Waters. It's a birder's paradise. It's also hot as blazes; on the day I visited, in August 2004, the temperature hit 113 degrees.

After World War II, developers saw the sea's recreational potential and fueled a building boom. Sportfishing was excellent. The sea supported an abundance of fish species, including tilapia, Gulf croaker, orangemouth corvina, and sargo. In the 1950s, developers imagined the Salton Sea as a potential winter resort. They built hotels, golf courses, and marinas, planned subdivisions, staged boat races that drew thousands of spectators, and sponsored golf tournaments with top PGA players. The fledgling communities of Salton City and Bombay Beach sprang up. In the 1960s, after a state park and recreation area opened, the number of visitors occasionally exceeded the number who visited Yosemite National Park.

By the late 1970s, the boom had gone bust. Visitors stopped coming. Hotels closed, and proposed subdivisions were never built. During the wet years of the 1970s, the Salton Sea began to rise and flooded out shoreline homes. Occasional die-offs of tilapia created a stink that put off tourists and property owners alike. Bombay Beach now consists of salt-encrusted mobile homes and decaying structures.

The other ambitious development on the west side of the sea was grandly named Salton City, rather than the town or village of Salton. A master-planned subdivision contains mile after mile of streets. Although few of the streets are paved, each has a street sign, and all signs bear nautical or waterfront names, such as Sea Vista, Marina Drive, Sea View, Dolphin, Sea Isle, Shore View, and Yacht Club Drive. Alas, most streets have no homes, nor did I see any yachts. There is a public boat ramp, but there was not a boat in sight. All that exists are widely scattered dwellings, each a modest bungalow or a mobile home; "upscale" is a double-wide. The local restaurant was closed for the summer. In 2000, the population of Salton City stood at 978.

The Salton Sea faces myriad challenges. Situated 227 feet below sea level, the sea has a major salinity problem. The cause is quite simple: the sun evaporates a considerable amount of water off the vast surface area of the sea, leaving behind the salts. As a consequence, salinity has marched steadily upward, from 33,000 parts per million (roughly the same as the Pacific Ocean) in the 1950s to more than 44,000 in 2000. Because the sea has no outlet, whatever flows into the sea stays there. And what flows in isn't so nice. The New River begins in Mexico and flows north through Mexicali, where more than a million residents contribute raw sewage to the river, then into the Salton Sea. Selenium, a nonmetallic element that can impair avian reproduction and deform or kill young birds, naturally occurs in Imperial Valley soil. Irrigation leaches selenium, fertilizers, and pesticides into the Imperial Irrigation District's drainage ditches, which empty into the Salton Sea.

Fertilizer in the runoff from the IID and domestic wastewater from Mexico contain large amounts of organic material that have turned the Salton Sea into a veritable nutrient soup, which supports large blooms of algae and has led to eutrophic conditions in the sea. In essence, the rate of oxygen consumption by the mass of algae and the decomposing organic material in the fertilizer and wastewater exceeds the rate of replenishment from photosynthesis and transfer from the atmosphere. The result has been anoxia—a severe lack of dissolved oxygen—which has caused a massive die-off of aquatic organisms and, in turn, enormous fish and bird kills. Although biologists maintain that the die-offs were part of a natural cycle, tourists had no interest in vacationing on beaches fouled by the smell of rotting fish. Critics dubbed the sea the "Salton Sewer."

Avian disease spread quickly in the 1990s. In 1996, tens of thousands of birds died, including 20 percent of the white pelicans in the West and more than 1,000 brown pelicans, the largest die-off of an endangered species ever reported. The pelicans had contracted botulism by eating sick tilapia floating on the sea's surface. In response, federal officials with the Salton Sea National Wildlife Refuge developed new management practices that improved the health of fish and reduced the number of bird deaths, but the damage to the sea's reputation was immense. Even some environmentalists began to question whether it was worth saving.

AS A TEENAGER, Sonny Bono water-skied on the Salton Sea. He later enjoyed considerable fame as the less glamorous half of Sonny and Cher. After his entertainment career headed south, he was elected mayor of Palm Springs and later, as a conservative Republican, to the U.S. House of Representatives. His district included the northern part of the Salton Sea. Protecting the sea became his pet cause during his years in Congress. After he died in a downhill skiing accident in 1998, Congress authorized the appropriation of funds to restore recreation and wildlife habitat at the sea and gave the wildlife refuge a new name—the Sonny Bono Salton Sea National Wildlife Refuge. His widow, Mary Bono, successfully ran for his congressional seat and has continued his fight to save the sea.

Advocates for saving the sea developed several Rube Goldberg–type plans with price tags upward of a billion dollars. Some solutions were far-fetched indeed. Sonny Bono favored building a dike to separate more than forty miles of the sea, where contaminated water and increased salinity would accumulate. The rest of the sea supposedly would slowly purify itself if relieved of these incursions of salty water. Another idea was to pump in clean water and pump out contaminated water, the latter of which conveniently would be sent to a giant salt sink in Mexico, thereby exporting our waste. No one asked the Mexican government what it thought of this idea.

Not everyone is enamored of saving the sea. The *Washington Post* called it "a fly-infested sinkhole consumed by a downright Biblical plague of botulism-carrying maggots, crazed fish, and drowning birds." Even some environmentalists oppose efforts to save the sea because it's an unnatural water body, sustained only by the agricultural return flows from the IID. They would prefer to protect the Colorado River Delta in Mexico, which has the benefit of being a natural water body. Mark Briggs of the Sonoran Institute in Tucson, Arizona, notes that the Salton Sea is "not self-sustaining and it was never meant to be. It's just a big evaporative basin that everything drains into. We're putting a lot of money into restoring an evaporative basin. It just doesn't make sense."

In 2003, the California legislature proclaimed that the state should "undertake the restoration of the Salton Sea ecosystem and the permanent protection of the wildlife dependent upon that ecosystem." How noble. Too bad the legislature didn't supply any money to make it happen. In 2007, after three years of studies and hearings, the California Department of Water Resources submitted its Salton Sea restoration plan to the state legislature. The plan calls

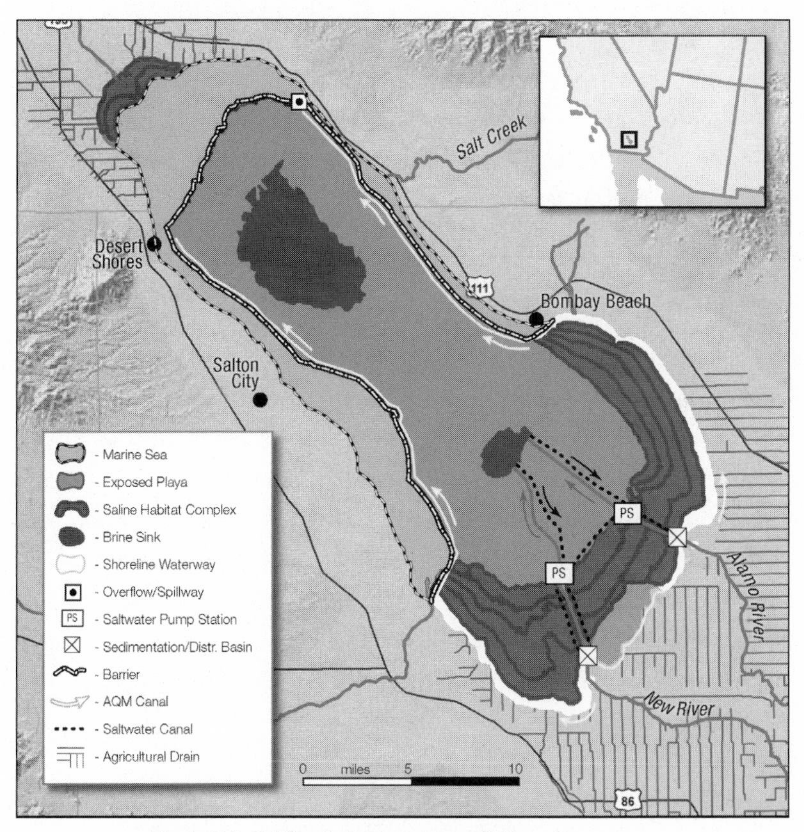

Figure E-1. California Department of Water Resources, Salton Sea Restoration Plan.

for using 52 miles of gravel and stone barriers and 158 miles of earthen berms to reduce the size of the sea by half—to 176 square miles. To reduce air pollution from blowing dust, a complex system of drip irrigation tubes would water plants on the 106,000 acres of lake bed that would lie exposed after the sea was reduced in size. As recently as 2003, the high end of estimates for the sea's restoration had been $3 billion. This new plan was estimated at an eye-popping $8.9 billion—$2 billion more than the total in the draft plan of only two months earlier. Even the plan's supporters concede that the price tag will make it a hard sell. With California's 2009–2010 state budget deficit projected to exceed $40 billion, Salton Sea supporters are competing for funds with education, health care, and other critical needs.

They are also competing with the urban water demands of San Diego. The water transfer agreement between the Imperial Irrigation District and San Diego exacerbates the Salton Sea's environmental challenges. As IID farmers fallow land, line ditches, and shift from flood irrigation to sprinklers, these conservation measures will reduce the excessive application of water and, in turn, produce less runoff. But if less tailwater flows into the Salton Sea, the sea will shrink further in size, increase further in salinity, and become virtually unsalvageable—a paradoxical and unintended consequence of efforts to conserve water. The reduced size of the sea would expose contaminants lying beneath the surface of the lake to the sun and wind, creating an air pollution problem. The adverse effects on pelicans and rails might violate both the federal and state endangered species acts. Lower inflows as a result of the water transfer will reduce the depth of the sea, which could affect shoreline habitat and expose rookeries located on islands. As the Salton Sea becomes more saline, the fish are less able to reproduce. The fishery will eventually collapse, and migratory birds will lose their food source.

As of January 2008, the birds supported by the sea were doing well. The wildlife refuge had constructed ponds that cater to wading birds and marshlands that provide excellent habitat for the Yuma clapper rail. The sea had experienced no recent outbreaks of avian botulism or cholera, but it had suffered a loss of all fish species except tilapia. And even the hardy tilapia suffer episodic die-offs during summer. Christian Schoneman of the wildlife refuge explained that salinity levels are steadily increasing and are now up to 51,000 parts per million—about 50 percent saltier than the Pacific Ocean. This rise has puzzled scientists, who worry about the ability of tilapia to survive. They believe tilapia can reproduce in water as saline as 60,000 parts per million, but if added stresses, such as high heat or lack of oxygen, compound the situation, no one knows what will happen. Should tilapia go the way of the croaker, corvina, and sargo, the birds will have lost their final food supply.

After meeting with Schoneman, I walked around the wildlife refuge. With binoculars, I spotted flocks of geese in the distance as they rose and circled in the air, preparing to head off to feed on other fields. They headed my way, coming ever closer until they began to pass directly over my head, so close that I could easily see their wing markings. Wave after wave of honking Ross's and snow geese whirred overhead in ever-shifting chevron formations. I watched spellbound for twenty minutes or more as upward of 20,000 geese passed over me.

Acknowledgments

I AM ENORMOUSLY grateful for the encouragement and support I received from friends, colleagues, and even perfect strangers. I want first to acknowledge Ben Smith for his remarkable generosity in spending untold hours reading and critiquing the entire manuscript. This book benefited immensely from his superb editing skills. He is a dear friend.

I owe special thanks to Peter Barnes, who hosted me as a writer-in-residence at the Mesa Refuge at Point Reyes, California, in the summer of 2004 and offered very helpful comments on the manuscript. To Terry Anderson, I am grateful for the time I spent at his Property and Environment Research Center (PERC) in Bozeman, Montana, as a Julian Simon Fellow in the summer of 2006. This center of free market environmentalism welcomed an advocate of a regulated market into their midst with open arms and critical comments. I benefited from PERC's stimulating atmosphere.

Several people generously read the entire manuscript. John Leshy's insightful suggestions helped me avoid a few blunders. Dennis Duggan's probing comments, and his literate and practical perspective, made this a better book. Katherine Adam's perceptive comments challenged me to sharpen my message.

For comments on drafts of individual chapters, I am grateful to George Abar, Peggy Barlett, Joan Card, Jim Cherry, Raymond Dougan, Paul Fisher, Dan Francis, Dustin Garrick, P. J. Hill, Kathy Jacobs, Gary Libecap, Val Little, Marc Miller, Barak Orbach, Chris Robinson, Carol Rose, Neal Shapiro, Kirsten Smolensky, and Laura Ziemer.

I received financial support from the National Science Foundation, Grant No. 0317375, which allowed me, with two economists, Gary Libecap and Alan Ker, to study water transfers in the American West. I also received a grant from Sharon Medgal, director of the University of Arizona Water Resources Research Center, to investigate irrigation districts in Arizona.

As I traveled around the country speaking about my book *Water Follies* and trying out new ideas, audiences' questions forced me to examine my assumptions and refine my thinking. Those exchanges planted the seeds of this book. I presented an early version of these ideas at the University of Texas. I also benefited from presentations at other universities, including Stanford University; the University of California, Berkeley, and the University of California, Davis; the University of Michigan; Lewis & Clark College; Boston College; Indiana University; the University of Illinois; the University of Georgia; the University of Toledo; the University of Nebraska; the University of Colorado; the University of Minnesota; the University of Nevada, Las Vegas; the University of New Mexico; and the University of Toronto.

At the University of Arizona (the world's best university for water research, I say shamelessly but accurately), presentations in a number of departments nurtured my ideas. At the James E. Rogers College of Law, a remarkably collegial environment with a talented and generous faculty, a work-in-progress seminar provided terrific suggestions for improving the book, as did an Environmental Breakfast Club colloquium. To Toni Massaro, a remarkable dean, colleague, scholar, and friend, I say thank you for your enthusiastic support of me and my work. I cherish our friendship.

For helpful references or assistance on specific issues or technical questions, I am indebted to Bill Alley, Mark Anderson, Colin Apse, Bruce Aylward, Erik Bakken, Eric Betterton, Paul Blowers, Diane Boyer, Janette Brimmer, Mark Brusseau, Pamela Bush, Jim Cherry, Rick Cobb, Jim Davenport, Barbara Dean, Kayti Didricksen, James O. Doyle, Len Drago, John Droz, Jodylee Duek, David Dykes, Darrol Eichner, Mark Ellsworth, Jay Gandolfi, Paul M. Heisig, Bob

Hirsch, Oscar Jackson, Dan Jensen, Michael Jess, Stephanie Johnson, Reland Kane, Don Kangas, Chris Kopach, Lea-Rachael Kosnik, Cathy Larson, Donna Marie Lisenby, Reina Maier, Bob Masters, Kelly McCaffrey, Michael McNulty, Junior Mukiibi, Nicholas Naylor-Leyland, David Nelson, Chuck Palmer, Richard Parachini, Amy Parker, Jennifer Pitt, Todd Rasmussen, Robert Renken, Jodi Rogstad, Tony Rossman, Steve Rypka, Neal Shapiro, Tina Shields, Jeff Silvertooth, Jack Simes, Chad Smith, Dana Smith, John Wells, Bill Wiley, Garry Woodard, and Connie Woodhouse.

Numerous friends, acquaintances, and strangers generously granted me interviews. I have noted these interviews in the Sources section. I owe special thanks to Bob Johnson, Pat Mulroy, and Lester Snow for carving out time from hellaciously busy schedules to meet with me. For stimulating lunchtime conversations that influenced my thinking, I am grateful to Dean Lueck and Sylvia Tesh.

Research assistants at the College of Law—Chris Brooks, David Harlow, and Cyndi Tuell—helped in the initial stages. For the past two years, Jean Morrill has borne the brunt of my innumerable requests to find something, usually after I tried and failed, ensuring that Jean's tasks would not be routine. Maureen Garmon of the College of Law library was remarkably resourceful in finding obscure reports and documents. I am also grateful to Meredith Marder, who took time from her own law review commitments to give me the benefit of her astute editing skills.

Jillian Aja and Jean West, College of Law class of 2009, took my extensive files and turned them into the Sources section. This job required patience, organizational skills, a sense of humor, and tact. We were aided by the word processing skills of Carol Ward, who brought a cheery disposition to a quite technical assignment.

I have been blessed with spectacular faculty support services, and I particularly want to note my thanks to Sandy Davis and Lucy Hoffman. But the heavy lifting has fallen on the shoulders of Kristie Gallardo and Barbara Lopez, who performed splendidly. Kristie transcribed many interviews, some recorded in such noisy venues that they were all but inaudible. Yet she worked her magic and produced faithful transcriptions. Kristie also typed multiple drafts of the early chapters. When Barbara Lopez took over, much work remained. Barbara typed, cut, clipped, retyped, and rearranged countless drafts. Even when I'd ask her to restore something I had deleted in the previous draft, Barbara did so cheerfully, accurately, and professionally. Her unflappable assistance makes my life easier. I cannot overstate Barbara's contribution to this book.

My literary agent, Deborah Grosvenor, offered her usual sage advice. I am also grateful to Bruce Babbitt for his support. I also want to thank Kyle Carpenter for creating the maps and helping with the other figures.

At Island Press, Todd Baldwin again served as my editor. I am grateful to Todd for taking on this project, even though his responsibilities at the press now include serving as vice president and associate publisher. Todd's vision profoundly shaped this book from conception to publication. I am extremely fortunate to have had his wise counsel. The book also benefited from Emily Davis' editorial input, from Milan Dozic's creativity in designing the cover, from Sharis Simonian's talents as production editor, and from Maureen Gately's interior design. Pat Harris, who served as the principal copyeditor, has my heartfelt thanks for her extraordinarily thorough review and comments. I am also grateful to Diane Ersepke for her help with copyediting, and for her numerous suggestions.

Karen Adam, my best friend and wife, is the most generous and caring person I have ever known. Her love and support made this book possible. I dedicate it to her.

List of Figures

Sources

A note on quotations. For sake of the flow of the text, I have changed some quotations. In a small number of instances, I corrected misspellings without using the patronizing "[sic]," changed a letter from lowercase to uppercase without noting the change in brackets, or eliminated words without using ellipsis marks (. . .). In no instance have I altered the meaning of any quotation.

General Background

Books

Ball, Philip. *Life's Matrix: A Biography of Water*. New York: Farrar, Straus and Giroux, 1999.

Barlow, Maude. *Blue Covenant: The Global Water Crisis and the Coming Battle for the Right to Water*. New York: New Press, 2007.

Barlow, Maude, and Tony Clarke. *Blue Gold: The Fight to Stop the Corporate Theft of the World's Water*. New York: New Press, 2002.

Childs, Craig. *The Secret Knowledge of Water: Discovering the Essence of the American Desert*. Seattle, WA: Sasquatch Books, 2000.

De Villiers, Marq. *Water: The Fate of Our Most Precious Resource*. New York: Houghton Mifflin, 2000.

Ehrlich, Gretel. *The Future of Ice: A Journey into Cold*. New York: Pantheon Books, 2004.

Fradkin, Philip L. *A River No More: The Colorado River and the West*. New York: Knopf, 1981.

Gleick, Peter H. *The World's Water, 2006–2007: The Biennial Report on Freshwater Resources*. Washington, DC: Island Press, 2006.

Glennon, Robert. *Water Follies: Groundwater Pumping and the Fate of America's Fresh Waters*. Washington, DC: Island Press, 2002.

Gosnell, Mariana. *Ice: The Nature, the History, and the Uses of an Astonishing Substance*. New York: Knopf, 2005.

Hundley, Norris, Jr. *The Great Thirst: Californians and Water, 1770s–1990s*. Berkeley: University of California Press, 1992.

Leopold, Luna B. *Water, Rivers, and Creeks*. Sausalito, CA: University Science Books, 1997.

Libecap, Gary D. *Owens Valley Revisited: A Reassessment of the West's First Great Water Transfer*. Stanford, CA: Stanford University Press, 2007.

McDonald, Bernadette, and Douglas Jehl. *Whose Water Is It? The Unquenchable Thirst of a Water-Hungry World*. Washington, DC: National Geographic Society, 2003.

Outwater, Alice. *Water: A Natural History*. New York: Basic Books, 1996.

Pearce, Fred. *Keepers of the Spring: Reclaiming Our Water in an Age of Globalization*. Washington, DC: Island Press, 2004.

Pearce, Fred. *When the Rivers Run Dry: Water, the Defining Crisis of the Twenty-First Century*. Boston: Beacon Press, 2006.

Pielou, E. C. *Fresh Water*. Chicago: University of Chicago Press, 1998.

Postel, Sandra. *Last Oasis: Facing Water Scarcity*. New York: Norton, 1992.

Postel, Sandra. *Pillar of Sand: Can the Irrigation Miracle Last?* New York: Norton, 1999.

Reisner, Marc. *Cadillac Desert: The American West and Its Disappearing Water*. New York: Viking Penguin, 1986.

Reisner, Marc, and Sarah Bates. *Overtapped Oasis: Reform or Revolution for Western Water*. Washington, DC: Island Press, 1990.

Royte, Elizabeth. *Bottlemania: How Water Went on Sale and Why We Bought It*. New York: Bloomsbury, 2008.

Stegner, Wallace. *Beyond the Hundredth Meridian: John Wesley Powell and the Second Opening of the West*. New York: Penguin, 1954.

Ward, Diane Raines. *Water Wars: Drought, Flood, Folly, and the Politics of Thirst*. New York: Riverhead Books, 2002.

Wiley, Peter, and Robert Gottlieb. *Empires in the Sun: The Rise of the New American West*. New York: Putnam, 1982.

Wood, Chris. *Dry Spring: The Coming Water Crisis of North America*. Vancouver, BC, Canada: Raincoast Books, 2008.

Worster, Donald. *Rivers of Empire: Water, Aridity, and the Growth of the American West*. New York: Pantheon Books, 1985.

Introduction

Books

Diamond, Jared. *Collapse: How Societies Choose to Fail or Succeed*. New York: Penguin, 2005.

Sax, Joseph L., Barton H. Thompson Jr., John D. Leshy, and Robert H. Abrams. Legal Control of Water Resources: Cases and Materials. 4th ed. St. Paul, MN: Thomson/West, 2006.

Articles and Reports

"Acquisitions by Southern Nevada Water Authority." *Water Strategist*, February 2007, p. 21.

"Acquisitions by Southern Nevada Water Authority." *Water Strategist*, July–August 2006, p. 7.

Audi, Tamara. "Vegas Plans a New Push to Attract More People." *Wall Street Journal*, 7 January 2008.

Baird, Joe. "Water Deal on Nevada Agenda: Spring Valley; The Battle Continues over Groundwater Along the Utah Border." *Salt Lake City Tribune*, 10 September 2006.

Baker, Dean. "Guest Opinion: Nevada's Ground Water Tug of War; State of Nevada's Interest Best Served by Protecting Environment." *Las Vegas Sun*, 10 September 2006.

Berkes, Howard. "Las Vegas Water Battle: 'Crops vs. Craps.'" Morning Edition, NPR, 12 June 2007. www.npr.org/templates/story/story.php?storyId=10953190.

Bowles, Jennifer, and Dan Lee. "Southern California Development Issues: Water Troubles Put Inland Developments in Limbo." *Riverside Press Enterprise*, 23 January 2008.

Boxall, Bettina. "Las Vegas Looks North to Slake Its Thirst," *Los Angeles Times*, 7 March 2007.

Brean, Henry. "Mulroy Says Pipeline Plan Nothing Like Owens Valley." *Ely Times*, 15 September 2006.

Brean, Henry. "Nevada Water Issues: New Colorado River Accord Makes Rural Groundwater More Valuable to SNWA." *Ely Times*, 19 January 2007.

Brean, Henry. "White Pine County: Crossing Waterline; 'We Don't Have a Choice,' Mulroy Bluntly Warns About Valley's Needs." *Las Vegas Review-Journal*, 18 August 2006.

Bredehoeft, John. "Review of the Evidence Provided by the Southern Nevada Water Authority of Proposed Development on the Hydrogeology of Spring Valley, White Pine and Lincoln County, Nevada." Report prepared for the Western Environmental Law Center, 30 June 2006.

"CityCenter News." www.vegastodayandtomorrow.com/citycenter.htm (accessed 13 May 2007).

Conaughton, Gig. "Leaders Call for Water Conservation Plan." *North County Times*, 29 September 2006.

Davenport, Paul. "Colorado River Agreement: Ariz. Lawmakers OK River Pact." Associated Press, 12 February 2007.

"Development Issues: Riverside: Conservation over Development; Drought and a Burgeoning Population May Force the Rest of Us to Choose: Lawns or Jobs?" *Los Angeles Times*, 17 January 2008.

Drape, Joe. "Setting Restaurant Records by Selling the Sizzle." *New York Times*, 22 July 2007.

Evans, K. J. "Howard Hughes: Sky Was No Limit." *Las Vegas Review-Journal*, 1999. www.1st100.com/part3/hughes.html.

Gable, Eryn. "Water: Nevada Looks at 'Crazy' Ideas to Meet Growing Demand." *Greenwire*, 12 April 2007.

Glennon, Robert, and Michael J. Pearce. "Transferring Mainstream Colorado River Water Rights: The Arizona Experience." *Arizona Law Review* 49, no. 2 (Summer 2007): 251.

GLS Research. "Las Vegas Visitor Profile." Annual report, calendar year 2006. Prepared for Las Vegas Convention and Visitors Authority.

Harlow, David. "Las Vegas' Gamble on White Pine County's Groundwater." University of Arizona, Rogers College of Law, unpublished student paper 2006.

Hopkins, A. D. "Steve Wynn: The Winner." *Las Vegas Review-Journal*, 1999. www.1st100.com/part3/wynn.html.

Hutchins-Cabibi, Taryn, Bart Miller, and Anita Schwartz. "Water in the Urban Southwest: An Updated Analysis of Water Use in Albuquerque, Las Vegas Valley, and Tucson." Boulder, CO: Western Resource Advocates, 2006.

Hutchinson, Alex. "Watering Holes: The Debate over a Plan to Pump Water out of the Nevada Desert Could Be the Next Battle in the War over the West's Most Vital Natural Resource." Popular Mechanics, February 2007. www.popularmechanics.com.

Kaplan, Michael. "Losing Your Shirt, but Not in the Casino." *New York Times*, 26 August 2007.

Koch, Ed. "Utah Turns Spigot Off for Nevada; Lawmakers: State Not Ready to Sign Away

Water Rights." *Las Vegas Sun*, 14 February 2007.

"Las Vegas Looks North to Slake Its Thirst: Vegas' Drinking Problem Is Nevada Ranchers' Headache as Officials Look to Tap Groundwater in Rural Counties to Slake a Thirst for Growth." *Los Angeles Times*, 7 March 2007.

"Legislation Seeks Increased Availability of Colorado River Supplies." *Arizona Water Resource*, February 2007.

Leslie, Jacques. "Running Dry." *Harper's*, July 2000, p. 51.

Marco, Lisa. "Nevada Water Issues: Southern Nevada Secures Water—for Now." *Las Vegas Sun*, 13 December 2006.

McKinnon, Shaun. "Ariz. to Share Water Duties with Nevada." *Arizona Republic*, 24 January 2007.

Moran, Terry, and Katie Hinman. "Water Wars: Quenching Las Vegas' Ferocious Thirst; Proposed Pipeline from Snake Valley to Sin City Meets Local Opposition." *ABC News* (national), 6 April 2007.

Moran, Terry, and Katie Hinman. "Water Wars: Quenching Las Vegas' Thirst." *ABC News*, 5 April 2007. http://abcnews.go.com/Nightline/ story?id=3012250&page=1.

Mulroy, Patricia. "A Fork in the River: Coping with Change on the Colorado." *Water Strategist*, May 2007.

"Nevada State Engineer Issues Spring Valley Ruling—the Largest Interbasin Transfer Request in Nevada's History." *Western Water Law and Policy Reporter*, August–September 2007, p. 278.

"New Reservoir NV/CA/AZ." *Water Report*, 15 January 2007, p. 17.

Rake, Launce. "'A Matter of Survival'; LV's Growth Will Stop in 2013 Without White Pine Water, Mulroy Says." *Las Vegas Sun*, 16 August 2006.

Rake, Launce. "Battle for Rural Water Heads to Capital." *Las Vegas Sun*, 10 September 2006.

Rake, Launce. "Feds Failed to Inform Tribes Before Pulling Water Protest." *Las Vegas Sun*, 3 October 2006.

Rake, Launce. "Fish Has Ally in Water War." *Las Vegas Sun*, 22 December 2006.

Riley, Brendon. "Thirsty Las Vegas Seeks OK to Pump Water from Rural Nevada Valley." *North County Times*, 12 September 2006.

Ritter, John. "Vegas Reaching for Rural Water: Pipeline Plan Creates Fear That City Will Pump Away Agricultural Livelihoods." *USA Today*, 19 October 2006.

Rivlin, Gary. "In Las Vegas, Too Many Hotels Are Never Enough." *New York Times*, 24 April 2007.

Robbins, Ted. "Stakes High for Las Vegas Water Czar." *Morning Edition*, NPR, 13 June 2007. www.npr.org/templates/story/story.php?storyId =10939792.

Roessler, Christina. "Las Vegas and the Groundwater Development Project: Where Does It Start? Where Will It End?" Reno, NV: Progressive Leadership Alliance of Nevada (PLAN), 2006.

Rothman, Hal. "Western Water: A Legend of Overallocation." *New West*, 31 January 2006.

Santos, Fernanda. "The Great Lakes Shrink, and Cargo Carriers Worry." *New York Times*, 22 October 2007.

Schoch, Deborah. "Water Laws May Throttle Growth: Statutes Force a District Near Lake Perris to Assess Whether Supply Will Be Available for Huge Warehouse Project, Which Is Now on Hold." *Los Angeles Times*, 14 January 2008.

Simon, Darren. "Reservoir Would Ease Drought." *Imperial Valley Press*, 18 April 2007.

"SNWA Approves One-Year Increase in Landscape Water Rebate Program." *U.S. Water News*, January 2007.

"Southern California Ag Issues: Water Cuts Slicing into Avocado Groves: Some Farmers Are 'Stumping' Hundreds of Healthy Trees to Keep Others Irrigated." *Los Angeles Times*, 26 February 2008.

Steinhauer, Jennifer. "Water-Starved California Slows Development." *New York Times*, 7 June 2008.

Tanner, Adam. "Las Vegas Growth Depends on Dwindling Water Supply." Reuters, 21 August 2007. www.reuters.com/article/domesticNews/ idUSN1335882320070821.

Tetreault, Steve. "Senator Cites Utah Farmers' Concerns." *Las Vegas Review-Journal*, 3 August 2007.

"Top Las Vegas Water Official Blasts Utah Request for Study." *Salt Lake Tribune*, 4 August 2007.

"Utah 'Optimistic' About Water Agreement with Nevada." *U.S. Water News Online*, January 2007. www.uswaternews.com/archives/arcpolicy/ 7utahopti1.html.

"Water Ruling Brings Hope: State Engineer's Decision Could Lead to a Second Major Source of Water for the Valley." Editorial. *Las Vegas Sun*, 19 April 2007.

Williams, Ted. "Sin City Goes Dry." *Audubon*, March–April 2007, p. 92.

Yancey, Kitty Bean. "Sun City Uncovered." *USA Today*, 11 May 2007.

Letters, Memoranda, Miscellaneous

CityCenter Las Vegas. Premiere 2007, published by MGM Grand. www.citycenter.com.

Colorado River Water Consultants. "Study of Long-Term Augmentation Options for the Water Supply of the Colorado River System." March 2008. http://waterplan.state.wy.us/BAG/green/Fact_Sheet.pdf.

Donnelly, David A. "Strategies for Augmenting the River: Evolving Issues and Solutions." CLE International presentation, Las Vegas, NV, 10 May 2007.

Howe, John, producer. Desert Wars: Water and the West. Salt Lake City: KUED 7, University of Utah, 2006. Documentary film featuring interviews with Jamie Cruz, director of energy and environmental services, MGM Mirage; Hal Rothman, history professor, University of Nevada, Las Vegas; Patricia Mulroy, general manager, Southern Nevada Water Authority; Cecil Garland, rancher, Snake Valley, Callao, UT; Dean Baker, rancher, Baker, NV; Don Duff, aquatic ecologist and president, Great Basin chapter of Trout Unlimited; Jay Banta, superintendent, Fish Springs National Wildlife Refuge; and others. www.kued.org/productions/desertwars/.

Las Vegas Convention and Visitors Authority. "Vegas FAQ: Top 25 Frequently Asked Questions." April 2007. www.lvcva.com/getfile/2007Top25Questions.pdf?fileID=106.

R & R Partners. "Puppy" and "The Neighbor" television advertisements. Las Vegas, NV: Southern Nevada Water Authority, 2007.

Interviews

Absher, Gordon. Vice president of public relations, MGM Mirage. 8 February 2008.

Minwegen, Tom. Deputy general manager, Las Vegas Valley Water District. 23 May 2007.

Mulroy, Pat. General manager, Southern Nevada Water Authority, Las Vegas. 10 May 2007.

Chapter 1. Atlanta's Prayer for Water

Articles and Reports

Associated Press. "Georgia: Governor Orders Cuts in Water Use." *New York Times*, 24 October 2007.

Blackwood, Harris. "Corps Cuts Lake Lanier Water Releases." *Gainesville Times*, 16 January 2007.

Bluestein, Greg. "Just a Little off the Bottom: Georgians Eyeing Scarce Water Want a Slice of Tennessee." Associated Press, 9 February 2008.

Bluestein, Greg. "Some Businesses Thrive in Georgia Drought." Associated Press, 25 October 2007.

Carroll, C. Ronald. "Mussels Can't Be Sacrificed for Poor Water Planning." *Atlanta Journal-Constitution*, 18 October 2007.

Chapman, Dan. "Water War Between the States Is for Real." *Atlanta Journal-Constitution*, 8 February 2008.

Dewan, Shaila. "Georgia Claims a Sliver of the Tennessee River." *New York Times*, 22 February 2008.

Dewan, Shaila, and Brenda Goodman. "New to Being Dry, the South Struggles to Adapt." *New York Times*, 23 October 2007.

Eberly, Tim, and Stacy Shelton. "Feds to Intervene in States' Water Feud: Georgia, Alabama, Florida Have Yet to Reach Agreement." *Atlanta Journal-Constitution*, 2 March 2008.

Evans, Ben. "Resolution to Water Feud Bleak." Associated Press, 27 February 2008.

Fanning, Julia L. "Water Use in Georgia, 2000; and Trends, 1950–2000." In *Proceedings of the 2003 Georgia Water Resources Conference, Held April 23–24, 2003, at the University of Georgia*, edited by Kathryn J. Hatcher. Athens: University of Georgia, Institute of Ecology, 2003. http://ga.water.usgs.gov/pubs/other/gwrc2003/pdf/Fanning-GWRC 2003.pdf (accessed 10 November 2007).

Fischer, Karlene. "TMDLs: The Basics." *ENRLS Update*, May 2007. State Bar of Arizona, Environmental and Natural Resources Law Section.

Fitzhugh, Thomas W., and Brian D. Richter. "Quenching Urban Thirst: Growing Cities and Their Impacts on Freshwater Ecosystems." *BioScience* 54, no. 8 August 2004. p. 751.

Garber, Kent. "High and Dry in the South: Mother Nature Started the Drought; How Man Made It Worse." *U.S. News and World Report*, 19 October 2007.

Georgia Department of Natural Resources, Environmental Protection Division. "Outdoor Water Use Schedules and Restrictions." www.gaepd.org/Documents/water_use_schedules.html (accessed 10 November 2007).

Goodman, Brenda. "Amid Drought, a Georgian Consumes a Niagara." *New York Times*, 15 November 2007.

Goodman, Brenda. "Drought-Stricken South Facing Tough Choices." *New York Times*, 16 October 2007.

Goodman, Brenda. "Georgia Loses Federal Case in a Dispute About Water." *New York Times*, 6 February 2008.

"Governor Undermines His Own Conservation Message." *Macon Editorial*, 8 February 2008.

Jarvie, Jenny. "Atlanta Water Use Is Called Shortsighted." *Los Angeles Times*, 4 November 2007.

Jarvie, Jenny. "Georgia Governor Leads Prayer for Rain." *Los Angeles Times*, 14 November 2007.

Melton, Harold D., and R. Todd Silliman. "Reflections on the A.C.F. and A.C.T. Basin Compacts." In *Proceedings of the 2005 Georgia Water Resources Conference, Held April 25–27, 2005, at the University of Georgia*, edited by Kathryn J. Hatcher. Athens: University of Georgia, Institute of Ecology, 2005.

Meyer, Christopher H. "Interstate Water Allocation." *Water Report*, no. 42 (15 August 2007).

Opdyke, Tom. "Cobb Top Water Guzzler Says He'll Try to Cut Back: Single-Family Home Used 440,000 Gallons in One Month." *Atlanta Journal-Constitution*, 13 November 2007.

Paumgarten, Nick. "There and Back Again." *New Yorker*, 16 April 2007, p. 58.

Pavey, Rob. "Water Must Be Weighed in Plan: Reactors' Usage Remains Concern." *Augusta Chronicle*, 9 February 2008.

Pearson, Michael. "Georgia's Water Crisis: Weather Permitting; Free-Flowing Debate: Should Government Restrict Rezoning During Drought? Will Rights Erode?" *Atlanta Journal-Constitution*, 31 October 2007.

"Rains Bring Relief, but Crisis Hasn't Ended." *New York Times*, 21 May 2008.

Shelton, Stacy. "Lanier's Eight-Inch Rise Useless." *Atlanta Journal-Constitution*, 27 January 2008.

Shirek, Jon. "Perdue Drops Lawsuit Against Corps." 7 November 2007. www.11alive.com/news/local/drought/story.aspx?storyid=106008&catid=219.

Starnes, Joe Samuel. "Not on Every Map, but a Desirable Location Anyway." *New York Times*, 25 May 2007.

U.S. Geological Survey. "Coastal Ground Water at Risk: Saltwater Contamination at Brunswick, Ga. and Hilton Head Island, S.C." n.d. USGS Water-Resources Investigations Report 01-4107. http://ga2.er. usgs.gov/coastal/coastalreport.cfm (accessed 7 June 2007).

U.S. Geological Survey. "Droughts in Georgia." USGS Open-File Report 00-380. October 2000.

WaterWebster. "Florida, Alabama, Georgia Water Sharing." www.waterwebster.com/FloridaAlabamaGeorgia.htm (accessed 7 November 2007).

"Water Use in Georgia: U.S.G.S. Trends in Water Use, 1950–2000." http://ga.water.usgs.gov/ (accessed 10 November 2007).

Watson, Jaye, and Jerry Carnes. "Atlanta Water Situation Called Dire." 11 October 2007. www.11Alive.com/news/article_news.aspx?storyid=104561.

Weiss, Mitch. "Southern Drought Could Shut Down Nuclear Plants." *USA Today*, 24 January 2008.

Whoriskey, Peter. "Three States Compete for Water from Shrinking Lake Lanier." *Washington Post*, 27 October 2007.

Letters, Memoranda, Miscellaneous

Perdue, Sonny. "Governor Perdue Declares Emergency; Requests Presidential Intervention." Press release, 20 October 2007.

The World Almanac. "Annual Rainfall for U.S. States." 1988.

Interviews

Baize, David. Assistant chief, Bureau of Water, State of South Carolina. 1 June 2007.

Bethea, Sally. Executive director, Upper Chattahoochee Riverkeeper, Atlanta, GA. 15 November 2007.

Cyr, Richard. General manager, Hilton Head Public Service District. 9 November 2007.

Hoard, Kathy. Commission of Athens-Clarke County, GA. 12 November 2007.

Jackson, Oscar. Physician. 8 November 2007.

Kennedy, Jim. State geologist, Georgia. 1 June 2007.

Pope, David. Director, Georgia/Alabama office, Southern Environmental Law Center. 24 March 2008.

Chapter 2. Wealth and the Culture of Water Consumption

Books

Engelman, Robert. *More: Population, Nature, and What Women Want*. Washington, DC: Island Press, 2008.

Royte, Elizabeth. *Bottlemania: How Water Went on Sale and Why We Bought It*. New York: Bloomsbury, 2008.

Articles and Reports

Acoba, Elena. "Fresh Trends for Showers: Functionality and Spalike Feel Really Clean Up." *Arizona Daily Star*, 8 April 2007.

Agrell, Siri. "The New Wave of H_2O." *Globe Life*, 18 July 2007.

Ammon, Christina. "Watershed Moment: A Former California Timber Town Becomes Ground Zero in the Battle over Bottled Water." *High Country News*, 25 June 2007, p. 4.

"An Outpouring of Statistics." *Daytona Beach News-Journal*, 10 May 2007.

"Bath Splash." Phoenix Home & Garden, February 2006, p. 52.

Benson, Reed D. "'Adequate Progress,' or Rivers Left Behind? Developments in Colorado and Wyoming Instream Flow Laws Since 2000." *Environmental Law* (Lewis & Clark Law School) 36, no. 4 (2006): 1283.

Benson, Reed D. "Rivers to Live By: Can Western Water Law Help Communities Embrace Their Streams?" *Journal of Land, Resources, and Environmental Law* 27, no. 1 (2007): 1.

Blevins, Jason. "Battle Looms over Kayak Parks." *Denver Post*, 2 June 2003.

Boerger, Paul. "McCloud Nestlé Water Bottling Plant Project." *Mt. Shasta News*, 5 July 2007.

"Bottled Water Makes a Splash with Flavors, Colors and Nutrients." *Daily Breeze*, 11 April 2007.

Brain, Marshall. "How Snow Making Works." www.jewishworldreview.com/0207/HowStuffWorks.php3 (accessed 21 June 2007).
Burros, Marian. "Fighting the Tide: A Few Restaurants Tilt to Tap Water." *New York Times*, 30 May 2007.
Chapman, Dan. "Ethanol Fever Could Produce Hangover." *Atlanta Journal-Constitution*, 14 January 2007.
Cioletti, Jeff. "CSD Report: Soft Drink Volume Drops for a Second Straight Year; State of the Industry Report '07." *Beverage World* 126, no. 4 (15 April 2007): S3.
Cohen, Nancy. "For Poland Spring Bottles, There Is an Afterlife." *Morning Edition*, NPR, 11 June 2007. www.npr.org/templates/story/story.php?storyId=10874230.
"Colorado Court Ruling Upholds Recreational Water Rights." *Water Strategist*, June 2003, p. 13.
"Colorado Water Court, Water Division 4 Awards Controversial 'Recreational In-Channel Diversion' for Whitewater Park on Gunnison River." *Western Water Law*, February 2004, p. 108.
"Consumers Prefer Bottled Water." *Food & Drink Weekly* 13, no. 16 (23 April 2007).
Courtney, Kevin. "Crystal Geyser Project: Crystal Geyser Loses Bid to Pump Mineral Water from Napa." *Napa Valley Register*, 13 September 2007.
Cronin, Mike. "Adventure Resort for Riverview?" *Arizona Republic*, 12 October 2006.
Deutsch, Claudia H. "A Spotlight on the Green Side of Bottled Water." *New York Times*, 3 November 2007.
Deutsch, Claudia H. "For Fiji Water, a Big List of Green Goals." *New York Times*, 7 November 2007.
Deutsch, Claudia H. "Two Growing Markets That Start at Your Tap." *New York Times*, 10 November 2007.
Dobbs, David. "Pass the Cocktail Sauce!" *Audubon*, January–February 2007, p. 18.
Enkoji, M. A. "American River Recreation: Water Policy Favors Rafters; Some American River Dams Must Boost Flow for Recreational Users." *Sacramento Bee*, 3 May 2007.
Fahrenthold, David A. "Bottlers, States, and the Public Slug It Out in Water War." *Washington Post*, 12 June 2006.
Howard, Brian. "Despite the Hype, Bottled Water Is Neither Cleaner nor Greener than Tap Water." *Environmental Magazine*, 9 December 2003.
Hutson, Susan S., et al. "Estimated Use of Water in the United States in 2000." USGS Circular 1268. Denver, CO: U.S. Department of the Interior, U.S. Geological Survey, 2004.
"In Praise of Tap Water." Editorial. *New York Times*, 1 August 2007.
"In-Channel Recreational Water Rights." *Water Report*, 15 February 2007.
Just-Drinks.com Editorial Team. "US: Bottled Water Sales Continue to Rise." *Just-Drinks*, 10 April 2007.
Lazarus, David. "How Water Bottlers Tap into All Sorts of Sources." *San Francisco Chronicle*, 19 January 2007.
Lazarus, David. "L.A. Business Tries to Make Fiji Water a Star." *San Francisco Chronicle*, 21 January 2007.
Lazarus, David. "Spin the (Water) Bottle: With $11 Billion in U.S. Sales, the Beverage's Marketers Have Become Clear Winners." *San Francisco Chronicle*, 17 January 2007.

"Letters to Business." *San Francisco Chronicle*, 21 January 2007.

Lipsher, Steve. "Snowmaking a Savior for Ski Areas' Late Start." *Denver Post*, 22 November 2007.

Lyttle, Bethany. "Living Here: Houses Near Water Skiing; Making Some Waves." *New York Times*, 18 May 2007.

Marsh, Bill. "A Battle Between the Bottle and the Faucet." *New York Times*, 15 July 2007.

Martin, Glen. "McCloud, Siskiyou County: Bottled Water War Heats Up Election." *San Francisco Chronicle*, 5 November 2006.

Massad, Jason. "Waveyard Water Use on Par with Golf Course." *East Valley Tribune*, 28 April 2007.

Mittelstaedt, Martin. "The Religious War on Bottled Water." *Environment Reporter*, 23 September 2006.

Mohl, Bruce. "Water Wars: In the Final Analysis, the Price Is All That Separates Them." *Boston Globe*, 18 September 2005.

Mooallem, Jon. "The Unintended Consequences of Hyperhydration." *New York Times Magazine*, 27 May 2007, p. 30.

Moskin, Julia. "Must Be Something in the Water." *New York Times*, 15 February 2006.

Nelson, Gary. "Waveyard Vows Low Water Use at Adventure Park." *Arizona Republic*, 27 April 2007.

Roberts, Sam. "Arizona Displaces Nevada as Fastest-Growing State." *New York Times*, 22 December 2006.

Ropp, Thomas. "Water Bottler Feeling Pressure." *Arizona Republic*, 7 October 2006.

Ropp, Thomas. "Water Bottler Told to Stop or Face Lawsuit." *Arizona Republic*, 7 October 2006.

Ross, Kimberly. "Water Bottling Plant: Nestlé Bottler Seeks More Public Input." *Redding Record Searchlight*, 27 August 2007.

"San Francisco Works to Curb Bottled Water Waste." *Online NewsHour*, PBS, 21 August 2007. www.pbs.org/newshour/bb/environment/july-dec07/bottle_08-21.html.

Scott, Jeffry. "Stone Mountain Cancels Snow Attraction." *Atlanta Journal-Constitution*, 11 October 2007.

Specter, Michael. "The Last Drop: Confronting the Possibility of a Global Catastrophe." *New Yorker*, 23 October 2006.

Spielman, Fran. "City to Pour On Pitch for Tap Water: It's Stainless Steel Containers Vs. Bottles." *Chicago Sun Times*, 16 November 2007.

Stone Mountain Park. "Stone Mountain Park Announces Opening of Coca-Cola Snow Mountain, America's Family Snow Park." Press release, 24 January 2007. http://stonemountainpark.com/press-room/archived_releases.aspx?id=171.

"Tap Water Goes Chic." *Modesto Bee*, 29 March 2007.

Vega, Cecilia M. "Restaurants Urged to Eschew Bottles in Favor of Tap Water." *San Francisco Chronicle*, 21 March 2008.

"Water: Canadian Churches Boycott Bottles for Moral Reasons." *Greenwire*, 26 September 2006. www.eenews.net/trial/.

Western States Water Council. "Addressing Water Needs and Strategies for a Sustainable Future." *Western States Water*, no. 1704 (12 January 2007).

Letters, Memoranda, Miscellaneous

Specter, Michael, and Amy Davidson. "Not a Drop to Drink." Q&A. *New Yorker Online*, 23 October 2006. http://www.newyorker.com/archive/2006/10/23/061023on_onlineonly021

Interviews

Lasgin, Jane. Nestlé Waters North America. April 2004.
Marsh, Lila. Attorney, Dallas. 7 August 2007.
Rustem, Bill. Public Sector Consultants. 11 September 2007.

Chapter 3. Our Thirst for Energy

Books

Bryce, Robert. *Gusher of Lies: The Dangerous Delusions of "Energy Independence."* New York: PublicAffairs, 2008.
Flannery, Tim. *The Weather Makers: How Man Is Changing the Climate and What It Means for Life on Earth*. New York: Atlantic Monthly Press, 2005.
Gore, Al. *An Inconvenient Truth: The Planetary Emergency of Global Warming and What We Can Do About It*. Emmaus, PA: Rodale Press, 2006.
Houghton, John. *Global Warming: The Complete Briefing*. Cambridge, England: Cambridge University Press, 1994.

Articles and Reports

Andrews, Edmund L., and Felicity Barringer. "Bush Seeks Vast, Mandatory Increase in Alternative Fuels and Greater Vehicle Efficiency." *New York Times*, 24 January 2007.
Barker, Brent. "Running Dry at the Power Plant." *EPRI Journal*, Summer 2007, p. 26.
Barringer, Felicity. "Fish Vs. California Water Supply." *New York Times*, 11 July 2008.
Barrionuevo, Alexei. "A Bet on Ethanol, with a Convert at the Helm." *New York Times*, 8 October 2006.
Begley, Sharon. "The Truth About Denial." *Newsweek*, 13 August 2007, p. 18.
Birger, Jon. "Fifty Most Powerful Women: The Outsider." *Fortune*, 16 October 2006.
Birger, Jon. "The Great Corn Gold Rush." *Fortune*, 16 April 2007, p. 75.
Boffey, Philip M. "Talking Points: The Evidence for Global Warming." *New York Times*, 4 July 2006.
Broder, John M. "House, 314–100, Passes Broad Energy Bill; Bush to Sign It Today." *New York Times*, 19 December 2007.
Cassman, Kenneth, Vernon Eidman, and Eugene Simpson. "Convergence of Agriculture and Energy: Implications for Research and Policy." CAST Commentary QTA2006-3. Ames, IA: Council for Agricultural Science and Technology, November 2006.
Chapman, Dan. "Ethanol Fever Could Produce Hangover." *Atlanta Journal-Constitution*, 14 January 2007.
Christensen, Niklas, and Dennis P. Lettenmaier. "A Multimodel Ensemble Approach to Assessment of Climate Change Impacts on the Hydrology and Water Resources of the

Colorado River Basin." Seattle, WA: University of Washington, Department of Civil and Environmental Engineering, n.d.
"Climate Change and Water in California." *Western Water Law*, December 2007, p. 37.
Cohen, Ronnie, Barry Nelson, and Gary Wolff. "Energy Down the Drain: The Hidden Costs of California's Water Supply." Oakland, CA: Natural Resources Defense Council and Pacific Institute, August 2004.
Cohen, Ronnie. "The Water-Energy Nexus." *Southwest Hydrology*, September–October 2007, p. 16.
Cole, Cyndy. "Power Generation Tied to Supplies of Water." *Arizona Daily Sun*, 14 August 2006.
Committee on the Scientific Bases of Colorado River Basin Water Management, National Research Council. "Colorado River Basin Water Management: Evaluating and Adjusting to Hydroclimatic Variability." Report in brief. Washington, DC: National Academies Press, February 2007.
Davey, Monica. "In Farm Belt, Ethanol Plants Hit Resistance." *New York Times*, 13 November 2007.
Davis, Tony, and Dan Sorenson. "Study Predicts Dust-Bowl Southwest." *Arizona Daily Star*, 6 April 2007.
Daykin, Tom, and Karen Herzog. "Water, Water Everywhere, and Every Drop to Drink." *Milwaukee Journal Sentinel*, 4 June 2007.
Dean, Cornelia. "That 'Drought' in Southwest May Be Normal, Report Says." *New York Times*, 22 February 2007.
Dewan, Shaila. "Google Is Reviving Hopes for Ex-Furniture Makers." *New York Times*, 15 March 2007.
Downing, Jim, and Matt Weiser. "Study to Affirm Climate Change Warnings." *Sacramento Bee*, 6 April 2007.
"Drought Could Shut Down Nuclear Power Plants: Southeast Water Shortage a Factor in Huge Cooling Requirements." Associated Press, 23 January 2008.
Egan, Timothy. "Life on the Ethanol-Guzzling Prairie." *New York Times*, 11 February 2007.
Ellis, John. "Judge Limits Delta Pumps: Effort to Save Fish Will Hurt Farmers and Others, the State Says." *Sacramento Bee*, 1 September 2007.
"Ethanol Boom Creates Dilemma over Water Use for Production, Irrigation." Editorial. *U.S. Water News*, www.uswaternews.com, April 2007.
Evatt, Robert. "Google to Unveil Facility's Plans." 29 April 2007. www.perimeterc enter.com/newsroom/.
Fitch, Eric J. "Exploring the Relationship Between Energy and Water." *Water Resources Impact* 9, no. 1 (January 2007), p. 4.
Gelt, Joe. "Interconnected Energy/Water Savings and Uses Worked into Conservation Planning." *Arizona Water Resource* 1, no. 5 (May–June 2006), p. 1.
Gordon, Greg. "Water Supply Can't Meet Thirst for New Industry." *Minneapolis–St. Paul Star Tribune*, 26 December 2005.
Granite Falls Energy. "How Ethanol Is Made." 16 May 2007. www.granitefallsenergy.com/index.cfm?show=10&mid=28.

Gross, Daniel. "With Help, Could Ethanol Be the Next Internet?" *New York Times*, 27 May 2007.

Hall, Noah, et al. "Climate Change and Freshwater Resources." *Natural Resources and Environment*, Winter 2008.

Herszenhorn, David M. "Farmers' Income Rises, as Do Food Prices, but It's Mostly Politics as Usual." *New York Times*, 24 April 2008.

Hightower, Mike, et al. "Emerging Energy Demands on Water Resources." *Water Resources Impact* 9, no. 1 January 2007, p. 8.

Hightower, Mike. "Energy Demands for Water Versus Water Availability." *Southwest Hydrology*, May–June 2007, p. 24.

Hill, Rachelle, and Tamim Younos. "The Water Cooler: The Intertwined Tale of Energy and Water." April 2008. Virginia Water Resources Research Center. www.vwrrc.vt.edu/watercooler_apr08.html.

Hoerling, Martin. "In the Southwest." *Southwest Hydrology*, January–February 2007.

House, Lon W. "Will Water Cause the Next Electricity Crisis?" *Water Resources Impact* 9, no. 1 (January 2007).

Hovey, Art. "Ethanol: An Uncertain Future." *Lincoln Journal Star*, 21 April 2007.

Hovey, Art. "Tamora Area Pops Up on Ethanol Watch List." *Lincoln Journal Star*, 1 November 2006.

Hymel, Mona L. and Roberta F. Mann. "Moonshine to Motorfuel: Tax Incentives for Fuel Ethanol." *Duke Environmental Law and Policy Forum*, Fall 2008.

Intergovernmental Panel on Climate Change. "Climate Change 2007: The Physical Science Basis; Summary for Policymakers." Geneva, Switzerland: IPCC Secretariat, February 2007.

Kanter, James, and Andrew C. Revkin. "Scientists Detail Climate Changes, Poles to Tropics." *New York Times*, 7 April 2007.

Keeney, Dennis. "Water Use by Ethanol Plants: Potential Challenges." Minneapolis, MN: Institute for Agriculture and Trade Policy, October 2006.

Kolbert, Elizabeth. "The Climate of Man—I. Disappearing Islands, Thawing Permafrost, Melting Polar Ice: How the Earth Is Changing." *New Yorker*, 25 April 2005.

Krauss, Clifford. "As a New Fuel Takes Its First Steps, Congress Proposes a Giant Leap." *New York Times*, 18 December 2007.

Krauss, Clifford. "Ethanol's Boom Stalling as Glut Depresses Price." *New York Times*, 30 September 2007.

Lambrecht, Bill. "States Across the Midwest Brace for Water-Guzzling Ethanol Plants." *St. Louis Post-Dispatch*, 15 April 2007.

Leatherman, Courtney. "An Unequivocal Change: Monumental Report Leaves Little Doubt That Humans Have Hand in Climate Change." *Nature Conservancy*, Summer 2007.

Lien, Dennis. "A Thirsty Fuel." *Twin Cities Pioneer Press*, 25 June 2006.

Lovett, Richard A. "Fuel's Gold: Turning Corn into Ethanol May Not Be Worth It." *San Diego Union-Tribune*, 3 August 2005.

Martin, Andrew. "The Price of Growing Fuel, Food, and Energy Compete for Land, Perhaps for Years." *New York Times*, 18 December 2007.

Mehta, Stephanie N. "Behold the Server Farm! Glorious Temple of the Information Age." *Fortune*, 7 August 2006, p. 69.

Mosier, Nathan S., and Klein Ileleji. "How Fuel Ethanol Is Made from Corn." n.d. Purdue Extension, BioEnergy. www.ces.purdue.edu// extmedia/ID/ID-328.pdf (accessed 3 September 2007).

National Research Council, Committee on Water Implications of Biofuels Production in the United States. "Water Implications of Biofuels Production in the United States." Washington, DC: National Academies Press, 2008.

Pasqualetti, Martin J., and Scott Kelley. "The Water Costs of Generating Electricity in Arizona." Tucson, AZ: Arizona Water Institute, 2008.

Paul, Jim. "Experts: Ethanol's Water Demands a Concern." 18 June 2006. *MSNBC*, www.msnbc.msn.com/id/13406453/.

Pimentel, David, et al. "Water Resources: Agricultural and Environmental Issues." *BioScience* 54, no. 10 (October 2004): 909–918.

"Power Hogs: Server Farm Electricity Use Soars." blogs.business2.com/greenwombat, 15 February 2007.

Renewable Fuels Association. "How Ethanol Is Made." n.d. www.ethanolrfa.org/resource/made/

Rosenthal, Elisabeth, and Andrew C. Revkin. "Science Panel Says Global Warming Is 'Unequivocal.'" *New York Times*, 3 February 2007.

Rosenthal, Elisabeth. "Studies Call Biofuels a Greenhouse Threat." *New York Times*, 8 February 2008.

Scott, Jeffry. "Stone Mountain Cancels Snow Attraction." *Atlanta Journal-Constitution*, 11 October 2007.

Seager, Richard. "An Imminent Transition to a More Arid Climate in Southwestern North America." Lamont-Doherty Earth Observatory of Columbia University, www.1deo.columbia.edu/res/div/ocp/drought/ science.shtml (accessed 4 April 2007).

Seager, Richard, et al. "Model Projections of an Imminent Transition to a More Arid Climate in Southwestern North America." Science Express, 5 April 2007. www.science xpress.org.

Sewall, Jacob O., and Lisa Cirbus Sloan. "Disappearing Arctic Sea Ice Reduces Available Water in the American West." *Geophysical Research Letters* 31 (2004): L06209.

Stachura, Sea. "Ethanol Vs. Water: Can Both Win?" 18 September 2006. Minnesota Public Radio, http://minnesota.publicradio.org/display/web/2006/09/07/ethanolnow/.

Stone Mountain Park. "Stone Mountain Park Announces Opening of Coca-Cola Snow Mountain, America's Family Snow Park." 24 January 2007. www.stonemountain park.com/press-room/archived_releases.aspx?id=171.

Streitfeld, David. "Food Prices Rise, Farmers Respond: Fewer Corn Acres, More Soybean." *New York Times*, 1 April 2008.

"Sun's Green Data Center." Green Wombat: Sun's Green Data Center. blogs.business2.com/greenwombat/2007/08/, 21 August 2007.

Syeed, Nafeesa, and David Pitt. "U.S. to Plant Most Corn Since 1944 but Costs Will Stay High, Report Says." *Arizona Daily Star*, 31 March 2007.

Tatko-Peterson, Ann. "Water Bottles: Health Vs. Environment Concerns." 11 September 2007. www.northjersey.com.

Taugher, Mike. "California's Changing Climate: Less Water, More Conflict." *Contra Costa Times*, 22 January 2007.

U.S. Department of Energy. "Energy Demands on Water Resources: Report to Congress on the Interdependency of Energy and Water." Albuquerque, NM: U.S. Department of Energy, Sandia National Laboratories, December 2006.

Wald, Matthew L. "Taming the Guzzlers That Power the World Wide Web." *New York Times*, 7 November 2007.

Wald, Matthew L., and Alexei Barrionuevo. "Chasing a Dream Made of Weeds." *New York Times*, 17 April 2007.

"Water for AG Use: Can't Drink Ethanol." Editorial. *Sacramento Bee*, 29 April 2007.

Williams, Ted. "Under the Influence of Ethanol: America's Corn-Based Ethanol Program Carries High Costs in Fish, Wildlife, and Tax Dollars." *Fly Rod and Reel*, April 2007, p. 18.

Woody, Todd. "Server Farm Goes Solar." 4 October 2007. http://money.cnn.com/2007/10/03/technology/solar_servers.biz2/index.htm.

Zarembo, Alan, and Bettina Boxall. "Permanent Drought Predicted for Southwest: Study Says Global Warming Threatens to Create a Dust Bowl–like Period." *Los Angeles Times*, 6 April 2007.

Letters, Memoranda, Miscellaneous

Bockelmann, Melissa. "Media Guide to the Electric Utility Industry." Edison Electric Institute, Washington, DC, undated.

Brown, Lester. "Distillery Demand for Grain to Fuel Cars Vastly Understated." 4 January 2007. www.earth-policy.org/Updates/2007/ Update63.htm.

Center for Sustainable Environments, Northern Arizona University. "The Interrelations of Water and Energy: Water and Energy Fact Sheet." September 2005. www.environment.nau.edu/water/WaterandEnergy Facts.htm (accessed 23 October 2007).

Institute for Agriculture and Trade Policy. "Greater Water Efficiency Crucial for Ethanol's Future, New Paper: State and Local Government, Industry Need to Make Water Conservation a Priority." Press release, 26 October 2006.

National Energy Foundation. "Generators and Transformers." Resources for Education, www.nef1.org.

"Planning and Constructing an Ethanol Plant in Minnesota: A Guidance Document." State of Minnesota, June 2006.

"Policy Keynote Address by Minnesota Governor Tim Pawlenty to the 10th Annual National Ethanol Conference." 9 February 2005.

Renewable Fuels Association. "Ethanol Production: 2007 Picks Up Where 2006 Left Off." Press release, 2 April 2007.

U.S. Department of Energy, Energy Efficiency and Renewable Energy, Biomass Program. "Ethanol." www1.eere.energy.gov/biomass/ (accessed 24 January 2007).

Udall, Brad. "Global Warming, the Hydrologic Cycle, and Water Management." CLE International, Law of the Colorado River, 2006–2007.

Interviews

Brimmer, Janette. Legal director, Minnesota Center for Environmental Advocacy. 7 June 2007.

McCloud, Michael. Project coordinator, XL Renewables, Chandler, AZ. 7 December 2007.

Chapter 4. Fouling Our Own Nests

Articles and Reports

Aamot, Gregg. "Wary of the Water in the Twin Cities." Yahoo Finance, 8 March 2007. http://biz.yahoo.com.

American Federation of State, County, and Municipal Employees, AFL-CIO, et al. "All Dried Up: How Clean Water Is Threatened by Budget Cuts." 22 September 2004 (www.win-water.org/reports).

American Institute of Chemical Engineers. "MTBE." www.aiche.org/government/prioritystatements/mtbe.htm (accessed 25 April 2005).

Artiola, Janick, and Mónica D. Ramírez. "Chlorinated Solvent Contaminants in Arizona Aquifers: Part I." *SciTransfer* (University of Arizona Superfund Basic Research Program), no. 001 (August 2006).

Artiola, Janick, and Mónica D. Ramírez. "Chlorinated Solvent Contaminants in Arizona Aquifers: Part II." *SciTransfer* (University of Arizona Superfund Basic Research Program), no. 002 (October 2007).

Bagwell, Keith. "EPA Writes Off Toxic Water Site as Uncleanable." *Arizona Daily Star*, 18 January 1998.

Boone, Rebecca. "Cryptosporidium Outbreak Hits the West." *Yahoo News*, 21 September 2007.

Bowman, Chris. "Groundwater Resources: Many Private Foothills Wells Tainted; The State Finds Widespread Contamination in Yuba and El Dorado Counties." *Sacramento Bee*, 23 January 2005.

Boxall, Bettina. "Colorado River Water Deal Is Reached: The Interior Secretary Calls It an 'Agreement to Share Adversity.' A Shortage Could Be Declared as Early as 2010." *Los Angeles Times*, 14 December 2007.

Boxall, Bettina. "Plan for Central Valley Farm Drainage Advances: Congress Must Still Give Final Approval and Could Significantly Alter the Controversial Proposal." *Los Angeles Times*, 10 March 2007.

"Brain-Eating Amoeba Kills Arizona Boy." *KPHO News* 5, 29 September 2007.

Bustillo, Miguel, and Kenneth R. Weiss. "Bush Plan Could Drain Effort to Clean Up Waters: Under His Budget, Funds for an Antipollution Program Would Be About Half the 2004 Level; Other Environmental Projects Also Face Cuts." *Los Angeles Times*, 9 February 2005.

Cavanaugh, Kerry. "Survey Reveals Contaminants in Most Wells." *Los Angeles Daily News*, 29 March 2007.

Center for Health Effects of Environmental Contamination. "Iowa—Nitrate Fact Sheet." n.d. www.cheec.uiowa.edu/nitrate/faq.html.

Clean Colorado River Alliance. "Recommendations to Address Colorado River Water Quality." Clean Colorado River Alliance, January 2006.

"Debate over the Effects of Perchlorate Jeopardizes the Future of Federal Legislation to Clean Up Drinking Water." *Western Water Law*, July 2007.

"Drinking Water Standards/Perchlorate: Perchlorate Detected in Drinking Water in Twenty-six States." *American Water Works Association*, 9 February 2005.

Environmental Working Group. "A National Assessment of Tap Water Quality." Executive summary. 20 December 2005. www.ewg.org/ tapwater/findings.php (accessed 21 December 2005).

Focazio, Michael J., et al. "The Chemical Quality of Self-Supplied Domestic Well Water in the United States." *Ground Water Monitoring and Remediation* 26, no. 3 (Summer 2006): 92–104.

Foley, Ryan J. "Study: Fertilizers Harm Freshwater Lakes." *San Francisco Chronicle*, 13 June 2005.

Force, Lisa, et al. "The Colorado: A River at Risk; Coping with Drought in the Colorado River Basin." Flagstaff, AZ: Grand Canyon Trust, November 2004.

Frohman, Jessica, Ananda Hirsch, and Ed Hopkins. "Communities at Risk: The Sierra Club Superfund Report; How the Bush Administration Is Failing to Protect People's Health at Superfund Sites." San Francisco, CA: Sierra Club, n.d.

Gelt, Joe. "Bioremediation—Water Treatment Tool to Fix Pollution Problems." *Arizona Water Resource* 15, no. 2 (November–December 2006).

Ghosh, Amlan, Muhammed Mukiibi, and Wendell Ela. "TCLP Underestimates Leaching of Arsenic from Solid Residuals Under Landfill Conditions." *Environmental Science and Technology* 38, no. 17 (2004): 4677–4682.

Goad, Ben, and David Danelski. "Perchlorate: Democrats Press for Federal Limits on Rocket-Fuel Chemical in Water Supplies." *Riverside Press Enterprise*, 25 April 2007.

Hasemyer, David. "Water Quality: A Lingering, Toxic Mess; Radioactive Pile Remains Threat near Colorado River." *San Diego Union-Tribune*, 13 February 2005.

Heilprin, John. "National Water Quality Reports: Public Data Shows Hundreds of Chemicals in Tap Water." *Contra Costa Times*, 20 December 2005.

Hernandez, Roberto. "Perchlorate Removal Offers Hope; Inland: A Carbon Filter Being Tested Is Seen as a Breakthrough in Removing the Pollutant from Water." *Riverside Press-Enterprise*, 25 April 2005.

Janofsky, Michael. "Federal Programs: Changes May Be Needed in Superfund, Chief Says." *New York Times*, 5 December 2004.

Juozapavicius, Justin. "Court Order on Ark. Chicken Waste Sought." *USA Today*, 20 February 2008. www.usatoday.com/news/nation/2008-02-20-453807717_x.htm (accessed 22 November 2008).

Kurt, Kelly. "Arkansas Poultry Farmers Say Lawsuit over Chicken Waste Could Doom Industry." Environmental News Network, 21 July 2005. www.enn.com/top_stories/article/16347 (accessed 22 November 2008).

Lambert, Lisa. "Chemical Found in Water May Hurt Thyroid, Study Says." *Yahoo News*, 5 October 2006.

Lee, Jennifer. "Second Thoughts on a Chemical: In Water, How Much Is Too Much?" *New York Times*, 2 March 2004.

Lee, Mike. "Groundwater Protection: Tests on Milk, Lettuce Find Perchlorate Is Widespread." *Sacramento Bee*, 30 November 2004.

Martin, Glen. "Experts Cast Doubt on Toxic Water Plan: Scientists Don't Think Pricey New Drainage Policy Will Succeed." *San Francisco Chronicle*, 18 January 2007.

McKinnon, Shaun. "Long-Term Project Will Reduce Tainted Water in Ariz., Calif." *Arizona Republic*, 4 September 2006.

Meersman, Tom. "Pollution Agency Is Getting Tougher on 3M." *Star Tribune*, 16 April 2007.

Meko, David M., et al. "Medieval Drought in the Upper Colorado River Basin." *Geophysical Research Letters* 34 (24 May 2007).

"Mississippi River Basin and Gulf of Mexico Hypoxia." www.epa.gov/ msbasin/ (accessed 19 June 2007).

Mouawad, Jad. "Oil Giants to Settle Water Suit." *New York Times*, 8 May 2008.

"New California Regulations Set Legal Limit for Perchlorate in Public Water Supplies." *Western Water Law*, November 2007, p. 12.

Novick, Samantha M. "Havasu's Goal: Curb Spring-Break 'Wildness.'" *Arizona Daily Star*, 12 March 2007.

O'Malley, Julia. "UAA Scientists Find Pollutant Increases Masculinity in Fish." *Anchorage Daily News*, 16 October 2006.

"Oklahoma Suit Against Arkansas Poultry Industry for Illinois River Pollution Continues After Supreme Court Refuses to Hear Arkansas' Attempt to Stop the Suit." *Western Water Law Policy Reporter*, May 2006, p. 198.

Ortega, Fred. "Debate over Safe Levels of Perchlorate Heats Up." *Pasadena Star News*, 3 May 2007.

Parker, Suzi. "Finger-Lickin' Bad: How Poultry Producers Are Ravaging the Rural South." *Grist*, 21 February 2006.

Rosner, Hillary. "At the Foot of the Rockies, a Radioactive Wasteland Is Transformed." *New York Times*, 7 June 2005.

"Settlement for Groundwater Contamination: PG&E to Pay $295 Million to Settle Water Contamination Suits." *Orange County Register*, 3 February 2006.

Squillace, Paul J., and Michael J. Moran. "Factors Associated with Sources, Transport, and Fate of Volatile Organic Compounds in Aquifers of the United States and Implications for Ground-Water Management and Assessments." U.S. Geological Survey Scientific Investigations Report 2005-5269. Washington, DC: U.S. Geological Survey, National Water-Quality Assessment Program, National Synthesis on Volatile Organic Compounds, 2006.

Taugher, Mike. "Streams Full of Pesticides from Runoff." *Contra Costa Times*, 5 August 2006.

"Toxin's Ill-Defined Threat Hits L.A. Suburb's Water." *Arizona Daily Star*, 24 April 2005.

U.S. Department of the Interior, U.S. Geological Survey. "USGS Scientists Evaluated a Range of Inorganic and Organic Contaminants in Domestic Wells from Every State and Puerto Rico." http://health.usgs.gov/dw_ contaminants/domestic_wells/ (accessed 26 December 2006).

U.S. Environmental Protection Agency. "Hypoxia in the Northern Gulf of Mexico: An Update by the EPA Science Advisory Board." Washington, DC: U.S. Environmental Protection Agency, December 2007.

U.S. Environmental Protection Agency, Office of Water. "Draft Wadeable Streams Assessment: A Collaborative Survey of the Nation's Streams." Washington, DC: U.S. Environmental Protection Agency, Office of Water, May 2006.

U.S. Geological Survey. "Agricultural Chemicals in Iowa's Ground Water, 1982–1995: What Are the Trends?" n.d. http://ia.water.usgs.gov/ nawqa/factsheets/fs-116-97.html.

U.S. Geological Survey, California Water Science Center Newsroom. "Study Finds Subsidence Continuing in the Coachella Valley." 17 December 2007. http://ca.water.usgs.gov/news/release071217.html (accessed 15 January 2008).

"UA Water Quality Center Builds Coalition of Research Interests." *Arizona Water Resource Supplement*, September–October 2005.

Vavricka, Emile, and Robert Morrison. "Perchlorate: An Emerging Contaminant of Environmental and Toxicological Concern." *Western Water Law*, June 2005, p. 207.

"Wastewater Treatment Needs Along the Lower Colorado River." U.S. Department of the Interior, Bureau of Reclamation, Office of Resources Management, 11 February 2005.

Weyer, Peter. "Nitrate in Drinking Water and Human Health." http://cheec.uiowa.edu/nitrate/health.html.

Wilkison, Donald, et al. "Effects of Wastewater and Combined Sewer Overflows on Water Quality in the Blue River Basin, Kansas City, Missouri and Kansas, July 1998–October 2000." U.S. Geological Survey Water-Resources Investigations Report 02-4107. Washington, DC: U.S. Department of the Interior, U.S. Geological Survey, 2002.

Wolowicz, Daniel. "Drinking Water Contamination: Toxins Found in Drinking Water Below Rocketdyne Test Site." *Simi Valley Acorn*, 26 August 2005.

Letters, Memoranda, Miscellaneous

"1, 4-Dioxane and Your Health. What is 1, 4-Dioxane?" Tucson: University of Arizona College of Pharmacy, Superfund Basic Research Program, and U.S.-Mexico Binational Center, n.d.

California Environmental Protection Agency. "State Water Resources Control Board." Press release, 23 January 2007.

TCBR, LLC. "Coachella Valley Information." www.tcbrllc.com/resources/coachella/ (accessed 22 January 2008).

Thomasberg, Kathleen. "Agricultural Ground Water Use and Related Ground Water Quality Impacts Salinas Valley, California." Water Quality and Conservation Program Manager, Monterey County Water Resources Agency, 2007 Annual Forum, Ground Water Protection Council, San Diego, 18 September 2007.

"What Is Arsenic?" Tucson: University of Arizona College of Pharmacy, University of Arizona College of Agriculture and Life Sciences, U.S.– Mexico Binational Center for Environmental Sciences and Toxicology, Superfund Basic Research Program, and Southwest Environmental Health Sciences Center, April 2007.

"What Is TCE?" Tucson: University of Arizona College of Pharmacy, Superfund Basic Research Program, and U.S.–Mexico Binational Center, May 2006.

Interviews

Nelson, David. Attorney, Phoenix, AZ. 7 June 2007.

Parker, Amy. U.S. Environmental Protection Agency. 19 June 2007.

Smith, Chad. American Rivers. 17 May 2007.

Chapter 5. The Crisis Masked

Books

Annin, Peter. *The Great Lakes Water Wars*. Washington, DC: Island Press, 2006.

August, Jack L., Jr. *Dividing Western Waters: Mark Wilmer and Arizona v. California*. Fort Worth, TX: TCU Press, 2007.

Barlow, Maude, and Tony Clarke. *Blue Gold: The Fight to Stop the Corporate Theft of the World's Water*. New York: New Press, 2002.

Barnett, Cynthia. *Mirage: Florida and the Vanishing Water of the Eastern U.S.* Ann Arbor: University of Michigan Press, 2007.

Galusha, Diane. *Liquid Assets: A History of New York City's Water System*. Fleischmanns, NY: Purple Mountain Press, 2002.

Koeppel, Gerard T. *Water for Gotham: A History*. Princeton, NJ: Princeton University Press, 2000.

Midkiff, Kenneth. *Not a Drop to Drink: America's Water Crisis (and What You Can Do)*. Novato, CA: New World Library, 2007.

Articles and Reports

Aasen, Eric. "Parched Texas Looks to Oklahoma for Water." *Dallas Morning News*, 5 August 2007.

Applebome, Peter. "In a Town Divided, a Wispy Boundary Between Land Use and Religion." *New York Times*, 23 October 2005.

Archibold, Randal C., and Kirk Johnson. "An Arid West No Longer Waits for Rain." *New York Times*, 4 April 2007.

Baker, Max B. "Proposal to Capture Water Has Oklahoma Steaming." *Star Telegram*, 21 January 2007.

Barringer, Felicity. "Growth Stirs a Battle to Draw More Water from the Great Lakes." *New York Times*, 12 August 2005.

Barringer, Felicity. "Lake Mead Could Be Within a Few Years of Going Dry, Study Finds." *New York Times*, 13 February 2008.

Barringer, Felicity. "Water Levels in Three Great Lakes Dip Far Below Normal." *New York Times*, 14 August 2007.

Becerra, Hector, and Sara Lin. "L.A. Water Supply Conditions: L.A. Facing Its Driest Year Ever; Downtown Has Received only 2.4 Inches of Rain Since July 1, Meteorologists Say." *Los Angeles Times*, 6 March 2007.

Bialkowski, Bill. "Not an Ordinary Day at the Cottage." *Georgian Bay*, Summer 2007.

Biette, David N. "Decision Time: Water Diversion Policy in the Great Lakes Basin." Washington, DC: Woodrow Wilson International Center for Scholars, Canada Institute, September 2004.

Breitler, Alex. "Farmers Face Off with Smelt Water Policy." *Stockton Record*, 14 June 2007.
Breitler, Alex. "Survey of Delta Smelt Produces Troubling Results." *Stockton Record*, 31 May 2007.
Bush, Pamela M. "Cooperative Water Resources Management: The Federal-Interstate Compact Approach in the Delaware River Basin." Toledo, OH: University of Toledo College of Law, Legal Institute of the Great Lakes, 12 November 2004.
"Canadian Critics Say Diversion Plan May Kill Great Lakes." *U.S. Water News Online*, October 2004.
"Certificate Authorizing the Cities of Concord and Kannapolis to Transfer Water from the Catawba River and Yadkin River Basins to the Rocky River Basin Under the Provisions of G.S. § 143-215.22I." Environmental Management Commission. Cities of Concord and Kannapolis Interbasin Transfer Certificate, 10 January 2007.
Citizens and Villages United to Save Ramapo. "Preserve Ramapo: The Not So Silent Majority." www.preserveramapo.org (accessed 17 May 2007).
"Colorado Releases Water to Fulfill Pact with Kansas, Nebraska." *Dodge City Globe*, 4 September 2006.
Conley, Mike. "Water Transfer Hearing Draws Crowd of 700, Boos, Sarcasm." *McDowell News*, 11 September 2006.
Dobuzinskis, Alex. "Court Could Devastate Water Supply: Half of Southland's Imported Resources from North at Risk." *Los Angeles Daily News*, 30 August 2007.
Earth Institute at Columbia University. "Water Shortages in Northeast Linked to Human Activity." www.earthinstitute.columbia.edu/new (accessed 9 January 2007).
Egan, Dan. "Great Lakes Circle the Drain." *Milwaukee Journal Sentinel*, 27 February 2005.
Ellis, John. "Judge Limits Delta Pumps: Effort to Save Fish Will Hurt Farmers and Others, the State Says." *Sacramento Bee*, 1 September 2007.
Fimrite, Peter. "Ruling to Protect Delta Smelt May Force Water Rationing in Bay Area." *San Francisco Chronicle*, 1 September 2007.
Flesher, John. "Turmoil over Great Lakes Water Pact." *Washington Post*, 7 April 2007.
Galloway, Gerald, and Ralph Pentland. "Securing the Future of Ground Water Resources in the Great Lakes Basin." Ground Water 43, no. 5 (September–October 2005): 737.
Gardner, Michael. "Governor to Issue Drought Decree, Press for More Conservation." *San Diego Union Tribune*, 4 June 2008.
"Georgian Bay Association Blames Man-Made Hole for Falling Water Levels in Great Lakes." Dredging News Online, www.sandandgravel.com/news/article.asp?v1=10285, 15 August 2007.
Gertner, Jon. "The Future Is Drying Up." *New York Times*, 21 October 2007.
"Give United Water Fee Hike, but Watch Closely." www.preserveramapo.org, 19 October 2006.
Grannemann, N. G., et al. "The Importance of Ground Water in the Great Lakes Region." Water-Resources Investigations Report 00-4008. Lansing, MI: U.S. Department of the Interior, U.S. Geological Survey, 2000.
"Great Lakes Panel Promises Quicker Study on Declining Water Levels." *CBS News*, 19 October 2007.
Hall, Noah D. "Toward a New Horizontal Federalism: Interstate Water Management in the

Great Lakes Region." University of Colorado Law Review 77, no. 2 (Spring 2006).
Hanna, John. "Kansas Makes Water Demands on Nebraska," http://www.usatoday.com/news/nation/2007-12-19-1892413517_x.htm
Hendee, David. "Hot Questions on Water Woes." *Omaha World Herald*, 11 November 2005.
Hendee, David. "Irrigation Empire, Part 5: Paring Back Irrigation on the Platte." *Omaha World Herald*, 21 May 2007.
Hendee, David. "Time's Up for Nebraska to Comply with a Republican River Ruling." *Omaha World Herald*, 22 January 2008.
Hendee, David. "Trouble on the River." *Omaha World Herald*, 9 November 2005.
Incalcaterra, Laura. "Rate Hike Approved for Rockland's United Water Customers." *Journal News* (Westchester County), 14 December 2006.
Incalcaterra, Laura. "Rockland Water Could Soon Be Supplying an Orange County Development." *Journal News* (Westchester County), 21 December 2006.
Incalcaterra, Laura. "United Water's Responsibilities Increase with Customers' Bills." *Journal News* (Westchester County), 15 December 2006.
Incalcaterra, Laura. "Water Conservation Requested in Rockland." *Journal News* (Westchester County), 6 August 2006.
"Irrigation Empire, Part 4: Republican River Valley's Prosperity Trickles Away." *Omaha World Herald*, 20 May 2007.
Jenkins, Nate. "Trouble Double for Nebraska as It Prepares for Water Meeting." *Omaha World Herald*, 29 July 2007.
Jones, Tim. "Great Lakes Key Front in Water Wars." *Chicago Tribune*, 28 October 2007.
"Kansas Attorney General Promises Action to Get More Water for Republican River." *Omaha World Herald*, 15 August 2007.
Keys, Lisa. "Buyer Be Wet." *New York Times*, 12 October 2007.
Koulouras, Jason. "Troubled Waters on Great Lakes." *Toronto Star*, 25 September 2004.
Lam, Tina. "Mich. Should Share Water, N.M. Governor Says." *Detroit Free Press*, 11 October 2007.
Lam, Tina. "Michigan Leaders Push U.S. for Fix in St. Clair River." *Detroit Free Press*, 28 October 2007.
Lam, Tina. "Richardson Changes Tune on Tapping Great Lakes for Dry West." *Detroit Free Press*, 15 October 2007.
Lam, Tina, and Ben Schmitt. "Superior Free Fall Struck So Fast." *Detroit Free Press*, 29 August 2007.
Lyon, Bradfield, et al. "Water Shortages, Development, and Drought in Rockland County, New York." *Journal of the American Water Resources Association* 41, no. 6 (December 2005): 1457–1469.
Mackie, Andrew. "Duke Energy in Hot Seat over Water Transfer." *Hickory Daily Record*, 4 October 2006.
Mackie, Andrew. "Opponents Raise Nearly $1 Million for Water Fight." *Hickory Daily Record*, 22 September 2006.
Mackie, Andrew. "Unquenchable Thirst." *Hickory Daily Record*, 5 September 2006.
Mayland, Kirt. "A Glass Half Full: The Future of Water in New England." Arlington, VA: Trout Unlimited, 2007.

McBrayer, Sharon, and Ragan Robinson. "Whose Water Is It?" *News Herald*, 28 September 2006.

McKinnon, Shaun. "Hauling Water Is Way of Rural Life." *Arizona Republic*, 27 June 2005.

McNichol, Phil. "Cottagers Ask for Help with Lake Levels." *Sun Times*, 29 August 2007.

Money, Jack. "Scenic Rivers' Water Quality Brings Debate." *Oklahoman*, 23 June 2002.

Munoz, Sara Schaefer. "Homeowners, Builders Tackle Water Scarcity." *Wall Street Journal*, 23 February 2007.

"Nebraska Governor Pleads for Local Help on Water Issues." *U.S. Water News Online*, February 2007.

"Oklahoma Asks Federal Court to Dismiss Suit Challenging State Moratorium on Water Exportation." *Western Water Law and Policy Reporter*, October 2007, p. 318.

Olinger, David, and Chuck Plunkett. "Liquid Assets: Turning Water into Gold. Part IV: Law Makes, Breaks Men." *Denver Post*, November 2005.

Owen, Dave. "Law, Environmental Dynamism, Reliability: The Rise and Fall of CALFED." *Environmental Law* (Lewis & Clark Law School) 37, no. 4 (2007).

Pentland, Ralph. "Great Lakes Compact—Water for Sale?" Washington, DC: Woodrow Wilson International Center for Scholars, Canada Institute, September 2004.

Pepperell, Penny. "The Drain Hole in the St. Clair Has Enlarged Alarmingly, Beating the Predictions." Georgian Bay Association. www.georgianbay.ca/env_drainhole.htm (accessed 8 September 2007).

Pollock, Dennis. "Feds Reduce Water to Valley Farms: Westland Water District Growers in Crisis, to Decide Which Crops to Abandon." *Fresno Bee*, 3 June 2008.

Porterfield, Katie. "Holding Water." *TN Magazine*, February 2007.

Potter, Lori. "Shut Down Wells, Kansas Says." *Kearney Hub*, 20 December 2007.

Quirmbach, Chuck. "Great Lakes Water Levels Drop." Environment Report, www.gkrc.org, 28 May 2007.

"S.C. Worries over Water Because N.C. Cities Can Drain River." *U.S. Water News Online*, April 2007.

Silva, Andrew. "Water Woes in Desert: State Acts to Ease Problem." *San Bernardino Sun*, 9 August 2007.

Sinykin, Jodi Habush, and Donna L. McGee. "Opportunities and Challenges for State Implementation of Water Conservation Under the Great Lakes Compact: A Report and Toolkit." *Michigan State Law Review* (2006): 1195.

Smith, Jerd. "Farmers Sweat Lack of Water." *Rocky Mountain News*, 10 May 2006.

Southern Environmental Law Center. "Catawba-Yadkin Interbasin Transfer." n.d. www.southernenvironment.org/water/catawba_ yadkin_interbasin_transfer (accessed 27 May 2007).

Squillace, Mark, and Sandra Zellmer. "Managing Interjurisdictional Waters Under the Great Lakes Charter Annex." NR&E, Fall 2003.

Squillace, Mark. "Rethinking the Great Lakes Compact." Michigan State Law Review 2006, no. 5 (2006): 1347.

"State Actions, NE: Governor Heineman Approves Water Policy Bill for Republican River." *Water Strategist*, May 2007, p. 17.

Steinhauer, Jennifer. "Governor Declares Drought in California and Warns of Rationing." *New York Times*, 5 June 2008.

Taugher, Mike. "Delta Woes May Whet Water War." *Contra Costa Times*, 26 February 2007.

Taugher, Mike. "Judge to Weigh Water Limits in Delta to Protect Salmon." *Contra Costa Times*, 2 October 2007.

"Texas Group Sues over Water Sale Ban." *U.S. Water News Online*, January 2007.

"The Great Lakes–St. Lawrence River Basin Compact and Agreement: International Law and Policy Crossroads." *Michigan State Law Review* 2006, no. 5 (2006).

"The Patrick Farm Sellout—Three Betrayals." www.preserveramapo.org (accessed 24 May 2007).

Van Der Voo, Lee. "A Showdown Looms on the Clackamas River over Drinking Water." *Lake Oswego Review*, 7 March 2007.

Vandegrift, Greg. "Great Lakes Water Wars." KARE11TV, 15 May 2007. www.kare11.com (accessed 15 May 2007).

Vellinga, Mary Lynne, and Todd Milbourn. "Sunrise Douglas Project Dealt Setback by Court." *Sacramento Bee*, 2 February 2007.

Wall Street Journal. "Homeowners, Builders Tackle Water Scarcity." http://online.wsj.com/article (accessed 5 March 2007).

"Watch the Storm Drains." *Journal News* (Westchester County), 16 September 2006.

"Water Appropriation Doctrine—So Easy in Theory . . . Much Tougher in Practice." Editorial. *U.S. Water News*, November 2006.

"Water Appropriation in Action: Water Woes Create Ill Will Among Colorado Farmers." Editorial. *U.S. Water News*, October 2006.

Water Supply. "United Water New York." www.co.rockland.ny (accessed 17 May 2007).

Weiser, Matt. "Sacramento Won't Fine Couple Who Let Lawn Die." *Sacramento Bee*, 3 July 2008.

W. F. Baird and Associates Coastal Engineers. "Regime Change (Man Made Intervention) and Ongoing Erosion in the St. Clair River and Impacts on Lake Michigan–Huron Lake Levels." Oakville, ON, Canada: W. F. Baird and Associates Coastal Engineers, June 2005.

Wilcox, Douglas, et al. "Lake-Level Variability and Water Availability in the Great Lakes." U.S. Geological Survey. http://pubs.usgs.gov/circ/2007/1311/ (accessed 9 September 2007).

Wood, Daniel B. "Ag Water Issues: Water Crisis Squeezes California's Economy." *Christian Science Monitor*, 12 September 2007.

Yamamura, Kevin. "Governor Endorses Canal: Delta-Circumventing Project Was Rejected by State's Voters in 1982." *Sacramento Bee*, 15 June 2007.

Yoder, Josh. "Dear Mr. Governor, Please Save Our River." *Hickory Daily Record*, 24 September 2006.

Young, Samantha. "California Resurrects Canal Proposal to Reroute Water Supplies." Associated Press, 18 June 2007.

Letters, Memoranda, Miscellaneous

Attorney General Henry McMaster, State of South Carolina. News release, 28 March 2007. "McMaster Details Plans to Fight Water Transfer."

"Catawba Riverkeeper Water Transfer Power Point." IBT Presentation for WKA Conference 2007.

"Council of Great Lake Governors, Great Lakes Water Management Initiative Request for Public Comment." 19 July 2004.

Lochhead, James S. "Inter-Jurisdictional Watershed Management an Opportunity in the Great Lakes Basin." Presentation to conference, "The National Water Crisis: A Great Lakes Response," at University of Toledo College of Law and University of Toledo College of Law Legal Institute of the Great Lakes. Undated manuscript.

Mayland, Kurt. "A Glass Half Full: The Future of Water in New England." Arlington, VA: Trout Unlimited, 2007.

"Motion of the State of South Carolina for Leave to File Complaint." Supreme Court of the United States, *State of South Carolina v. State of North Carolina*, 7 June 2007.

North Carolina Department of Environment and Natural Resources. "Environmental Panel Grants Interbasin Water Transfer." Press release, 10 January 2007.

Rocky Mountain News. "Colorado Water: Dividing the Waters." 2005. www.rockymountainnews.com.

Rocky Mountain News. "Colorado Water: Running Dry." 2005. www.rockymountainnews.com.

Rocky Mountain News. "Colorado Water: The Last Drop." 2005. www.rockymountainnews.com.

"The Interbasin Transfer of Water—Why Is It Important to You?" www.ci.concord.nc.us.

U.S. Department of the Interior, U.S. Geological Survey. "Ground Water in the Great Lakes Basin." n.d. http://wi.water.usgs.gov/glpf/.

U.S. Geological Survey, "Assessment of the Water Resources of Rockland County, NY, with Emphasis on the Sedimentary Bedrock Aquifer," http://ny.water.usgs.gov/projects/rockland/rockland.htm (accessed 17 May 2007).

Interviews

Annin, Peter. Author. 20 September 2007.

Bleed, Ann. Director, Nebraska Department of Natural Resources. 7 May 2008.

Bush, Pamela. Attorney, Delaware River Basin Commission, 11 June 2007.

Injerd, Dan. Chief, Office of Water Resources, Illinois Department of Natural Resources. 28 March 2008.

Lisenby, Donna. Catawba Riverkeeper. 11 June 2007.

Miller, Daniel. Geologist, Rockland County, NY; bureau head for water supply, Rockland County Department of Health. 24 March 2008.

Privette, Annette. Public relations spokesperson, Concord, NC. 11 June 2007.

Smith, Duane. Director, Oklahoma Water Resources Board. 2007.

Chapter 6. Business as Usual

Books

Ashworth, William. *Ogallala Blue: Water and Life on the High Plains*. New York: Norton, 2006.

Bear, J., A. Cheng, S. Sorek, D. Ouazar, and I. Herrera, eds. *Seawater Intrusion in Coastal Aquifers: Concepts, Methods, and Practices*. Dordrecht, Netherlands: Kluwer, 1999.

Billington, David P., and Donald C. Jackson. *Big Dams of the New Deal Era: A Confluence of Engineering and Politics*. Norman: University of Oklahoma Press, 2006.

Cheng, Alexander H.-D., and Driss Ouazar, eds. *Coastal Aquifer Management: Monitoring, Modeling, and Case Studies*. Boca Raton, FL: Lewis, 2004.

Doremus, Holly, and A. Dan Tarlock. *Water War in the Klamath Basin: Macho Law, Combat Biology, and Dirty Politics*. Washington, DC: Island Press, 2008.

Glennon, Robert. *Water Follies: Groundwater Pumping and the Fate of America's Fresh Waters*. Washington, DC: Island Press, 2002.

Grossman, Elizabeth. *Watershed: The Undamming of America*. Washington, DC: Counterpoint, 2002.

Leslie, Jacques. *Deep Water: The Epic Struggle over Dams, Displaced People, and the Environment*. New York: Farrar, Straus and Giroux, 2005.

Lowry, William R. *Dam Politics: Restoring America's Rivers*. Washington, DC: Georgetown University Press, 2003.

McCully, Patrick. *Silenced Rivers: The Ecology and Politics of Large Dams*. London: Zed Books, 2001.

National Research Council. *Confronting the Nation's Water Problems: The Role of Research*. Washington, DC: National Academies Press, 2004.

Postel, Sandra, and Brian Richter. *Rivers for Life: Managing Water for People and Nature*. Washington, DC: Island Press, 2003.

Webb, Robert H., Stanley A. Leake, and Raymond M. Turner. *The Ribbon of Green: Change in Riparian Vegetation in the Southwestern United States*. Tucson: University of Arizona Press, 2007.

Wohl, Ellen. *Disconnected Rivers: Linking Rivers to Landscapes*. New Haven: Yale University Press, 2004.

Articles and Reports

"A Dry Legacy: The Challenge for Colorado's Rivers." Trout Unlimited, *Colorado Water Project*, January 2002.

Alley, W. M., et al. "Sustainability of Ground-Water Resources." USGS Circular 1186. Denver, CO. U.S. Department of the Interior, U.S. Geological Survey, 1999.

Alley, William M. "Tracking U.S. Groundwater." *Environment* 48, no. 3 (April 2006): p. 11.

Alley, William M., and Stanley A. Leake. "The Journey from Safe Yield to Sustainability." *Ground Water* 42, no. 1 (January–February 2004): 12.

Allison, M. Lee, and Todd C. Shipman, Arizona Geological Survey. "Earth Fissure Mapping Program." 2006 Progress Report. Open-file Report 07-01. June 2007.

American Rivers. "America's Most Endangered Rivers of 2007." www.americanrivers.org, 17 April 2007.

"America's Most Endangered Rivers of 2005." *#3 Fraser River*, http://www.americanrivers.org/site?News2?page=NewsArticle&id=7193 (Undated).

Amos, Adell Louise. "Hydropower Reform and the Impact of the Energy Policy Act of 2005 on the Klamath Basin: Renewed Optimism or Same Old Song?" *Journal of Environment Law and Litigation* 22, University of Oregon, (2007): 1.

Anderson, Mark T., and Lloyd H. Woosley Jr. "Water Availability for the Western United States—Key Scientific Challenges." U.S. Geological Survey, 2005.

"Another Sign of Long-Term Water Worries." www.journalstar.com, 6 October 2006.

Associated Press. "Oklahoma's Underground Water Resources Might Be Disappearing." *Examiner-Enterprise.com*, www.examiner-enterprise.com (accessed 23 April 2007), 23 April 2007.

Babbitt, Bruce. "A River Runs Against It: America's Evolving View of Dams." *Open Spaces Magazine*, www.open-spaces.com, 11 June 2007.

Bailey, Eric. "Fate of Klamath River Dams in Play: Federal Officials Call for Upgrades to Four of Them to Help Salmon Get Upriver; but It May Be Cheaper to Take the Barriers Down." *Los Angeles Times*, 31 January 2007.

Bailey, Eric. "Removing Four Klamath River Dams May Save Money, Report Finds: The Federal Study Says Pulling the Plug Could Cost $100 Million Less Than Keeping Them" *Los Angeles Times*, 2 December 2006.

Barbassa, Juliana. "State Halts Key Water Pump to Protect Endangered Delta Smelt." Associated Press, 31 May 2007.

Barnard, Jeff. "Feds Require Fish Ladders at Ore. Dams." Associated Press, 30 January 2007.

Barringer, Felicity. "On the Snake River, Dam's Natural Allies Seem to Have a Change of Heart." *New York Times*, 13 May 2007.

Barringer, Felicity. "Pact Would Open River, Removing Four Dams." *New York Times*, 14 November 2008.

"Black Rock Follies." *Columbia Institute for Water Policy*, www.columbia-institute.org (accessed 6 June 2007).

"Black Rock Will Not Solve Yakima River Fishery Problems." *Columbia Institute for Water Policy*, www.columbia-institute.org (accessed 6 June 2007).

Bodi, F. Lorraine, and Dulcy Mahar, Bonneville Power Administration. "Hydropower & Fish, Northwest Challenge: Keeping Fish and Clean Hydro." *Water Report*, no. 47 (15 January 2008).

Bowles, Jennifer. "Coachella Valley Sinking as Aquifer Is Over-Pumped, Report Says." *Press-Enterprise*, 18 December 2007.

Bowles, Jennifer. "EPA Pans Plan for Power Plant; Project: The $1.3 Billion Hydroelectric Idea Draws Concerns over the Flooding of a Canyon." *Riverside Press-Enterprise*, 26 July 2006.

Bowman, Margaret, et al. "Exploring Dam Removal." *American Rivers and Trout Unlimited*, August 2002.

"Breaching Snake River Dams Is Action of Last Resort." *Walla Walla Union-Bulletin*, 28 May 2007.

Breitler, Alex. "Controversial Canal's Impact on Delta Remains Largely Unknown." *Stockton Record*, 30 August 2007.

Bruckner, Robin. "NOAA Open Rivers Initiative." *NOAA Restoration Center*, www.nmfs.noaa.gov/habitat/restoration (accessed 27 October 2005).

Chang, Muncel. "Salt Water Vs. Fresh Water—Ghyben-Herzberg Lens." Groundwater, http://sparce.eucc.ou.edu (accessed 19 September 2000).

Chorneau, Tom. "Governor in Hot Water over Dam Plans." *San Francisco Chronicle*, 6 October 2007.

Chorneau, Tom. "Governor's Delta Plan Reignites Peripheral Canal Debate." *San Francisco Chronicle*, 27 September 2007.

Chorneau, Tom. "Water More Precious as State Grows: Strong Feelings over Dams; Not Everyone Agrees with Governor That New Reservoirs Are Best Way to Prepare for Dry Years—Some Experts Wonder if They'll Even Be Needed." *San Francisco Chronicle*, 14 April 2007.

Cole, Nancy. "Shrinking Aquifer Looms as Big Problem for Farms." *Arkansas Democrat Gazette*, 24 September 2006.

Collier, Michael, et al. "Dams and Rivers a Primer on the Downstream Effects of Dams." USGS Circular 1126. Tucson, AZ: U.S. Department of the Interior, U.S. Geological Survey, 1996.

"Congressional Update/Energy Bureau of Reclamation/Hydropower." *Western States Water*, 18 November 2005.

Cook, Jim. "Guest Opinion: Behind the Dams on the Klamath River." *Sacramento Bee*, 18 September 2007.

"Corps Opposes Breaching Snake River Dams." *Environment News Service*, 21 February 2002.

Curwen, Thomas. "Hetch Hetchy: Parting the Waters of What Once Was; Revisiting the Ceaseless Dream of the Hetch Hetchy Valley Moves Us Closer to the Lost Sanctuaries of the World." *Los Angeles Times*, 5 April 2005.

Custodio, Emilio. "Intensive Use of Ground Water and Sustainability." *Ground Water* 43, no. 3 (May–June 2005): 291.

Davis, Tony. "Pumping of Groundwater Spurs Surge in Earth Fissures." *Arizona Daily Star*, 5 June 2007.

Davis, Tony. "The Battle for the Verde." *High Country News* 39, no. 9 (14 May 2007): 10.

"Detection and Measurement of Land Subsidence Using Global Positioning System Surveying and Interferometric Synthetic Aperture Radar, Coachella Valley, California, 1996–2005." Scientific Investigations Report 2007-5251. U.S. Geological Survey.

Dininny, Shannon. "Proposed Reservoir Could Seep Toward Hanford." *Seattle Post-Intelligencer*, 18 September 2007.

Doyle, Michael. "Hetch Hetchy Study a No-Go: House Panel Rejects Bush Proposal for Funds to Look at Draining the Reservoir." *Sacramento Bee*, 27 June 2007.

"Editorial: Can He Build a Dam? Water Projects May Elude." *Sacramento Bee*, 13 January 2007.

"Editorial: Spending More Now to Study Auburn Dam Would Be Wasteful." *Auburn Journal*, 30 January 2007.

"Editorial: We'll Need Dams if We're to Address Water-Supply Woes." *Modesto Bee*, 16 January 2007.

"Executive Summary Yakima River Basin Storage Alternatives Appraisal Assessment." *United States Bureau of Reclamation*, May 2006.

"Feds Recommend Keeping Klamath Dams." *Los Angeles Times*, 17 November 2007.

"Fish Ladders Required for Klamath Dams." *News and Review*, www.oregon news.com, 30 January 2007.

Fisher, Douglas. "Bush Budget Proposes Hetch Hetchy Study; Restoration Advocates Surprised, Pleased." *San Jose Mercury News*, 7 February 2007.

Gardner, Michael. "Battle Brews over Plans for Three Reservoirs: Governor's Pitch for New Dams Faces Resistance in Legislature." *San Diego Union-Tribune*, 8 October 2007.

Gardner, Michael. "Watershed Moment in Reservoir Campaign: Coinciding Warnings Driving Push for Dams." *San Diego Union-Tribune*, 20 May 2007.

Geissinger, Steve. "Dams at Center of Conflict Between Governor, Dems: 'Any Time You Build a New Dam, One Thing You Can Count on Is a Lawsuit.'" *Oakland Tribune*, 29 January 2007.

Gelt, Joe. "Study Says Northern Arizona's Water Supplies Unsustainable." *Arizona Water Resource*, College of Agriculture and Life Sciences, University of Arizona, 15, no. 3 (January–February 2007).

Geluso, James. "Water Shortage to Keep River Dry This Holiday." *Drought*, 30 June 2007.

Glennon, Robert. "Water Scarcity, Marketing, and Privatization." Texas Law Review 83 (2005): 1873.

"Ground Water Facts." www.ngwa.org/public/gw_use/faqs.aspx, February 2006.

"Ground Water Report to the Nation: A Call to Action." *Ground Water Protection Council*, 2007.

"Ground-Water Depletion Across the Nation." USGS Fact Sheet-103-03, November 2003.

Hagengruber, James. "Reservoir Could Spread Radiation." *Spokesman-Review*, 19 September 2007.

Harden, Blaine. "U.S. Orders Modification of Klamath River Dams Removal May Prove More Cost-Effective." *Washington Post*, 31 January 2007.

"How Do We Use Ground Water Resources in the U.S.?" www.waterbank.com/faq1.html, n.d.

Huston, Susan S., et al. "Estimated Use of Water in the United States in 2000." USGS Circular 1268. Reston, VA. U.S. Department of the Interior, U.S. Geological Survey, 2004.

"Idaho Governor Is Hoping to Keep More Snake Water." *RedOrbit News*, 19 February 2007.

"Is Seawater Intrusion Affecting Ground Water on Lopez Island, Washington?" USGS, Fact Sheet. pubs.usgs.gov (accessed 6 January 2008).

Jenkins, Matt. "Into Thin Air?" *High Country News*, 30 April 2007.

Jenkins, Matt. "Peace on the Klamath." *High Country News*, 23 June 2008.

Johns, Norman D. "Bays in Peril: A Forecast for Freshwater Flows to Texas Estuaries." *National Wildlife Federation*, October 2004.

Johnson, Ted. "Battling Seawater Intrusion in the Central and West Coast Basins." Water Replenishment District of Southern California, Technical Bulletin 13 (Fall 2007).

Kendy, Eloise. "The False Promise of Sustainable Pumping Rates." *Ground Water* 41, no. 1 (January–February 2003): 2.

Kent, Christopher A. "Water Resource Planning in the Yakima River Basin: Development Vs. Sustainability." University of Hawaii Press, Yearbook of the Association of Pacific Coast Geographers, 2004.

Kosnik, Lea-Rachel. "Hydropower's Contribution to Climate Change Reduction." University of Missouri, St. Louis, Unpublished Manuscript (Draft), December 2007.

Lane-Miller, Chelsea, and Brad DeVries. "America's Most Endangered Rivers of 2007." *American Rivers*, 2007.

Leslie, Jacques. "Before the Flood." *New York Times*, 22 January 2007.

Lester, David. "Black Rock Benefit? A $2.4 Billion Question." *Yakima Herald-Republic*, 9 January 2007.

Lester, David. "Reservoir Could Spread Hanford Pollution." *Yakima Herald-Republic*, 19 September 2007.

Lin, Judy. "Defeated Dams Still Supported: Governor Isn't Backing Away from $4 Billion in Bonds After Negative Vote by Senate Panel." *Sacramento Bee*, 25 April 2007.

Lucas, Greg. "GOP Cites Warming in Bid for New Dams." *San Francisco Chronicle*, 1 February 2007.

Maclin, Elizabeth, and Matt Sicchio, eds. "Dam Removal Success Stories." *Prepared by Friends of the Earth, American Rivers, and Trout Unlimited*, December 1999.

Magilligan, Francis J., and Keith H. Nislow. "Long-Term Changes in Regional Hydrologic Regime Following Impoundment in a Humid-Climate Watershed." Reprinted from *Journal of the American Water Resources Association* 37, no. 6 (December 2001).

Marder, Meredith K. "The Battle to Save the Verde River." *Arizona Law Review* 50 (2008).

Martin, Mark. "Governor's New State Water Plan to Include Two Dams." *San Francisco Chronicle*, 9 January 2007.

Maupin, Molly A., and Nancy L. Barber. "Estimated Withdrawals from Principal Aquifers in the United States, 2000." U.S. Geological Survey, 2005.

McCool, Daniel. "The River Commons: A New Era in U.S. Water Policy." *Texas Law Review* 83 (2005): 1903.

Milstein, Michael. "Klamath Wells Deep in Trouble." *Klamath Forest Alliance*, www.klamathforestalliance.org, 3 May 2004.

Nash, J. Madeleine/Hetch Hetchy. "Is This Worth a Dam?" *Time*, 11 July 2005.

Needham, Jerry. "Stored Water Is Focus of Fight." www.mysanantonio.com, 7 August 2007.

O'Brien, Maria. "Lessons Learned Resolving the Unresolvable: Rio Grande Silvery Minnow Vs. City of Albuquerque." *Thirteenth Institute for Natural Resources Law Teachers*, 4 June 2005.

Olsen, Ken. "Nevada Stakes Its Salmon Claim, Snake River Dams Run Up Against a Powerful Alliance in an Unlikely Place." *High Country News*, 4 February 2008.

Osborn, Rachael Paschal. "Black Rock Follies." Columbia Institute for Water Policy. http://columbia-institute.org/ci/cihome/Home.html (accessed 6 June 2007).

Osborn, Rachael Paschal. "Odessa Aquifers: Crisis in Sustainability." Columbia Institute for Water Policy, 10 November 2006. http://columbia-institute.org/oa/odessa/Home.html (accessed 18 December 2008).

Otto, Betsy, et al. "Paving Our Way to Water Shortages: How Sprawl Aggravates the Effects of Drought." American Rivers, Natural Resources Defense Council and Smart Growth America, 2002.

Philip, Tom. "A Time to Deliver Water Solutions." *Sacramento Bee*, 14 January 2007.
Pollard, Vic. "Governor's Plan to Back Dams Faces Hitches." *Bakersfield Californian*, 8 January 2007.
"Proposed Klamath River Basin Restoration Agreement for the Sustainability of Public and Trust Resources and Affected Communities." www.edsheets.com/klamathdocs.html, 15 January 2008 (Draft 11).
"Reclamation, Inventory of Reclamation Water Surface Storage Studies with Hydropower Components." U.S. Department of the Interior, October 2005.
Reichard, Eric G., and Theodore A. Johnson. "Assessment of Regional Management Strategies for Controlling Seawater Intrusion." *Journal of Water Resources Planning and Management*, August 2005.
Reisner, Marc. "Coming Undammed." *Audubon*, September–October 1998.
"Rivers Unplugged—Removing Dams That Don't Make Sense." American Rivers, n.d.
"Salmon Undammed." *Trout*, Spring 2007.
Sarris, Greg. "After the Fall." *Los Angeles Times*, 5 April 2005.
Schultz, E. J. "Governor Launches New Water Battle: In Fresno, He Pushes $4 Billion Bond in 2008 for New Dams." *Fresno Bee*, 27 March 2007.
"Schwarzenegger Promotes Dams as Way to Boost Water Reserves." Department of Water Resources, *California Water News*, 26 March 2007.
"Science and Technology to Support Fresh Water Availability in the United States." Report of the National Science and Technology Council Committee on Environment and Natural Resources, November 2004.
Shaw, Hank. "Latinos Ally with Governor on Water." *Stockton Record*, 24 April 2007.
Shaw, Hank. "New Dams: Necessity, or Billion-Dollar Mistakes?" *Stockton Record*, 21 January 2007.
Squatriglia, Chuck. "Hetch Hetchy Line in Bush's Budget Blasted." *San Francisco Chronicle*, 7 February 2007.
"Surficial and Northern Atlantic Coastal Plain Aquifer Systems, Long Island." U.S. Geological Survey, capp.water.usgs.gov/gwa/ch_m/M-text3.html (accessed 10 June 2007).
"Ten Reasons Why Dams Damage Rivers." American Rivers, www.amrivers.org, n.d.
"The Black Rock Dam Proposal." Columbia Institute for Water Policy, www.columbia-institute.org (accessed 6 June 2007).
"The Ground Water Supply and Its Use." www.wellowner.org/agroundwater/gwsupplyanduse.shtml, 2007.
Thompson, Don. "Democrats Oppose New Dams in California, Favor Conservation." *San Jose Mercury News*, 25 January 2007.
Thomson, Gus. "Dam Costs Skyrocket: Bureau Projects $9.6 Billion Needed for Construction." *Auburn Journal*, 30 January 2007.
Thomson, Gus. "Water Storage Issues: Doolittle Backing Plan for New Dams; Water Storage Structures One of Governor's Proposal." *Auburn Journal*, 14 January 2007.
"Two Weeks in the West." *High Country News*, 2007.
Walters, Dan. "Democrats Shunning Pat Brown." *Sacramento Bee*, 15 January 2007.
"Water 2025: Preventing Crises and Conflict in the West." Water 2025 Status Report. U.S. Department of the Interior. August 2005.

Weiser, Matt. "Auburn Dam Price Tag Soars: Study Puts Cost at Twice Earlier Estimates, Says It Wouldn't Protect Capital in Giant Flood." *Sacramento Bee*, 31 January 2007.

Whitney, David. "Discord Threatens Klamath River Water Talks." *Sacramento Bee*, www.sacbee.com, 12 August 2007 (accessed 14 December 2007).

Yardley, William. "Climate Change Adds New Twist to Debate over Dams." *New York Times*, 23 April 2007.

Young, Samantha. "A Look at Where Californians Get Their Water for Farms and Cities." *Associated Press*, 7 April 2007.

Young, Samantha. "Cost of Reviving California Dam Project Soars Toward $10 Billion." Associated Press, 30 January 2007.

Young, Samantha. "In California, Warming Trend Renews Water Debate." Associated Press, 7 April 2007.

Young, Samantha. "Senate Panel Rejects Governor's Plan to Build More Dams." Associated Press, 24 April 2007.

Yuskavitch, Jim. "Klamath Dams." *Fly Fisherman*, May 2007.

Letters, Memoranda, Miscellaneous

Galloway, Devin, et al. "Land Subsidence in the United States." U.S. Geological Survey, (Undated).

Hirsch, Robert M. "Science and Water Availability." U.S. Department of the Interior, U.S. Geological Survey, 7 June 2005.

Interviews

Barlow, Paul. U.S. Geological Survey. 1 June 2007.

Clark, John. U.S. Geological Survey. 1 June 2007.

Guenther, Herb. Director, Arizona Department of Water Resources. 17 July 2006.

Heisig, Paul. U.S. Geological Survey. 17 May 2007.

Hetrick, John. Salt River Project. July 2005.

Jess, Michael. Associate director, University of Nebraska–Lincoln Water Center. 31 May 2007.

Johns, Norman. National Wildlife Federation. 14 December 2007.

Johnson, Robert. Commissioner, U.S. Bureau of Reclamation, Tucson, AZ. 31 August 2007.

Johnson, Ted. Chief hydrologist, Water Replenishment District of Southern California. December 2007.

Reichard, Eric. U.S. Geological Survey. 19 June 2007.

Schleicher, Gloria. Tenants Harbor, ME. July 2006.

Snow, Lester. Director, California Department of Water Resources. 30 August 2007.

Chapter 7. Water Alchemists

Books

Reisner, Marc. *Cadillac Desert: The American West and Its Disappearing Water*. New York: Penguin Books, 1986.

Articles and Reports

Bauman, Joe. "Western Water Issues." *Deseret News*, 13 April 2007.

Best, Allen. "No Mere Pipe Dream." *Colorado Biz*, www.cobizmag.com, 1 October 2006.

Breitler, Alex. "Sierra Snowpack: Cloud Seeding Begins in the Sierra." *Stockton Record*, www.recordnet.com, 10 February 2007.

Cotton, William R. "Basic Cloud Seeding Concepts." *Southwest Hydrology* 6, no. 2 (March–April 2007): 16.

"Critical Issues in Weather Modification Research." Committee on the Status of and Future Directions in U.S. Weather Modification Research and Operations. National Research Council of the National Academies, www.nap.edu 2003.

Davidge, Ric. "Water Exports." *New Letter V1 #12 Bulk Water*, www.waterbank.com, June 1994.

Fountain, Henry. "The Science of Rain-Making Still Produces Cloudy Results." *New York Times*, 19 October 2003.

Frazier, Deborah. "Water Plan Faces Familiar Foes, $4 Billion Pipeline from Wyo. Would Supply Front Range." *Rocky Mountain News*, 7 October 2006.

Garstang, Michael, et al. "Weather Modification." *American Meteorological Society*, May 2004.

Geluso, James. "Water Supply Options." www.bakersfield.com, 30 May 2007.

Gochis, David J. "Emergent Precipitation Enhancement Techniques and the Rights to Developed Water." Undated, University of Arizona, Department of Hydrology & Water Resources, Research Assistant.

Griffith, Don A., and Mark E. Solak. "The Potential Use of Winter Cloud Seeding Programs to Augment the Flow of the Colorado River." Prepared for the Colorado River Commission. March 2006.

Hull, Jeff. "The Modern Rain Dance." *New York Times*, 2 July 2006.

Hunter, Steven M. "Answers to Frequently Asked Questions About Cloud Seeding to Augment Mountain Snowpacks." *North American Interstate Weather Modification Council*, www.naiwmc.org, May 2006.

Klymchuk, Daniel. "Water Exports—a Manitoba Bonanza?" *Frontier Backgrounder*, Published by the Frontier Centre for Public Policy, September 2001.

Kohlhoff, Karl, and David Roberts. "Beyond the Colorado River: Is an International Water Augmentation Consortium in Arizona's Future?" *Arizona Law Review* 49, no. 2, p. 257–295.

Lopez, Daniel. "Cloud Seeding Decision Delayed: Water Board Wants More Data." *Monterey Herald*, www.montereyherald.com, 11 December 2007.

Meyer, Jeremy P. "Water Lifeline or Pipe Dream?" *Denver Post*, 3 June 2007.

Miller, Jared. "Pipeline Could Water Wyo., Too." *Casper Star-Tribune*, 1 November 2007.

Owens, Dennis. "Water, Water Everywhere, but Canada Won't Sell It." *Wall Street Journal*, 31 August 2001.

Rossi, Jim. "Is the Water Bag Proposal a Trojan Horse?" *North Coast Journal*, 6 February 2003.

Ryan, Tom. "Weather Modification for Precipitation Augmentation and Its Potential Usefulness to the Colorado River Basin States." cwcb.state.co.us/flood, October 2005.

U.S. Congress, Office of Technology Assessment, *Alaskan Water for California? The Subsea*

Pipeline Option Background Paper, OTA-BP-O-92 (Washington, DC: U.S. Government Printing Office, January 1992).

"Weather Modification-Mitigation/Climate Change." *Western States Water Council*, www.westgov.org, 20 July 2007.

"Western Governors/Water Resources Evolving Technologies." *Western States Water*, 20 November 2006.

Woodka, Chris. "Water Proposals Could Hit Snags in Compacts." *Pueblo Chieftain*, 23 May 2007.

Woodka, Chris. "Water Users Want Brakes on Pipeline Project." *Pueblo Chieftain*, 28 October 2007.

"Wyoming State Officials 'Skeptical' of Colo. Man's Water Pipeline." *U.S. Water News Online*, www.uswaternews.com, October 2006.

Letters, Memoranda, Miscellaneous

California Coastal Commission, North Coast District Office. Memorandum from Peter Douglas, Executive Director, et al. *Re: Protest to State Water Resources Control Board Concerning Applications of Alaska Water Exports to Divert Water from Gualala and Albion Rivers (for Commission Meeting of December 13, 2002, Item F3a)*, Eureka, California, 22 November 2002.

Guenther, Herb, et al. Letter to the Honorable Gale A. Norton, Secretary, Department of the Interior. 3 February 2006.

Leedom, John N. Letter to Stan Sokol, General Counsel, Office of Sciences and Technology Policy. 28 April 2006.

Marburger, John H. III, Director, Office of Science and Technology Policy. Letter to the Honorable Kay Bailey Hutchison, U.S. Senator. 13 December 2005.

Interviews

Betterton, Eric. Professor of atmospheric sciences, University of Arizona. 26 January 2007.

Chapter 8. The Ancient Mariner's Lament

Books

Barnett, Cynthia. *Mirage: Florida and the Vanishing Water of the Eastern U.S.* Ann Arbor: University of Michigan Press, 2007.

Gleick, Peter H. *The World's Water, 2006–2007: The Biennial Report on Freshwater Resources*. Washington, DC: Island Press, 2006.

National Research Council of the National Academies, Committee on Advancing Desalination Technology. *Desalination: A National Perspective*. Washington, DC: National Academies Press, 2008.

Articles and Reports

"Balancing Water Needs on the Lower Colorado River: Recommendations of the Yuma Desalting Plant/Ciénega de Santa Clara Workgroup." 22 April 2005.

Bond, James H. "Water Authority Had to Drop Desal." *North County Times*, 8 August 2006.

Burge, Michael. "Desalination: GE Water, Planned Desalination Plant Team Up." *San Diego Union-Tribune*, 25 May 2007.
Burge, Michael. "Desalination Plant Needs Back-up Plan." *San Diego Union-Tribune*, 27 September 2007.
Burge, Michael. "Environmental Groups Sue over Desalination." *San Diego Union-Tribune*, 15 January 2008.
Burge, Michael. "Poseidon Plant Nears OK: Desalination Project Needs Another Panel's Approval." *San Diego Union-Tribune*, 16 November 2007.
Burge, Michael. "Rough Sea Ahead for Plan to Desalinate North County Areas." *San Diego Union-Tribune*, 3 September 2006.
Caldwell, Alicia A. "El Paso Desalination Plant Opens Up After Fifteen Years of Planning." *Daily Texan*, 9 August 2007.
"California Department of Water Resources Awards $25 Million in Grants for Water Desalination Projects." *Western Water Law* (July 2005): 246.
"California Desalination Report with More Than a Grain of Subjectivity." *Water Conditioning and Purification*, January 2007.
"Chlorine Resistant Polyamide Reverse Osmosis Membranes." Presented at American Membrane Technology Association 2007 Conference and Exposition, Las Vegas, NV, July 2007.
Conaughton, Gig. "Coastal Commission Approves Carlsbad Desal Plant." *North County Times*, 16 November 2007.
Conaughton, Gig. "Coastal Group, Company Dicker over Carlsbad Desal Info." *North County Times*, 28 July 2007.
Conaughton, Gig. "Water Authority: Still Not Interested in Carlsbad Plant." *North County Times*, 26 October 2006.
Cooley, Heather, et al. "Desalination, with a Grain of Salt: A California Perspective." *Pacific Institute*, June 2006.
Davis, Tony. "Ultimate Solution?" *High Country News*, 24 November 2008, p. 14.
"Desal Plant Repair Estimated at $29 Million." *Water News* 21, no. 10 (October 2004).
"Desalination: A National Perspective." Committee on Advancing Desalination Technology, National Research Council of the National Academies. Washington, DC: National Academies Press, 2008.
"Desalination by Forward Osmosis." Presentation at American Membrane Technology Association 2007 Conference and Exposition, Las Vegas, NV, July 2007.
"Desalination of Ground Water: Earth Science Perspectives." U.S. Department of the Interior, USGS Fact Sheet 075-03, October 2003.
Dibble, Sandra. "Desalination Plant: Desalination Test a Model of Cooperation: Conservationists Work with Water Managers." *San Diego Union-Tribune*, 26 March 2007.
Downey, Dave. "Desalination Plant Clears Hurdle." *North County Times*, 7 June 2007.
"Evaluating Yuma-Area Groundwater as a Feed Water Supply for the Yuma Desalting Plant." Presented at the American Membrane Technology Association Biennial Conference, 29 July–2 August 2006.
"Evaluation of Slowsand Filtration for Reducing Costs of Desalination by Reverse Osmosis."

Presented at American Membrane Technology Association 2007 Conference and Exposition, Las Vegas, NV, July 2007.

Fimrite, Peter. "Bay Area Desal Plant: Desalination Viable for Drinking Water, Engineers Report; but Plant Would Be Expensive—at Least $111 Million." *San Francisco Chronicle*, 7 February 2007.

Foskett, Ken. "Water Woes: Desalination Is Costly, Challenging." *Atlanta Journal-Constitution*, 17 February 2008.

Hart, Chris. "Desalination Dilemma—Lessons Learned from Tampa Bay." *U.S. Water News*, November 2003.

Howe, Jeff. "The Great Southwest Salt Saga." *Wired*, 12 November 2004.

Howe, Kevin. "Desal Company Beats Petition: San Diego Plant Back on Track." *Monterey Herald*, 7 June 2007.

Howe, Kevin. "Desal Panel Raises Red Flag: Environmental Concerns Persist." *Monterey Herald*, 13 October 2006.

Howe, Kevin. "Desal Plans Go Before Coastal Panel: Staff Urges Denial of Cal Am Plant Permits." *Monterey Herald*, 8 December 2006.

Howe, Kevin. "Desalination Debate Rages On; Monterey County: No Action Taken; Lawyers Will Study Report." *Monterey Herald*, 21 March 2007.

Ireland, Philip K. "Desalination Plant Project Moving Ahead Despite Agency's Withdrawal." *North County Times*, 29 July 2006.

Johnson, Jim. "Cal Am Gets Rate Hike, OK for Desal Test; Coastal Panel Rejects Staff's Advice on Plant." *Monterey Herald*, 15 December 2006.

Karajeh, Fawzi, and Fethi BenJemaa. "A Legislative Boost for Water Desalination as an Alternative Water Supply Strategy in California." *Water Conservation News*, Fall–Winter 2004–2005, p. 7.

Landry, Clay J., and Christina Quinn. "The New Economy of Water." *Water Resources Impact*, November 2006.

Lobeck, Joyce. "Yuma Desalting Plant: Yuma Desalting Plant Draws Global Visitors." *Yuma Sun*, 15 May 2007.

McCord, Shanna. "Desalination Project: Water Bills to Jump 40 Percent with Desalt Plant." *Santa Cruz Sentinel*, 20 November 2005.

McKinnon, Shaun. "Rivals Come Together on Desalter amid Water Debate." *Arizona Republic*, 28 January 2007.

"Modifications to Projects of Title 1 of the Colorado River Basin Salinity Control Act (Public Law 93-320, As Amended, 43 U.S.C. § 1571)." Report to the Congress by the Secretary of the Interior. Draft 11 February 2003.

Overly, Jeff. "Desalination Plant Clears Hurdle: Huntington Beach Council Approves Desalination Plant, the First Step on a Long Road for the Project." *Orange County Register*, 1 March 2006.

Perry, Tony. "Coastal Commission Approves Desalination Plant in Carlsbad." *Los Angeles Times*, 7 August 2008.

Peterson, Lee. "Water Desalination Proposals Draw Out Fans and Foes." *Torrance Daily Breeze*, 14 August 2006.

"Report to the Congress, The Yuma Desalting Plant and Other Actions to Address

Alternatives." August 2005. (Attachment to letter to Senator Pete Domenici from Secretary of the Interior Gayle Norton, 26 October 2005.)

"Reverse Osmosis." Department of Health, Education, and Welfare Public Health Service, *Food and Drug Administration*, 21 October 1980.

Royte, Elizabeth. "A Tall, Cool Drink of . . . Sewage?" *New York Times Sunday Magazine*, 10 August 2008.

"Seawater Desalination and the California Coastal Act." California Coastal Commission, March 2004.

"Seawater Desalination Gaining Momentum in Texas." *Western Water Law*, August 2005, p. 278.

Service, Robert F. "Desalination Freshens Up." *Science* 313 (25 August 2006): 1088.

Shea, Andrew L., and Nikolay Voutchkov. "Large-Scale Seawater Desalination and Alternative Project Delivery." *Design-Build Dateline*, February 2005.

Skoloff, Brian. "Many States Seen Facing Water Shortages." *Boston Globe*, 26 October 2007.

"Small Business Gold Mine in Water." By Entrepreneur.com www.thestreet.com, 10 January 2007.

"Tampa Bay Seawater Desalination Plant, Florida, USA." www.water-technology.net/projects/tampa/, 2007.

Voutchkov, Nikolay. "Advances in Membrane Seawater Desalination Technology." *Asian Water*, July–August 2006.

Voutchkov, Nikolay. "Plants for Seawater Desalination in California." *Journal of the Australian Water Association*, November 2006.

Voutchkov, Nikolay. "Poseidon Advances Seawater Desalination in California." U.S. Water News 23, no. 11 (November 2006).

"Water Quality Improvement Center, Advanced Water Treatment Research." U.S. Department of the Interior, Bureau of Reclamation, 2005.

Water Recycling and Desalination Staff. "DWR Announces Second Round of Funding for Water Desalination." *Water Conservation News*, Fall–Winter 2005–2006, p. 5.

Waymer, Jim. "Tampa Troubles Raise Desalination Concerns." *Florida Today*, 10 January 2005.

Yeung, Bernice. "Sea Change." *California Lawyer*, October 2005, p. 24.

"Yuma Desalting Plant: Evaluation of Alternative Treatment Processes and Reverse Osmosis Membrane Types." Presented at the American Membrane Technology Association Biennial Conference. 29 July–2 August 2006.

Letters, Memoranda, Miscellaneous

"A Solution to the YDP/Cienega Controversy." Central Arizona Project, 2 May 2005.

"Aquasis" www.aquasis.com, 10 May 2007.

"Consolidated Appropriations Act, 2005," PL 108-447, 8 December 2004.

"Energy and Water Development Appropriations Bill," Report 108-554. 18 June 2004.

"Letter from P. Lynn Scarlett, Asst. Secretary for Policy, Management & Budget to the Honorable Pete V. Domenici, Chairman, 26 October 2005.

"Letter from Susan Culp, et al. to The Honorable Janet Napolitano Re: Imminent Threat to the Ciénega de Santa Clara and the Colorado River Delta." 5 November 2003.

Mickley, Michael. "New Techniques in Desalination and Brine Management." Presented by Michael Mickley, PE, PhD, President Mickley & Associates, Boulder, CO. CLE Internation, Law of the Colorado River. (Undated)

"Reclamation: Managing Water in the West," Yuma Desalting Plant Demonstration Run (Video). U.S. Department of the Interior, Bureau of Reclamation.

"Report to the Congress by the Secretary of the Interior." Modifications to Projects of Title 1 of the Colorado River Basin Salinity Control Act (Public Law 93-320, As Amended, 43 U.S.C. § 1571). Draft 11 February 2003.

"Subject: Modifications to Projects of Title 1 of the Colorado River Basin Salinity Control Act (Public Law 93-320, As Amended, 43 U.S.C. § 1571)"

"Yuma Desalting Plant." Bureau of Reclamation: Lower Colorado Region. www.usbr.gov/lc/yuma (accessed 9 December 2006).

Interviews

Cherry, Jim. Area manager, Yuma office, U.S. Department of the Interior. March 2007.

"Yuma Desalting Plant Demonstration Run." U.S. Department of the Interior, Bureau of Reclamation.

Chapter 9. Shall We Drink Pee?

Articles and Reports

"American Water Receives Approval for 'Green' Water Recycling Project." *Water Online*, www.wateronline.com, 30 January 2007.

Archibold, Randal C. "From Sewage, Added Water for Drinking." *New York Times*, 27 November 2007.

Archibold, Randal C. "Los Angeles Eyes Sewage as a Source of Water." *New York Times*, 16 May 2008.

"Arizona Town Sells Wastewater Effluent for $67 Million." *U.S. Water News* 24, no. 12 (December 2007).

Baker, L. A., et al. "Environmental Consequences of Rapid Urbanization in Warm, Arid Lands: Case Study of Phoenix, Arizona (USA)." In *The Sustainable City III*. Proceedings of the Sienna Conference, held June 2004. Southampton, UK: WIT Press, 2004.

Chambers, Douglas B., and Thomas J. Leiker. "A Reconnaissance for Emerging Contaminants in the South Branch Potomac River, Cacapon River, and Williams River Basins, West Virginia, April–October 2004." U.S. Department of the Interior, U.S. Geological Survey, Open-File Report 2006-1393.

Chapman, Ginette. "From Toilet to Tap: The Growing Use of Reclaimed Water and the Legal System's Response." *Arizona Law Review* 47 (2005): 773.

"Common Personal Care Products, Pharmaceuticals, and Genetic Fingerprinting Used to Trace Wastewater in Streams." U.S. Department of the Interior, U.S. Geological Survey, 24 July 2002.

"Crossing the Line: Water Quality Shared by Missouri and Kansas." U.S. Department of the Interior, U.S. Geological Survey, www.usgs.gov/ newsroom/article, 7 September 2006.

Dean, Cornelia. "Drugs Are in the Water: Does It Matter?" *New York Times*, 3 April 2007.
Donn, Jeff, et al. "Report Doubtful on Safety of Water: Pharmaceuticals in What We Drink Could Have Long-Term Consequences." *California Water News*, 10 March 2008.
"Drug Traces Found in Water Pose Problem for Wildlife: Pharmaceuticals Passing Unaltered from Humans into Nation's Waterways." *U.S. Water News*, November 2005.
Friederici, Peter. "Facing the Yuck Factor." *High Country News*, 17 September 2007.
Gelt, Joe. "Prescott Valley's Effluent Water-Rights Auction Is Innovative, Profitable." *Arizona Water Resource*, January–February 2008.
Glennon, Robert. "Water Scarcity, Marketing, and Privitization." *Texas Law Review* 83 (2005): 1873.
"Ground Water Monitoring and Remediation." *National Groundwater Association Publication* 24, no. 2 (Spring 2004).
Herrmann, Jerry. "PV Stands to Make $67.1 Million off Its Effluent Water." *Daily Courier*, 30 October 2007.
Ingram, Bruce. "Trouble on the Potomac." *Fly Fisherman*, March 2007.
"Intersex Fish Are Found at High Rate in a Region." *New York Times*, 7 September 2006.
"Intersex Sunfish Found for First Time in Potomac Basin." *U.S. Water News Online*, www.uswaternews.com/archives/arcquality/, February 2007.
Kolpin, Dana W., et al. "Pharmaceuticals, Hormones, and Other Organic Wastewater Contaminants in U.S. Streams, 1999–2000: A National Reconnaissance." *American Chemical Society: Environmental Science and Technology*, pubs.acs.org/hotartcl, 15 March 2002.
Lee, Mike. "Marketing Recycled Water: S.D. Looks North for Help Marketing Recycled Water; Orange County Staged Successful PR Campaign." *San Diego Union-Tribune*, 12 September 2005.
Lee, Mike. "Recycled Water Use: Recycled Tap Water's 'Unsettled Question.'" *San Diego Union-Tribune*, 26 September 2006.
Lee, Mike. "Recycling Wastewater: Sanders Against Sending Treated Wastewater to Tap." *San Diego Union-Tribune*, 20 July 2006.
Montoya, Thierry R., and Robert Orozco. "Endocrine Disruptors—a Water Quality Concern to Monitor." *Western Water Law*, October 2003, p. 335.
Mozilo, Ralph. "Cave Creek Water Company." Editorial. *Sonoran News*, 7–13 February 2007.
"Pharmaceuticals, Hormones, and Other Organic Wastewater Contaminants in U.S. Streams." USGS Fact Sheet, toxics.usgs.gov/ pubs/, June 2002.
"Pollution Link Eyed in River's 'Intersex' Fish." *Arizona Daily Star*, 20 January 2007.
"Population Trends Along the Coastal United States: 1980–2008." National Oceanic and Atmospheric Administration, www.oceanservice.noaa.gov/programs (accessed 25 June 2007).
"Purchase of Up to 2,724 AFA of Effluent." *Water Strategist*, November 2007.
Rake, Launce. "Chemicals Cause Changes in Fish and Raise Concerns for Humans." *Las Vegas Sun*, 20 October 2006.
"Reclaimed Water System Status Report—2007." Tucson Water Department, Tucson, Arizona.
"Reclaimed Water." Tucson Water Department, www.ci.tucson.az.us/water/reclaimed.htm (accessed 24 June 2007).

"Recycled Drinking Water: Yuck! San Diego Should Flush 'Toilet to Tap' Plan." Editorial. *San Diego Union-Tribune*, 24 July 2006.
Rogers, Paul. "Recycling Water: Making Sewage Water Good to Drink; Valley District, San Jose, Look to Ensure Adequate Future Supply." *San Jose Mercury News*, 25 September 2007.
Rose, Bleys. "Farmers, Vintners Cool to Prospect of Recycled Water for Irrigation." *Santa Rosa Press Democrat*, 16 May 2007.
Royte, Elizabeth. "Drugging Our Waters." NRDC: *On Earth*, www.nrdc.org/onearth/06, Fall 2006.
"Scientists Use Wastewater for Community Drug Testing." SignOnSan Diego.com, signonsandiego.printthis.clickability.com, 24 August 2007.
Scruggs, Caroline, et al. "EDCs in Wastewater: What's the Next Step?" *Update: The Official Newsletter of Weftec*, Summer 2006.
"Study Finds Antibiotics in Waterway: Antibiotics Used in Livestock Feeding Operations Discovered in Water Supplies." *The Aquifer: A Publication of the Groundwater Foundation* 20, no. 1 (Summer 2005).
"This Is Your River on Drugs." *American Rivers*, Winter–Spring 2007.
"Transactions." *Water Strategist*, November 2007, p. 2.
Underwood, Anne. "Rivers of Doubt: Minute Quantities of Everyday Contaminants in Our Drinking Supply Could Add Up to Big Trouble." *Newsweek*, 4 June 2007.
"What Is Reclaimed Water?" Tucson Water Department, www.ci.tucson.az.us/water/what_is_reclaim.htm (accessed 24 June 2007).
Woodhouse, Betsy. "Pharmaceuticals and Other Wastewater Products in Our Waters—a New Can of Worms?" *Southwest Hydrology*, November–December 2003, p. 12.
Letters, Memoranda, Miscellaneous
Arnold, Robert G., et al. "Water Quality: Estrogens and Effluent. Total Estrogenic Activity in Reclaimed Water and Stormwater." *WaterSustainability*, The University of Arizona, n.d.
"Ina Road Water Pollution Control Facility—Pima County." Wastewater Management Department, Tucson, AZ. Brochure.
Redding, Melanie. "Pharmaceuticals and Personal Care Products in Water." Sixth Annual Hydrogeology Symposium, Tacoma, WA, 2 May 2007.

Interviews

Avery, Chris. Attorney, City Attorney's Office, Tucson, AZ. 6 July 2005.
Basefsky, Mitch. Tucson Water. 30 July 2007.
Markus, Mike. Orange County Water District. 30 November 2007.

Chapter 10. Creative Conservation

Books

Davis, Scott. *Microhydro: Clean Power from Water*. Gabriola Island, BC, Canada: New Society, 2003.
Flores, Heather C. *Food Not Lawns: How to Turn Your Yard into a Garden and Your Neighborhood into a Community*. White River Junction, VT: Chelsea Green, 2006.

Steinberg, Theodore. *American Green: The Obsessive Quest for the Perfect Lawn*. New York: Norton, 2006.

Articles and Reports

"Albuquerque Progress Report." 2004.
"Albuquerque Water Use at Record Lows." *New Mexico Business Weekly*, 19 June 2007.
Bennett, Darryn. "Water Officials Mull Meter Restrictions." *North County Times*, www.nctimes.com/articles/2007, 28 December 2007.
Branom, Mike. "Fountain Hills Park Overhaul Kicks Off." *East Valley Tribune*, 24 June 2007.
"CAGRD Baseline Brochure Effectiveness Study: Final Analysis." Water Conservation Alliance of Southern Arizona, 1 September 2007.
Conaughton, Gig. "Water Conservation: Water Sales Hit Historic Peak; Calls for Conservation Appear to Be Failing to Soak In." *North County Times*, 11 December 2007.
Dickinson, Mary Ann. "Rinse and Save: How Saving Water Saved Energy." *Water Conservation News*, Fall–Winter 2004–2005.
Dicum, Gregory. "The Dirty Water Underground." *New York Times*, 31 May 2007.
"Fighting Turf War with Polka Dots." *CBS News*, www.cbsnews.com/stories/2005, 5 January 2005.
Gleick, Peter. "Billions of Drops in the Bucket: Just Rethinking How We Use Water Can Be as Effective as Huge Infrastructure Projects." *Los Angeles Times*, 6 January 2008.
Gleick, Peter H. "Flushing Water and Money Down the Drain." *San Francisco Chronicle*, 12 October 2006.
Gleick, Peter H., et al. "Waste Not, Want Not: The Potential for Urban Water Conservation in California." Pacific Institute, November 2003.
Goodnough, Abby. "Florida Is Slow to See the Need to Save Water." *New York Times*, 19 June 2007.
Hall, Ron. "San Antonio: Water Conservation's Future?" *Golfdom Insider*, www.golfdom.com/golfdom/article/articleDetail, 23 June 2007.
Hewitt, Alison. "County Plan Cuts Greenery to Save Water." *Whittier Daily News*, www.whittierdailynews.com (accessed 22 October 2007).
Kolbert, Elizabeth. "Turf War." *New Yorker*, 21 July 2008.
Klotz, Eric. "Utah Water Conservation Report." 2005.
Lafferty, Mike. "Grass." *Columbus Dispatch*, 5 June 2007.
Laster, Jennifer Roolf. "Dual-Flush Toilets Provide Water-Saving Option." www.mysanantonio.com/salife/gardening/stories/, 26 January 2007.
Lee, Mike. "Water-Conservation Appeals Sink In—but Slowly." *San Diego Union-Tribune*, 29 September 2007.
Little, Val L., and Rebecca Gallup. "Evaluation and Cost Benefit Analysis of Municipal Water Conservation Programs." Water Conservation Alliance of Southern Arizona, n.d.
Little, Val L., and Rebecca Gallup. "Municipal Water Conservation." Water Conservation Alliance of Southern Arizona. *Water Report*, no. 30 (15 August 2006).
Navarro, Mireya. "Flora with a Star in Its Corner." *New York Times*, 7 July 2005.
"New California Water Legislation: From Fish Farming to Low-Flush Toilets—Senate Bill-201 and Assembly Bill 2496." *Western Water Law*, July 2006, p. 244.

Rehfeld, Barry. "Steps to Take Before the Collective Well Runs Dry." *New York Times*, 13 November 2005.

Sangree, Hudson. "Prison Goal: Flush Control; New Electronic Toilet Device Can Prevent Misuse, Save Water." *Sacramento Bee*, 16 June 2007.

Schoch, Deborah. "Irrigation Issues: Irrigation Irritations Plague Cities; as Drought-Conscious Southern California Cities Urge Water Conservation; Too Many City- and State-Owned Watering Systems Waste Millions of Gallons on Roadways." *Los Angeles Times*, 27 October 2007.

Shelton, Shelley. "'Smart' Irrigation." *Arizona Daily Star*, 28 June 2007.

"Water CASA Research." *Water Conservation Alliance of Southern Arizona*, www.watercasa.org/research (accessed 10 January 2006).

"Water-Saving Rules for New Homes in Albuquerque Area." *U.S. Water News Online*, www.uswaternews.com (accessed 18 February 2008), February 2008.

"WaterSense: Efficiency Made Easy; High-Efficiency Toilets (HETs)." U.S. Environmental Protection Agency, www.epa.gov/watersense/pubs/ het.htm (accessed 27 June 2007).

Letters, Memoranda, Miscellaneous

Arakawa, PE, Stephen N. and Mark J. Graham, PE. "Changing Values—Recognizing the Value of Conservation." Twenty-Fifth Annual Water Law Conference, Coronado, CA, 22–23 February 2007.

"Per Capita Water Use in Long Beach Reduced 12% Since 2000." Press Release. Long Beach Water Department, 2 November 2006.

"Town of Chino Valley Announces Water Sustainability Plan." Chino Valley, AZ, 11 May 2007.

"Utilizing Center Pivot Sprinkler Irrigation Systems to Maximize Water Savings." United States Department of Agriculture, Natural Resources Conservation Service, n.d.

Water Conservation Strategy Resolution. City of Albuquerque, 27 June 2007.

Interviews

Carr, Chuck. Pasadena, CA. 4 January 2006.

Guz, Karen. Director of conservation programs, San Antonio Water System. 27 June 2007.

Chapter 11. Water Harvesting

Books

Lancaster, Brad. *Rainwater Harvesting for Drylands and Beyond: Guiding Principles to Welcome Rain into Your Life and Landscape*. White River Junction, VT: Chelsea Green, 2006.

Articles and Reports

Berman, John, and Shani Meewella. "Greywater Guerrillas: No Water Down the Drain in Vain." *ABC News*, abcnews.go.com, 21 August 2007.

Boren, Rebecca. "State Offers a Tax Credit: Reuse That Household H_2O." *Arizona Daily Star*, 8 April 2007.

"Clean Water: What Price Clean Water? Our View: Runoff Rules May Affect Your Wallet More Than Water Quality." Editorial. *North County Times*, 25 January 2007.

Davis, Richard S. "Green Infrastructure: A Storm Perfected?" *Western Water Law*, November 2007, p. 3.

Downery, Dave. "Storm Water Runoff: Rain in Drain Biggest Pain for Ocean." *North County Times*, 14 May 2007.

Dunn, Alexandra Dapolito, and Nancy Stoner. "Green Light for Green Infrastructure." *Environmental Law Institute*, Washington, DC, May–June 2007.

"Green Infrastructure." U.S. Environmental Protection Agency, cfpub.epa.gov/npdes/home.cfm (accessed 20 December 2007).

Kaye, Adam. "Clean-Water Initiative Flushed in Encinitas." *North County Times*, 10 March 2006.

Kaye, Adam. "Encinitas Seeks Proposals to Refund Clean-Water Fee." *North County Times*, 30 March 2006.

Kaye, Adam. "Encinitas Sued over Clean-Water Fee." *North County Times*, 23 October 2004.

Kray, Jeff. "A Perfect Storm(water)." *Environmental News*, 19 March 2008.

Lancaster, Brad. "Rainwater Harvesting for Drylands." www.harvesting rainwater.com (accessed 30 June 2007).

"New Homes and Lots in River Pointe: Manchester Township, NJ." *New Home Source*, www.newhomesource.com/search/community (accessed 2 February 2007).

Opdahl, Cristina. "Harvesting the Sky: Thirsty Santa Fe Catches On to Catching Rainwater." *High Country News*, 2 April 2007.

Payne, Troy L., and Janet Neuman. "Remembering Rain." *Environmental Law* 37, no. 1 (2007).

Price, Benjamin. "Scientist Sees Gray Water Risk." www.onlineathens.com, 25 November 2007.

"Rainwater Harvesting." www.portlandonline.com (accessed 30 June 2007).

"Residential Rainwater Collection and Use, San Juan Island." www.psat.wa.gov/Publications/LID (accessed 30 June 2007).

Rogers, Terry. "Regulation: State: Clean Up Coastal Waters; Local Governments Told to Curb Bacterial Pollution." *San Diego Union-Tribune*, 26 April 2007.

"Rooftop Rainwater Harvesting." King Street Center, City of Seattle. www.psat.wa.gov/Publications (accessed 30 June 2007).

"Rooftops to Rivers: Green Strategies for Controlling Stormwater and Combined Sewer Overflows." Natural Resources Defense Council (NRDC), May 2006.

Tweit, Susan J. "Raising the Roof." *Audubon*, March–April 2008, p. 40.

Weiss, Kenneth R. "Water Runoff Issues: Slowing a Tide of Pollutants; Runoff from Land Dwellers—Urban and Agricultural—Harms Coastal Waters, but There Are Solutions." *Los Angeles Times*, 25 December 2006.

Letters, Memoranda, Miscellaneous

Glanzberg, Joel, Speaker. "Water Harvesting Traditions in the Desert Southwest." Xeriscape Conference & Trade Fair 2005, Adapting to Our Changing Reality: Plants, Climate, Water, Technology. 2005.

Lancaster, Brad, Speaker. "Turning Drains into Sponges." ARCSA—SW Conference 2006, Tucson, Arizona, 27–28 October 2006.

Little, Val L. "Graywater Guidelines." The Water Conservation Alliance of Southern Arizona.

Phillips, Ann Audrey. "City of Tucson Water Harvesting Guidance Manual." Prepared for the City of Tucson, Department of Transportation, Stormwater Section, March 2003.

"Pollution Prompts Record Number of Beach Closings Nationwide: No-Swim Days Due to Stormwater Doubled from Previous Year, Says New Report." Press Release. Natural Resources Defense Council, 7 August 2007.

Shapiro, Neal, City of Santa Monica. "Managing Stormwater Through LID to Restore an Urbanized Watershed. Wisely Managing Our Urban Water Resources." ARCSA-SW Conference 2006, Tucson, Arizona, 27–28 October 2006.

"Using Gray Water at Home." The Arizona Department of Environmental Quality's Guide to Complying with the New, Simplified Type 1 General Permit, Publication No. C 01-06, April 2005.

Interviews

Brandon, Tony. Consultant, Athens, GA. 26 March 2007.

Dean, Teddy, and Alice Dean. Homeowners, San Juan Island, WA. 22 August 2007.

Mugavin, Walter J. New Jersey. 2 February 2007.

Pope, Tim. Northwest Water Source. 9 August 2007.

Chapter 12. Moore's Law

Articles and Reports

Beard, Betty. "Another Intel Plant in City Was Too Good to Pass Up." *Arizona Republic*, 25 October 2007.

"Corporate Responsibility Report 2005. Corporate Responsibility Report—Performance Indicators." Intel, www.intel.com/intel/finance (accessed 1 July 2007).

"Global Environmental Management Initiative (GEMI)," www.gemi.org/water/intel.htm (accessed 1 July 2007).

Fehr-Snyder, Kerry. "Intel's Fab Future: Ever-Smaller Chip Creates Huge Upside." *Arizona Republic*, 25 October 2007.

Flynn, Laurie J. "Google and Intel Lead Technology Effort to Cut Power Wasted by Computers." *New York Times*, 13 June 2007.

"Fun Facts: Exactly How Small (and Powerful) Is 45 Nanometers?" Intel Corporation Fact Sheet, n.d.

"Intel Corporation: Engaging Corporate-Level Support for Plant-Level Water Initiatives." Global Environmental Management Initiative. www.gemi.org/water/intel.htm. *Gemi*, 1 July 2007.

Jarman, Max. "Company Has No Plans to Close Older Chandler Factory. Chips Still in Demand for Graphics." *Arizona Republic*, 25 October 2007.

Jensen, Edythe. "Water Doesn't Get Wasted After Facilities Use It." *Arizona Republic*, 25 October 2007.

Markoff, John. "Intel Says Chips Will Run Faster, Using Less Power." *New York Times*, 27 January 2007.

"Moore's Law, Made Real by Intel Innovation." www.intel.com/technology/ mooreslaw (accessed 21 December 2007).

Pemberton, John. "Construction of Plant Exhilarated Manager." *Arizona Republic*, 25 October 2007.

Robertson, Jordan. "Intel Opening New Chip Plant in Arizona." Marketplace from American Public Media, marketplace.publicradio.org/ apheadline, 14 November 2007.

"Water Efficiency Leader Awards: 2007 WEL Winners." www.epa.gov (accessed 8 December 2007).

Letters, Memoranda, Miscellaneous

"Chipmaking." Brochure. Intel, (Undated).

"Intel Announces Investment in Rio Rancho, New Mexico Site." Press Release. 26 February 2007.

"Intel Eliminates Use of Lead from Future Microprocessors." Intel News Release. www.intel.com/pressroom/archive/releases (accessed 1 July 2007).

Shadman, Farhang, Director, NSF Engineering Research Center for Environmentally Benign Semiconductor Manufacturing. "Water Use and Reuse in High-Technology Industries." UA Water Forum 2006, 2 November 2006.

Interviews

Drago, Len, and Matthew Brandy. Intel Corporation, Chandler, AZ. 7 December 2007.

Robinson, Frank. Intel Corporation, Albuquerque, NM. 13 August 2007.

Shadman, Farhang. Professor, University of Arizona, Department of Hydrology. 25 January 2007.

Chapter 13. The Enigma of the Water Closet

Books

Benidickson, Jamie. *The Culture of Flushing: A Social and Legal History of Sewage*. Vancouver: University of British Columbia Press, 2007.

Coulter, Ann. *Godless: The Church of Liberalism*. New York: Crown Forum, 2006.

Goodland, Robert, Laura Orlando, and Jeff Anhang, eds. *Toward Sustainable Sanitation: What Is Sustainability in Sanitation?* Fargo, ND: International Association of Impact Assessment, 2001.

Lupton, Ellen, and J. Abbott Miller. *The Bathroom, the Kitchen, and the Aesthetics of Waste: A Process of Elimination*. Cambridge, MA: The Center; distributed by Princeton Architectural Press, 1996.

Morris, Robert D. *The Blue Death: Disease, Disaster, and the Water We Drink*. New York: HarperCollins, 2007.

Ogle, Maureen. *All the Modern Conveniences: American Household Plumbing, 1840–1890*. Baltimore: Johns Hopkins University Press, 2000.

Palmer, Roy. *The Water Closet: A New History*. Newton Abbot, Devon, England: David and Charles, 1973.
Snitow, Alan, Deborah Kaufmann, and Michael Fox. *Thirst: Fighting the Corporate Theft of Our Water*. San Francisco: Wiley, 2007.
Tomes, Nancy. *The Gospel of Germs: Men, Women, and the Microbe in American Life*. Cambridge, MA: Harvard University Press, 1998.
Woelfle-Erskine, Cleo, July Oskar Cole, and Laura Allen, eds. *Dam Nation: Dispatches from the Water Underground*. New York: Soft Skull Press, 2007.

Articles and Reports

Andreen, William. L. "The Evolution of Water Pollution Control in the United States: State, Local, and Federal Efforts, 1789–1972: Part I." *Stanford Environmental Law Journal* 22, no. 1 (January 2003): 215.
Andreen, William L. "The Evolution of Water Pollution Control in the United States—State, Local, and Federal Efforts, 1789–1972: Part II." *Stanford Environmental Law Journal* 22, no. 2 (June 2003): 145.
Barringer, Felicity. "Clean Water Fund Facing Major Cuts." *New York Times*, 8 February 2005.
Carlton, Jim. "Water Isn't a Fixture in Some Cutting-Edge Urinals." *Wall Street Journal*, 11 November 2003.
"Clean Water: How States Allocate Revolving Loan Funds and Measure Their Benefits." U.S. Government Accountability Office, June 2006.
Copeland, Claudia. "Clean Water Act Issues in the 108th Congress." Congressional Research Service, Order Code IB10108. 27 January 2003.
Copeland, Claudia, and Mary Tiemann. "Water Infrastructure Needs and Investment: Review and Analysis of Key Issues. Congressional Research Service, Order Code RL31116. 19 March 2007.
"Efficient Use: Against the Flow. Opposition to No-Water Urinals Shows Union Power." Editorial. *Los Angeles Daily News*, 2 March 2004.
Flaherty, Carol. "Biodegradable Portable Toilets a Hit." *Billings Gazette*, 27 June 2006.
Gefter, Amanda. "The Ugly Underneath." *Philadelphia Citypaper*, 13 July 2006.
Gleick, Peter H., et al. "Waste Not, Want Not: The Potential for Urban Water Conservation in California." Pacific Institute, November 2003, p. 42.
Horsley, Scott. "California Plumbers Stall Plans for No-Flush Urinals." National Public Radio, 1 March 2006.
Lavelle, Marianne, and Joshua Kurlantzick. "The Coming Water Crisis." *U.S. News and World Report*, 12 August 2002.
Morris, Robert D. "Pipe Dreams." *New York Times*, 3 October 2007.
Phoenix, Laurel E. "Aging Infrastructure: Coming Soon to a City Near You—Introduction." *Water Resources Impact*, September 2005, p. 3.
Reuters. "Biggest Threat to U.S. Drinking Water? Rust." *New York Times*, 24 January 2007.
Rockefeller, Abby A. "Civilization and Sludge: Notes on the History of the Management of Human Excreta." *Current World Leaders* 39, no. 6 (1997).
Stone, May N. "The Plumbing Paradox." *Winterthur Portfolio* 14 (Autumn 1979): 283.
Tarr, Joel A., et al. "Water and Wastes: A Retrospective Assessment of Wastewater

Technology in the United States, 1800–1932." *Technology and Culture* 25 (1984): 226.
"The Clean Water and Drinking Water Infrastructure Gap Analysis." *U.S. Environmental Protection Agency*, EPA-816-R-02-020, September 2002.
"The Dry Argument in the Men's Room." *World Plumbing Review* 1 (2006).
"Water Efficiency Technology Fact Sheet: Composting Toilets." *U.S. Environmental Protection Agency*, EPA 832-F-99-066, September 1999.
"Water Infrastructure." *U.S. General Accounting Office*, November 2001.
Williams, Florence. "Waste Not." *New York Times Magazine*, Spring 2007, p. 38; http://www.nytimes.com/2007/03/18/realestate/keymagazine/318GREEN.t.html?_r=1&em&ex=1174276800&en=b964c8bc01002fe8&ei=5087&oref=slogin
"World Toilet Expo & Forum 2006, 16th–18th November 2006, Bangkok, Thailand." World Toilet Organization, www.worldtoilet.org/hp/wto hp.htm (accessed 19 November 2006).
Yardley, William. "Gaping Reminders of Aging and Crumbling Pipes." *New York Times*, 8 February 2007.

Letters, Memoranda, Miscellaneous

"Composting Toilets." *Envirolet*, Envirolets Waterless & Low Water System Catalog, www.envirolet.com, 2004.
"Composting Toilet Process." *Clivus Multrum System Processes*, www.clivus multru.com/about_system_process.shtml (accessed 27 July 2007).
"Drinking Water [D–]." American Society of Civil Engineers, www.asce. org (accessed 5 April 2007).
"How Do We Use Drinking Water in Our Homes?" www.epa.gov/ safewater.
Kavanaugh, Michael C. "Unregulated and Emerging Chemical Contaminants: Technical and Institutional Challenges." Abstract. National Ground Water Association, Chicago, 2004.
Letter from Charles V. Dinges, American Society of Civil Engineers, to The Honorable John J. Duncan, 28 April 2004.
Missouri v. Illinois, 200 U.S. 496, 1906.
Popenoe, Chris. "Poop at the Zoo: The Bronx Zoo's Eco Restroom." Videos on Green and Eco Living, www.linkedin.com/in/popenoe and dawn-productions.blip.tv/.
"SunMar Self Contained Composters and Clivus Multrum Composting Toilet System Sales, Toilet Composter Installation, and Waterless Toilet Service." *Natural Home Building Source*, www.thenaturalhome.com/ compost.html (accessed 27 July 2007).
"Toilets." *Flex Your* Power, www.fypower.org/com/tools/products_results. html?id=100139 (accessed 27 July 2007).
"Why Is It Better?" *Falcon Waterfree Technologies*, www.falconwaterfree.com/ why/environmental.htm (accessed 3 November 2007).

Chapter 14. The Diamond-Water Paradox

Books

Barlow, Maude, and Tony Clarke. *Blue Gold: The Fight to Stop the Corporate Theft of the World's Water*. New York: New Press, 2002.

Dinar, Ariel, ed. *The Political Economy of Water Pricing Reforms*. New York: Oxford University Press, 2000.

Shirley, Mary M., ed. *Thirsting for Efficiency: The Economics and Politics of Urban Water System Reform*. New York: Pergamon Press, 2002.

Articles and Reports

"Case Studies of Sustainable Water and Wastewater Pricing." United States Environmental Protection Agency, December 2005.

Casselman, Ben. "Desperate Sprinklers." *Wall Street Journal*, 20 July 2007.

"CBO's Comments on H.R. 1071, a Bill on Subsidizing New Desalination Facilities." Congressional Budget Office, www.cbo.gov/ftpdoc.cfm, 24 May 2005.

"Current Water Rate Schedules." Tucson Water Department, www.ci.tucson.az.us/water/rates.htm, 7 August 2006.

Gleick, Peter. "The Human Right to Water." *Water Policy* 1 (1998): 487.

Gleick, Peter H., et al. "California Water 2030: An Efficient Future." Pacific Institute, September 2005, p. 7.

Glennon, Robert. "The Price of Water." *Journal of Land, Resources, and Environmental Law* 24 (2004): 337.

Glennon, Robert. "Water Scarcity, Marketing, and Privatization." *Texas Law Review* 83, no. 7 (June 2005): 1873.

Hardy, Terri. "Water Metering in Sacramento: Sacramento Water Wasters, Beware; Council Clears the Way to Phase In Meters over Twenty Years." *Sacramento Bee*, 2 November 2005.

Hendee, David. "NRDs Differ on Requiring Irrigation Well Meters." www.omaha.com, 1 August 2007.

"How Federal Policies Affect the Allocation of Water." Congress of the United States Congressional Budget Office, August 2006.

Kenny, Joan F. "Public Water-Supply Use in Kansas, 1987–97." Prepared in cooperation with the Kansas Water Office. U.S. Geological Survey, ks.water.usgs.gov/Kansas/pubs/fact-sheets/fs.187-99.html, January 2000.

Kolbert, Elizabeth. "Comment: Changing Lanes." *New Yorker*, August 11, 18, 2008.

Leavenworth, Stuart. "Conservation Efforts, Water Meters on the Way: New Law Requires Them in Sacramento by 2025." *Sacramento Bee*, 30 September 2004.

Maxwell, Steve. "Water Is Cheap—Ridiculously Cheap!" *Journal of the American Water Works Association* 97, no. 6 (June 2005): 4.

"NUS Consulting Survey Reports U.S. Water Costs Climbing Even Higher." *U.S. Water News*, September 2006.

Schoengold, Karina, et al. "Price Elasticity Reconsidered: Panel Estimation of an Agricultural Water Demand Function." *Water Resources Research* 42 (2006): 1029.

Shaw, Hank. "Groundwater Issues: Legislation to Measure Aquifer Faces Opposition." *Stockton Record*, 19 February 2007.

"Water Rate Structures in Utah: How Utah Cities Compare Using This Important Water Use Efficiency Tool." *Western Resource Advocates*, January 2005.

"Water Rates: Conserving Water and Protecting Revenues." Southwest Florida Water

Management District, www.swfwmd.state.fl.us/ conservation/waterrates/ (accessed 31 July 2007).

Letters, Memoranda, Miscellaneous

Kotis, Greg, and Mark Hollmann. Urinetown: The Musical. Directed by Spiro Veloudos.

Chapter 15. The Steel Deal

Articles and Reports

"Acequias." Published by New Mexico State Engineer Office. July 1997, Last Modified: 29 October 1998.
Anderton, Dave. "New Horizon at Geneva Steel." *Deseret Morning News*, 30 June 2005.
Brockman, James C. "Municipal Water Rights in New Mexico. Various Approaches—Various Development Impacts." *Water Report*, September 2006, pp. 1, 15.
"Elderly Woman Arrested in Clash over Dry Lawn Calls Experience 'Nightmare.'" *Fox News*, 9 July 2007.
Harwood, Kyle. "Santa Fe Water: Resources and Policy; Evolving "Wet Growth Regulations." *Water Report*, 15 February 2007, p. 22.
"History of Geneva Steel, Utah." www.onlineutah.com/steelhistory.shtml (accessed 3 August 2007).
"Jerry Olds, Utah State Engineer, Division of Water Rights." *Desert Wars: Water and the West*, kued.org/productions/desertwars (accessed 3 August 2007), 25 September 2006.
Matlock, Staci. "Farmers Feel Squeezed In as Water Becomes Scarce." *New Mexican*, 11 June 2006.
Tarlock, A. Dan, and Sarah B. Van de Wetering. "Western Growth and Sustainable Water Use: If There Are No "Natural Limits," Should We Worry About Water Supplies?" *Public Land and Resources Law Review* 27 (2006): 33.
Zellmer, Sandra. "Anti-Speculation and Water Law." *Water Report*, 15 April 2008, p. 1.

Letters, Memoranda, Miscellaneous

Hollenhorst, John. "Official Warns of Shrinking Groundwater Supplies." KSL.com, 5KSLTV, 14 September 2003.
"Order of the State Engineer." State of Utah, Department of Natural Resources, Division of Water Rights. 8 February 2007.

Interviews

Anderson, Oliver. 20 September 2005.
Astill, Dennis. Anderson Development Company, Adrian, MI. 7 August 2007.
Brockman, James C. Attorney, Santa Fe, NM. 13 August 2007.
Fowler, Bill. Anderson Development Company, Adrian, MI. 7 August 2007.
Jensen, Daniel. Attorney, Salt Lake City, UT. 3 August 2007.
Martinez, Eluid. Former state engineer, New Mexico, and commissioner, U.S. Bureau of Reclamation. 17 August 2007.

Olds, Jerry D. State engineer, UT. 27 December 2007.
Robinson, Christopher F. The Ensign Group, Salt Lake City, UT. 27 December 2007.
Ruhlman, Mark. Branch Design Development, Salt Lake City, UT. 17 August 2007.
Stein, Jay. Attorney, Santa Fe, NM. 9 August 2007.
Tullis, Rich. Assistant general manager, Central Utah Water Conservation District. 13 August 2007.
Waltz, Fred. Attorney, Taos, NM, 10 August 2007.
White, Bill. Attorney, water broker, Salt Lake City, UT. 8 August 2007.

Chapter 16. Privatization of Water

Books

Barlow, Maude. *Blue Covenant: The Global Water Crisis and the Coming Battle for the Right to Water*. New York: New Press, 2007.
Barlow, Maude, and Tony Clarke. *Blue Gold: The Fight to Stop the Corporate Theft of the World's Water*. New York: New Press, 2002.
Midkiff, Kenneth. *Not a Drop to Drink: America's Water Crisis (and What You Can Do)*. Novato, CA: New World Library, 2007.
Rothfeder, Jeffrey. *Every Drop for Sale: Our Desperate Battle over Water in a World About to Run Out*. New York: Tarcher/Putnam, 2001.
Scarborough, Brandon, and Hertha L. Lund. *Saving Our Streams: Harnessing Water Markets; A Practical Guide*. Bozeman, MT: Property and Environment Research Center, 2007.
Segerfeldt, Fredrik. *Water for Sale: How Business and the Market Can Resolve the World's Water Crisis*. Washington, DC: Cato Institute, 2005.
Shiva, Vandana. *Water Wars: Privatization, Pollution, and Profit*. Cambridge, MA: South End Press, 2002.
Snitow, Alan, Deborah Kaufmann, and Michael Fox. *Thirst: Fighting the Corporate Theft of Our Water*. San Francisco: Wiley, 2007.
Zinnbauer, Dieter, and Rebecca Dobson. *Global Corruption Report 2008: Corruption in the Water Sector*. New York: Cambridge University Press, 2008.

Articles and Reports

"Annual Report." *Oregon Water Trust*, 2005.
Arnold, Craig Anthony. "Privatization of Public Water Services: The States' Role in Ensuring Public Accountability." *Pepperdine Law Review* 32 (2005): 561.
Deutsch, Claudia H. "There's Money in Thirst: Global Demand for Clean Water Attracts Companies Big and Small." *New York Times*, 10 August 2006.
Finnegan, William. "Leasing the Rain: The World Is Running Out of Fresh Water, and the Fight to Control It Has Begun." *New Yorker*, 8 April 2002.
Forero, Juan. "Who Will Bring Water to the Bolivian Poor? Multinational Is Ousted, but Local Ills Persist." *New York Times*, 15 December 2005.
Glennon, Robert. "Water Scarcity, Marketing, and Privatization." *Texas Law Review* 83 (June 2005): 1873.

Koller, Frank. "No Silver Bullet: Water Privatization in Atlanta, Georgia—a Cautionary Tale." *CBCNews*, 5 February 2003.
Larson, Elizabeth. "Watershed Planning: Scott River Water Trust Seeks to Help Fish, Farmers; Program Leases Water Rights from Growers." *Capital Press*, 9 February 2007.
"National Land Trust Census Report." *Land Trust Alliance*, 2005.
Neuman, Janet C. "Have We Got a Deal for You: Can the East Borrow from the Western Water Marketing Experience?" *Georgia State University Law Review* 21 (2004): 449.
Neuman, Janet C. "The Good, the Bad, and the Ugly: The First Ten Years of the Oregon Water Trust." *University of Nebraska Law Review* 83 (2004): 432.
Purkey, Andrew. "Blue Ribbon Management for Blue Ribbon Streams." *PERC Reports*, June 2007.
"Rivers of Progress Built on Partnerships." *Oregonian*, 22 March 2007.
Rose, Carol M. "Privatization—the Road to Democracy." *St. Louis University Law Journal* 50 (2006): 691.
Scarborough, Brandon. "Buy That Fish a Drink." *PERC Reports*, Summer 2007, p. 5.
Schalch, Kathleen. "Profile: Difficulties of Providing Safe Drinking Water to Developing Nations." National Public Radio, 7 January 2003.
"Stories from the Field." Columbia Basin Water Transactions Program, www.cbwtp.org/jsp/cbwtp/stories/stories.jsp (accessed 4 August 2007).
"Teanaway River." Washington Water Trust, Winter 2002.
"Water Privatization Fiascos: Broken Promises and Social Turmoil." A special report by Public Citizen's Water for All program.
Wolff, Gary, and Eric Hallstein. "Beyond Privatization: Lessons for Restructuring Water Systems to Improve Performance." *Water Report*, 15 May 2006, p. 7.

Letters, Memoranda, Miscellaneous

Gersh, Jeff. "Balance in the Basin. CD." National Fish & Wildlife Foundation.
Kasic, Kathy. "Against the Current." A Metamorph films production in association with Trout Unlimited. January 2007.
Pagel, Martha O. "Presentation: Oregon Options for Market-Based Instream Flow Restoration." Instream Use and Changing Values (ESA, Urban Streams, Recreation), American Bar Association Section of Environment, Energy, and Resources, 25th Annual Water Law Conference, Coronado, CA, 22–23 February 2007.

Interview

Purkey, Andrew. Program director, Columbia Basin Water Transactions Program, Western Partnership Office, Portland, OR. 11 April 2007.

Chapter 17. Take the Money and Run

Books

Erie, Steven P. *Beyond Chinatown: The Metropolitan Water District, Growth, and the Environment in Southern California*. Stanford, CA: Stanford University Press, 2006.

Green, Dorothy. *Managing Water: Avoiding Crisis in California*. Berkeley: University of California Press, 2007.

Haddad, Brent M. *Rivers of Gold: Designing Markets to Allocate Water in California*. Washington, DC: Island Press, 2000.

Hoffman, Abraham. *Vision or Villainy: Origins of the Owens Valley–Los Angeles Water Controversy*. College Station: Texas A&M University Press, 1981.

Articles and Reports

"Background on California's Quantification Settlement Agreement (QSA)." Imperial Irrigation District, March 2003.

"California Company Plans Yuma Paper Mill." *Arizona Republic*, Associated Press, 10 September 2007.

"California Strikes a Water Truce." *High Country News*, October 2003, p. 93.

"California's Imperial Irrigation District Announces Intent to Halt New Water Transfers." *Western Water Law*, January 2006, p. 64.

Conaughton, Gig. "Water Authority Agrees to Pay $40 Million to Imperial Valley." *North County Times*, 9 May 2007.

Conaughton, Gig. "Water Authority Wants to Rework Imperial Deal's Price." *North County Times*, 10 September 2007.

Gardner, Michael. "Water Authority OKs Larger Payment: Millions Will Go to Imperial Valley." *San Diego Union-Tribune*, 9 May 2007.

Hanak, Ellen. "Who Should Be Allowed to Sell Water in California? Third-Party Issues and the Water Market." San Francisco: Public Policy Institute of California, 2003.

Hettena, Seth. "Cowboy-Turned-Lawyer Is Bush Administration's Water Czar." *San Diego Union-Tribune*, 19 June 2004.

"Imperial Irrigation District." www.iid.com/aboutiid (accessed 3 May 2004).

Leshy, John D. "Special Project: Irrigation Districts. Irrigation Districts in a Changing West—an Overview." *Arizona State Law Journal* (1982): 345.

Lusk, Brianna. "Judge Dismisses QSA-IID Lawsuit." *Imperial Valley Press*, 15 February 2008.

"Official Says Water-Using Plant Won't Hurt Yuma." *Arizona Daily Star*, 13 November 2007.

Pope, Don. "Conversions of Water from Irrigation Use to Municipal and Industrial (Domestic) Use." Yuma County Water Users' Association, n.d.

Sax, Joseph L. "Understanding Transfers: Community Rights and the Privatization of Water." *Hastings West-Northwest Journal of Environmental Law and Policy* 1 (1993): 13.

Schwartz, Nelson D. "Far from the Reservation, but Still Sacred?" *New York Times*, 12 August 2007.

"SD Water Authority and Imperial Irrigation District Settle Arbitration." *U.S. Water News* June 2007, p. 30.

Simon, Darren. "IID Dismantles Local Entity." *Imperial Valley Press*, 27 July 2006.

Simon, Darren. "New Imperial Irrigation District Directors: Hanks, Sanchez, and Abatti Sworn In at IID." *Imperial Valley Press*, 9 December 2006.

Taylor, James J. "The Colorado River Quantification Settlement Agreement on the Brink—a View from San Diego." *California Water Law and Policy*, April 2003, p. 191.

"Will Morgan v. Imperial Irrigation District Unravel the Quantification Settlement Agreement?" *California Water Law and Policy*, December 2003, p. 85.

Letters, Memoranda, Miscellaneous

"Colorado River Water Delivery Agreement: Federal Quantification Settlement Agreement." 10 October 2003.

Gastelum, Ronald R., Esq., Chief Executive Officer, The Metropolitan Water District of Southern California, Los Angeles. "Implementation of the Quantification Settlement Agreement Plus 2004 Status Report." Presentation. May 2004.

"Quantification Settlement Agreement (QSA Cases) Coordinated Civil Case." Superior Court of California, County of Sacramento, www.saccourt.com/CoordCases (accessed 13 December 2004).

Rossmann, Antonio, Lecturer in Water Law, U.C. Berkeley (Boalt Hall). "Peace on the River in California? Not Quite. Legal and Institutional Issue Remain after the QSA." Presentation, delivered at Colorado River Water Users Assn., Caesar's Palace, Law Vegas. 11 December 2003.

Interviews

Aladjem, David. Attorney, Sacramento, CA. August 2003.

Cox, Don. Farmer and former board member, Imperial Irrigation District. 19 May 2004.

Dishlip, Herb. Central Arizona Water Conservation District. 28 May 2004.

Ellsworth, Mark. 19 March 2007.

Fulton, Mark. Blythe, CA. 20 May 2004.

Gastelum, Ron. Attorney, chief executive officer, Metropolitan Water District of Southern California. 13 May 2004.

Horne, Andy. Board member, Imperial Irrigation District. 18 May 2004.

Howitt, Richard. Economist, University of California, Davis. 5 February 2005.

Maruca, Joe. Member, Imperial County Board of Supervisors. 19 May 2004.

McKeith, Malissa. Attorney, Los Angeles. 18 June 2004.

Morgan, Deirdre. Imperial Irrigation District. 18 May 2004.

Pope, Donald. Manager, Yuma County Water Users' Association. 16 December 2003.

Porelli, Jack. Coachella Valley Water District, Coachella Valley, CA. 23 August 2007.

Raley, Bennett. Attorney, assistant secretary for water and science, U.S. Department of the Interior, Las Vegas, NV. 13 May 2004.

Robbins, Steven. General manager, Coachella Valley Water District, Coachella Valley, CA. 10 January 2008.

Rossman, Tony. Special counsel, Imperial County. 10 June 2004.

Scoonover, Mary. Resources Law Group. 11 October 2004.

Shields, Tina. Assistant manager, Water Department, Imperial Irrigation District. 3 July 2007.

Slocum, Charlie, and Gary Langford. Wellton-Mohawk Irrigation and Drainage District. 15 December 2003.

Smith, Ed. General manager, Palo Verde Irrigation District. 20 May 2004.

Smith, Rodney T. Economist, Austin, TX. 4 February 2005.

Swan, William. Attorney, Imperial Irrigation District. 30 April 2004.
Underwood, Dennis. Vice president, Metropolitan Water District of Southern California, Las Vegas, NV. 14 May 2004.

Chapter 18. The Future of Farming

Books

Pollan, Michael. *The Omnivore's Dilemma: A Natural History of Four Meals*. New York: Penguin Press, 2006.

Articles and Reports

"Americans Spend Less Than 10 Percent of Disposable Income on Food." *Salem-News*, 19 July 2006.
Barber, Dan. "Amber Fields of Bland." *New York Times*, 14 January 2007.
Barrionuevo, Alexei. "More Farmers Seek Subsidies as U.S. Eats Imported Produce." *New York Times*, 3 December 2006.
"Cultivating Waste." www.washingtonpost.com, 10 July 2006.
Egan, Timothy. "For Farmers, Subsidies Are a Matter of What Kind of Row You Hoe." *New York Times*, 18 February 2005.
Egan, Timothy. "Red State Welfare." *New York Times*, 28 June 2007.
"Finding Free Money." www.washingtonpost.com, 2 July 2006.
Gleick, Peter. "Charting California's Water Future: Panic Makes Poor Policy." *San Francisco Chronicle*, 22 July 2007.
Hesser, Amanda. "Salad in Sealed Bags Isn't So Simple, It Seems." *New York Times*, 14 January 2007.
Horovitz, Bruce. "Salads Grow into Profitable Garden of Eatin.'" *USA Today*, 24 February 2005.
"How Federal Policies Affect the Allocation of Water." *Congress of the United States Congressional Budget Office*, August 2006.
Kristof, Nicholas D. "Dems Ram Through Grotesque Farm Bill." *New York Times*, August 2007.
"Lessons from the Green Revolution: Do We Need New Technology to End Hunger?" www.foodfirst.org, March–April 2000.
"Lettuce: In and Out of the Bag." *Agricultural Outlook*, April 2001, p. 10.
McKinnon, Shaun. "Farmers Fear Dry Future." Arizona Republic, 25 July 2004.
Molden, David. "Water for Food, Water for Life." A comprehensive assessment of water management in agriculture. International Water Management Institute, 2007.
Morgan, Dan, et al. "Growers Reap Benefits Even in Good Years: Crops That Sell High Qualify for Payments." www.washingtonpost.com, 3 July 2006.
"Number of Farms, Land in Farms, and Average-Size Farm: United States, 1990–2005." www.inforplease.com (accessed 21 March 2008).
Philp, Tom. "Increased Demand, Unreliable Supply: Taste for California Almonds Grows with Farmers' Quest for Water." *Sacramento Bee*, 18 March 2007.

Pollick, Dennis. "Ag Loses Ground in State 90,000 Farm Acres Changed to Urban Uses in 1998–2000." *Fresno Bee*, 4 June 2003.
Putnam, Dan, et al. "Alfalfa, Wildlife, and the Environment: The Importance and Benefits of Alfalfa in the Twenty-First Century." Executive Summary, California Alfalfa and Forage Association, 2001.
Sirekis, Cyndie. "Ten Years of Celebrating Safe, Affordable, Abundant Food." *The Voice of Agriculture—American Farm Bureau*, 29 January 2007.
Sokolow, Alvin D. "Protecting Farmland in the United States: An Outline of Optional Policy Strategies and Techniques." University of California-Cooperative Extension, Revised, January 2004.
Sorenson, Dan. "Pampered Tomatoes." *Arizona Daily Star*, 23 April 2006.
"State's Farmland Disappearing at a Faster Rate: More Than 90,000 Acres Urbanized from 1998–2000." California Department of Conservation News Room, 2 June 2003.
Tronstad, Russell, et al. "Marginal Value of Water for End-of-Season Cotton Production." *Arizona Review*, Spring 2005, p. 6.
"Value-Added Products Cropping Up Everywhere." *Fresh Cut Magazine*, December 2004.
Western, Sam. "A New Green Revolution." *High Country News*, 26 December 2005.

Interviews

Anderson, Oliver. Farmer, Maricopa-Stanfield Irrigation and Drainage District. 20 September 2005.
Ellsworth, Mark. Regional manager, Earthbound Farm, Yuma, AZ. 19 March 2007.
Hetrick, John. Salt River Project. 5 July 2005.
Muggli, Roger. General manager, Tongue and Yellowstone Irrigation District, Miles City, MT. 17 August 2007.
Norton, John. Cotton Norton Stevenson Consulting, Phoenix, AZ. 20 September 2005.
Sigg, Joe. Arizona Farm Bureau, Phoenix. 5 July 2005.

Chapter 19. Environmental Transfers

Books

Haddad, Brent M. *Rivers of Gold: Designing Markets to Allocate Water in California*. Washington, DC: Island Press, 2000.

Articles and Reports

"All American Canal: Canal Bill Could Be Final Answer." Editorial. *Imperial Valley Press*, 13 December 2006.
"Appeals Court OKs Concrete Lining for Canal on U.S.–Mexico Border." 7 April 2007.
Bradshaw, Stan, and Laura Ziemer. "Water Leasing in Montana Through Trout Unlimited's Eyes." *PERC Reports*, Summer 2007, p. 12.
Dibble, Sandra. "Wetlands Become a Focus in Debate over Canal Lining." *San Diego Union-Tribune*, 6 June 2005.
Eilperin, Juliet. "Tide of Sentiment Shifts in Water War." *Washington Post*, 15 January 2006.
Glennon, Robert, and Peter Culp. "The Last Green Lagoon: How and Why the Bush

Administration Should Save the Colorado River Delta." *Ecology Law Quarterly* 28 (2002): 903.
Hendricks, Tyche. "In Bone-Dry Region, a Battle for Leaking Water." *San Francisco Chronicle*, 13 April 2007.
Hinojosa-Huerta, Osvel, et al. "Andrade Mesa Wetlands of the All-American Canal." *Natural Resources Journal* 42, no. 4 (Fall 2002).
"Irrigated Land Area in US for 1970 Through 2000 by System Type." Compiled by Freddie Lamm and Vicki Brown. Summarized by Freddy Lamm, Kansas State University. *Irrigation Journal*, 20 January 2004. www.oznet.ksu.edu/pn_irrigate/news/ILandarea.htm.
Jenkins, Matt. "The Efficiency Paradox." *High Country News*, 5 February 2007, p. 8.
King, Mary Ann. "Getting Our Feet Wet: An Introduction to Water Trusts." *Harvard Environmental Law Review* 28 (2004): 495.
Malloch, Steven. "Liquid Assets: Protecting and Restoring the West's Rivers and Wetlands Through Environmental Water Transactions." Trout Unlimited, Inc., March 2005.
Neuman, Janet. "The Good, the Bad, and the Ugly: The First Ten Years of the Oregon Water Trust." *Nebraska Law Review* 83 (2004): 432.
Neuman, Janet. "Wading into the Water Market: The First Five Years of the Oregon Water Trust." *Journal of Environmental Law and Litigation* (University of Oregon) 14 (1999): 135.
Nolde, Haley. "Mexicali: Living on Borrowed Water." www.journalism. berkeley.edu (accessed 3 July 2007).
"Oregon Water Trust." *Water Strategist*, March 2006.
Perry, Tony. "Court Rules Imperial Valley Canal to Be Lined: The Project Will Decrease Seepage of Colorado River Water Flowing to San Diego County." *Los Angeles Times*, 7 April 2007.
"Private Water Leasing a Montana Approach." A report on the 10-year history of a unique Montana program. Trout Unlimited, 2004.
Schmidt, Lisa. "Partners Spawn Ideal Trout Stream." Natural Resources Conservation Service,, www.mt.nrcs.usda.gov/new/project/tu_ bulltrout.html (accessed 6 August 2007).
Spillman, Benjamin. "Paving the Way for Water Savings." *Desert Sun*, 5 April 2006.
Zamora-Arroyo, Francisco, et al. "Conservation Priorities in the Colorado River Delta Mexico and the United States." 2005.

Letters, Memoranda, Miscellaneous

"All American Canal." Fact Sheet. Imperial Irrigation District, March 2003.
"Canal Lining Projects." Fact Sheet. San Diego Country Water Authority, January 2007.
Gheleta, Michael A., Esq. "International Water Conflicts." Presentation, Colorado River CLE, Las Vegas, May 2007.
Ziemer, Laura S., Director, Trout Unlimited's Montana Water Project. "Addressing River and Stream Flow Depletions Through TMDLs: The Montana Experience." American Bar Association Section of Environment, Energy, and Resources, 24th Annual Water Law Conference, Coronado, CA, 23–24 February 2006.

Interviews

Randall, Brianna. Montana Water Trust, Bozeman. June 2006.

Ziemer, Laura. Attorney, director of Trout Unlimited's Montana Water Project, Bozeman. 9 July 2006.

Chapter 20. The Buffalo's Lament

Books

Anderson, Terry L., and Pamela Snyder. *Water Markets: Priming the Invisible Pump*. Washington, DC: Cato Institute, 1997.

Committee on Western Water Management, National Research Council. Water Transfers in the West: Efficiency, Equity, and the Environment. Washington, DC: National Academy Press, 1992.

Ellis, Richard. *The Empty Ocean: Plundering the World's Marine Life*. Washington, DC: Island Press, Shearwater Books, 2003.

Foley, Duncan K. *Adam's Fallacy: A Guide to Economic Theology*. Cambridge, MA: Harvard University Press, 2006.

Isenberg, Andrew C. *The Destruction of the Bison: An Environmental History, 1750–1920*. Cambridge, England: Cambridge University Press, 2000.

Rinella, Steven. *American Buffalo: In Search of a Lost Icon*. New York: Spiegel and Grau, 2008.

Roberts, Callum. *The Unnatural History of the Sea*. Washington, DC: Island Press, 2007.

Articles and Reports

Carey, Janis M., and David L. Sunding. "Emerging Markets in Water: A Comparative Institutional Analysis of the Central Valley and Colorado–Big Thompson Projects." *Natural Resources Journal* 41 (2001): 283.

Carter, Harold O., et al. "Sharing Scarcity: Gainers and Losers in Water Marketing." Davis: University of California, Agricultural Issues Center, 1994.

Dellapenna, Joseph W. "The Importance of Getting Names Right: The Myth of Markets for Water." *William and Mary Environment Law and Policy Review* 25, no. 2 (Winter 2000): 317.

Gleick, Peter H. "California's 'Economic Productivity' of Water Use: Jobs, Income, and Water Use in California." Pacific Institute, Version 1, 22 October 2004,

Glennon, Robert, and Michael J. Pearce. "Transferring Mainstem Colorado River Water Rights: The Arizona Experience." Arizona Law Review 49 (2007): 235.

Glennon, Robert, et al. "Law and the New Institutional Economics: Water Markets and Legal Change in California, 1987–2005." *Washington University Journal of Law and Policy* 26 (2008).

Glennon, Robert, et al. "Transferring Water in the American West: 1987–2005." University of Michigan Journal of Law Reform 40 (2007): 1021.

Glennon, Robert, et al. "Water Markets in the West: Prices, Trading, and Contractual Forms." Economic Inquiry 46 (2008): 91.

Glennon, Robert. "Water Scarcity, Marketing, and Privatization." Texas Law Review 83 (2005): 1873.

Graff, Thomas J., and David Yardas. "Reforming Western Water Policy: Markets and Regulation." National Resources and Environment 12 (1998): 165.
Howe, Charles W. "Water Transfers and Their Impacts: Lessons from Three Colorado Water Markets." *American Water Resources Association* 39, no. 5 (October 2003): 1055.
Loy, Wesley. "Halibut Greeted at Dock with Price Offer over $5, Demand: The First Fish Are Favored in Restaurants, Command Premium Dollars." *Anchorage Daily News*, 13 March 2007.
MacDonnell, Lawrence J. "Commentary: Public Water; Anti-Speculation, Water Reallocation, and High Plains A & M, LLC v. Southeastern Co Water Consv. Disctrict." University of Denver Law Review 10, no. 1 (Fall 2006): 1.
Montoya, Maria E. "The Decline of the Great Plains." *Reviews in American History* 25, no. 4 (2001): 610.
Rose, Carol M. "Privatization—the Road to Democracy?" *St. Louis University Law Journal* 50 (2006): 691.
Steinfels, Peter. "Economics: The Invisible Hand of the Market." *New York Times*, 25 November 2006.
"Sustaining America's Fisheries and Fishing Communities: An Evaluation of Incentive-Based Management." *Environmental Defense*, 2008.
Thompson, Barton H., Jr. "Institutional Perspectives on Water Policy and Markets." California Law Review 81 (1993): 671.
Tierney, John. "World's Fish Supply Not as Dire as Claimed." *Arizona Daily Star*, 7 November 2006.
"Water Heist, How Corporations Are Cashing In on California's Water." *Public Citizen*, December 2003.
Yolles, Peter L. "Update 2000: Progress and Limitations in Developing a Water Market in California." *Water Resources Update*, January 2001.
Letters, Memoranda, Miscellaneous
"Landmark Study Shows Clear Pathway to Restoring Imperiled Fish Populations." *Environmental Defense*, press release, 28 March 2007.

Interviews

Fereday, Jeff. Attorney, Boise, ID. 9 August 2007.
MacDonnell, Larry. Attorney, Boulder, CO.
McNulty, Michael. Attorney, Tucson, AZ. 7 June 2004.
Muggli, Roger. General manager, Tongue and Yellowstone Irrigation District, Miles City, MT. 17 August 2007.

Conclusion. A Blueprint for Reform

Articles and Reports

Albright, David. "House Eyes Groundwater Effects of Bottled Water Production." *Inside EPA*, newsletter, 16 January 2008.

Alley, William M., and Charles J. Taylor. "The Value of Long-Term Ground Water Level Monitoring." *Ground Water* 39, no. 6 (November–December 2001): 801.
Barringer, Felicity. "West Coast Enigma as Salmon Vanish Without a Trace." *New York Times*, 17 March 2008.
"California Water Code § 386."—Findings Prior to transfer; fees of petitioner for transfer, 1982.
"Conservation." www.ci.tucson.az.us/water/conservation.htm (accessed 20 March 2008).
Davies, Lincoln L. "Just a Big 'Hot Fuss'? Assessing the Value of Connecting Suburban Sprawl, Land Use, and Water Rights Through Assured Supply Laws." *Ecology Law Quarterly* 34 (2007): 1217.
"Estimating Water Use in the United States—a New Paradigm for the National Water-Use Information Program." *National Research Council*, National Academy Press, Washington, D.C., 176 p. 2002.
Fimrite, Peter. "Threat of Closing Jolts Fishing Industry." *San Francisco Chronicle*, 13 March 2008.
Hanak, Ellen. "Linking Housing Growth to Water Supply: New Planning Frontiers in the American West." *Journal of the American Planning Association* 72, no. 2 (Spring 2006): 164.
"House Farm Bill Includes Regional Water Enhancement Program." *Infrastructure Watch*, 19 September 2007.
Leshy, John D. "Interstate Groundwater Resources: The Federal Role." *Journal of Environmental Law and Policy* 14 (2008): 1475.
Leshy, John D. "The Federal Role in Managing the Nation's Groundwater." *Environmental Law and Policy* 11 (2004–2005): 1.
"Members of Congress Create House Water Caucus." *Water Strategist*, May 2007, p. 17.
"San Antonio Programs and Rebates." www.saws.org/conservation/programs/ (accessed 20 March 2008).
Taylor, Charles J., and William M. Alley. "Ground-Water-Level Monitoring and the Importance of Long-Term Water-Level Data." USGS Circular 1217. Denver, CO: U.S. Department of the Interior, U.S. Geological Survey, 2001.
"Threat of Closing Jolts Fishing Industry." *San Francisco Chronicle*, 13 March 2008.
"Water Information for Albuquerque." www.cabq.gov/water/ (accessed 20 March 2008).
"Water Saving Tips." *Water Conservation Alliance of Southern Arizona*, www.watercasa.org (accessed 20 March 2008).
Weiser, Matt. "Officials Shut Salmon Fishing in Seven Coastal Areas of California, Oregon." *Sacramento Bee*, 13 March 2008.

Epilogue. The Salton Sea

Books

Mulholland, Catherine. *William Mulholland and the Rise of Los Angeles*. Berkeley: University of California Press, 2000.
Stringfellow, Kim. *Greetings from the Salton Sea: Folly and Intervention in the Southern California Landscape, 1905–2005*. Santa Fe, NM: Center for American Places, 2005.

Articles and Reports

Almeida, Christina. "Salton Sea: State Releases Salton Sea Restoration Plan; Lawmakers to Debate." *Associated Press*, 25 May 2007.

Bowles, Jennifer, and Jim Miller. "Expensive Salton Sea Plan Unveiled: $8.9 Billion; Even the Proposal's Supporters Say the Price Tag Will Make It a Hard Sell." *Riverside Press Enterprise*, 25 May 2007.

"Coachella Valley Water District 2006–07 Annual Review and Water Quality Report."

Kelly, David. "$8.9-Billion Salton Sea Plan Proposed: The 75-Year Effort to Save the Polluted Resource Would Cut It to About a Fifth Its Current Size and Add Sections of Wildlife Habitat." *Los Angeles Times*, 26 May 2007.

"Sea Timetable Is Ridiculous." Editorial. *Imperial Valley Press*, 31 May 2007.

Simon, Darren. "Salton Sea: 11-Member Local Entity Approves Mitigation Plan." *Imperial Valley Press*, 13 October 2005.

Sterling, Terry Greene. "The People of the Sea." *High Country News*, 3 March 2008, p. 10.

Letters, Memoranda, Miscellaneous

"Plagues & Pleasures." DVD. Directors Chris Metzler and Jeff Springer.

Index

Note: Page numbers in italics refer to illustrations.

About Island Press

Since 1984, the nonprofit Island Press has been stimulating, shaping, and communicating the ideas that are essential for solving environmental problems worldwide. With more than 800 titles in print and some 40 new releases each year, we are the nation's leading publisher on environmental issues. We identify innovative thinkers and emerging trends in the environmental field. We work with world-renowned experts and authors to develop cross-disciplinary solutions to environmental challenges.

Island Press designs and implements coordinated book publication campaigns in order to communicate our critical messages in print, in person, and online using the latest technologies, programs, and the media. Our goal: to reach targeted audiences—scientists, policymakers, environmental advocates, the media, and concerned citizens—who can and will take action to protect the plants and animals that enrich our world, the ecosystems we need to survive, the water we drink, and the air we breathe.

Island Press gratefully acknowledges the support of its work by the Agua Fund, Inc., Annenberg Foundation, The Christensen Fund, The Nathan Cummings Foundation, The Geraldine R. Dodge Foundation, Doris Duke Charitable Foundation, The Educational Foundation of America, Betsy and Jesse Fink Foundation, The William and Flora Hewlett Foundation, The Kendeda Fund, The Andrew W. Mellon Foundation, The Curtis and Edith Munson Foundation, Oak Foundation, The Overbrook Foundation, the David and Lucile Packard Foundation, The Summit Fund of Washington, Trust for Architectural Easements, Wallace Global Fund, The Winslow Foundation, and other generous donors.

The opinions expressed in this book are those of the author(s) and do not necessarily reflect the views of our donors.